Materials Science and Technologies

nova
science publishers
www.novapublishers.com

Materials Science and Technologies

Aerogels: Properties and Applications
Alina Iuliana Pruna, PhD (Editor)
2022. ISBN: 978-1-68507-788-4 (Hardcover)
2022. ISBN: 978-1-68507-895-9 (eBook)

High-Performance Calcium-Carbonate Concrete
Natt Makul
2022. ISBN: 978-1-68507-412-8 (eBook)

Analytical Models of Interstitial-Atom-Induced Stresses in Isotropic Metallic Materials
Ladislav Ceniga (Editor)
2022. ISBN: 978-1-68507-429-6 (eBook)

Evaluation of Reserves and Resources in Unconventional Reservoirs
Cenk Temizel, Cengiz Yegin, PhD
and Shah Kabir
2022. ISBN: 978-1-68507-841-6 (eBook)

Analytical Models of Coherent-Interface-Induced Stresses in Composite Materials II
Ladislav Ceniga
2021. ISBN: 978-1-68507-003-8 (eBook)

Analytical Models of Coherent-Interface-Induced Stresses in Composite Materials III
Ladislav Ceniga
2021. ISBN: 978-1-53619-996-3 (eBook)

More information about this series can be found at
https://novapublishers.com/product-category/series/materials-science-and-technologies/

Davor Margetic, PhD
and Renjith Thomas
Editors

Photosensitizers and Their Applications

nova
science publishers
www.novapublishers.com

DOI: https://doi.org/10.52305/KWAV1221

Library of Congress Cataloging-in-Publication Data

ISBN: 978-1-68507-880-5

Published by Nova Science Publishers, Inc. † New York

Contents

Preface

Application of light in chemistry presents an active field of research with a very promising outlook, particularly in technology and medicine. Molecules which mediate the transfer of light to another molecule (photosensitizers) are the cornerstone of exploitation of light. The current state of research and recent progress in applications of photosensitizers in diverse fields are documented in twelve chapters written by scientists from around the world.

Chapter 1 gives a general overview on the applications of photosensitizers in synthetic organic chemistry. This review is followed by Chapters 2-7, which deal with the various applications of photosensitizers in medicine, in particular for photodynamic cancer therapy. The subsequent Chapters 8-12 cover different aspects of applications of photosensitizers in material science, such as for dye-sensitized solar cells and photo-responsive materials.

December 20, 2021

Davor Margetić
Renjith Thomas

Chapter 1

Photosensitizers in Organic Synthesis

Nikola Basarić[*]
Department of Organic Chemistry and Biochemistry, Ruđer Bošković Institute, Zagreb, Croatia

Abstract

Photochemical reactions taking place via triplet excited state often require the use of photosensitizers. The processes which can be sensitized are numerous and the review is intended to cover only some examples showing the principles of how triplet sensitization can be used to construct different synthons in organic synthesis. Triplet excited states behave as biradical species and undergo typical radical processes including, just to name some, additions to a double bond, H-abstractions or electron transfer. The variety of structures which can be constructed is endless and there are continuous efforts of scientists to develop new triplet excited state reactions.

Keywords: triplet excited state, photochemistry, photosensitizers, organic synthesis

Introduction

Photosensitizers and Energy Transfer

Photosensitization is a process in which a molecule exhibits electronic excitation as a result of interaction with another molecule, called

[*] Corresponding Author's Email: nbasaric@irb.hr.

In: Photosensitizers and Their Applications
Editors: Davor Margetic and Renjith Thomas
ISBN: 978-1-68507-880-5

photosensitizer, which was excited by the use of light [1]. Usually the sensitizer in its triplet excited state transfers its excitation energy to another molecule (quencher), whereupon the sensitizer is deactivated to the ground state and the quencher is excited to the triplet excited state. Sensitization is a very useful process in organic synthesis or in different photochemical and photophysical applications since it allows for the use of light of higher wavelength (photons of lower energy), for the excitation of molecules which do not absorb that light. Note also that photosensitization refers to energy transfer processes [2], and not to photocatalysis [3] where photocatalysts are excited and undergo H-transfer or photoinduced electron transfer (PET) with the substrate, which are covered by a number of excellent review articles [4-10]. This chapter aims to point to a selected number of different important applications of photosensitizers in organic synthesis and cannot serve as a thorough review providing all examples reported, due to a vast number of publications in the field. The text is organized in retrosynthetic way [11] providing information of pathways how to construct different molecular architectures.

Sensitization involves the process of energy transfer from an excited donor to an acceptor. The energy transfer can be radiative, or nonradiative. The radiative energy transfer involves deactivation of the donor by emission of light, which is absorbed by an acceptor whereupon it gets excited. Such a mechanism has no significance in the singlet-triplet energy transfer since the singlet-triplet excitation is spin forbidden, very unlikely and with very small values of oscillator strengths and molar absorption coefficients. The nonradiative energy transfer can take place via two distinctly different mechanisms. The first mechanism involves interaction of transition dipole moments of the donor and acceptor (Scheme 1) [12].

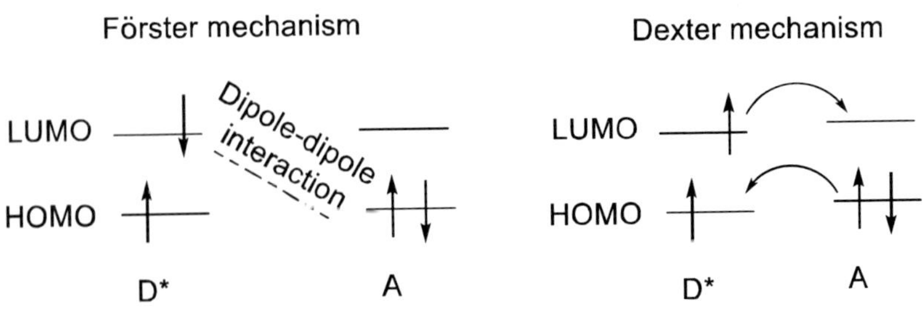

Scheme 1. Schematic representation of Förster (left) and Dexter mechanism of energy transfer for the process D* + A → D + A*.

The rate and efficiency of the transfer is described by the Förster theory and depends on the distance between the donor and the acceptor. It usually takes place at 10-100 Å, decreasing with the sixth power of the distance between the donor and the acceptor. The dipole-dipole energy transfer is also called Förster resonance energy transfer (FRET). Both radiative and nonradiative energy transfer can take place if there is an overlap of the emission band of the donor with the absorption band of the acceptor. FRET is important mechanism for singlet-singlet energy transfer only, since singlet-singlet transitions are characterized by large transition dipole moments, whereas singlet-triplet transitions are not [12, 13].

The other mechanism for the nonradiative energy transfer, which is important for the sensitization and the triplet-singlet energy transfer, requires a close contact between the donor and the acceptor whereupon mixing of the wave functions takes place allowing for the electron exchange. Such a mechanism, called Dexter mechanism (Scheme 1), depends on the distance between the donor and acceptor and takes place at short distances only, < 10 Å. The close contact between the donor and the acceptor is accomplished by diffusion of the donor and the acceptor. After the diffusion, the donor and the acceptor form an encounter complex, followed by the formation of a collision complex in which the energy transfer actually takes place. After the transfer, the excited acceptor has to move away from the donor, also taking place via an encounter complex and the diffusion. Therefore, the rate of the energy transfer in a solution often depends on the rate of the diffusion in that solvent [12].

The rate for the energy transfer by the Dexter mechanism (k_{EnT}) is given by Eq. 1, it drops exponentially with the increase of the donor and acceptor distance.

$$k_{\mathrm{EnT}} = K\,J\,e^{-\frac{2R_{\mathrm{DA}}}{\mathrm{L}}} \qquad \text{(Eq. 1)}$$

where $\frac{2R_{\mathrm{DA}}}{\mathrm{L}}$ is a measure of the distance between the donor and the acceptor, and J is the spectral overlap integral defined by Eq. 2:

$$J = \int_0^{\infty} f_{\mathrm{D}}(\nu)\,\epsilon_{\mathrm{A}}(\nu)\mathrm{d}\nu \qquad \text{(Eq. 2)}$$

where $f_{\mathrm{D}}(\nu)$ is the normalized donor emission spectrum (in case of triplet-singlet energy transfer, the phosphorescence spectrum) and $\epsilon_{\mathrm{A}}(\nu)$ is the normalized acceptor (singlet-to-triplet) absorption spectrum. Since both spectra are normalized (to an integral of unity), the rate of Dexter energy transfer does not depend on the oscillator strengths (and therefore transition dipole moments) for the transitions [12, 14]. Contrary to the estimation of the spectral overlap for FRET where fluorescence and absorption spectra are easily experimentally available, estimation of the spectral overlap for the Dexter energy transfer is difficult. However, it can be safely assumed that the energy transfer is fast if the triplet excited state energy of the donor is higher than the triplet excited state energy of the acceptor, and usually it is defined by the rate of diffusion.

The sensitizers used in organic synthesis are usually ketones, such as acetone, benzophenone, acetophenone or thioxanthone, which are characterized by high quantum yield of the intersystem crossing (ISC) and triplet population; however, other molecules are also used (Figure 1). The energies of the triplet excited states can be found in a number of publications [15].

Photosensitized Reactions of Alkenes, Isomerization and Addition

Alkenes without aromatic substituents and not in conjugation with a carbonyl group are generally characterized by a large energy gap between the singlet and the triplet excited state. Therefore, ISC in alkenes is generally inefficient and the use of sensitizers is essential for photochemical reactions proceeding via triplet excited states [12].

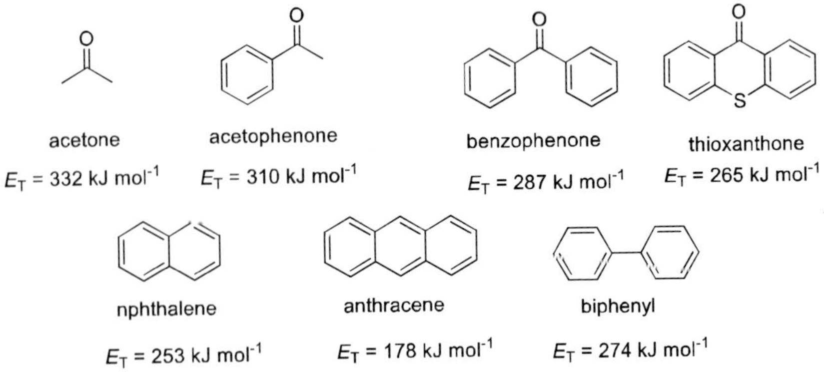

Figure 1. Common triplet sensitizers and their triplet excited state energies [15].

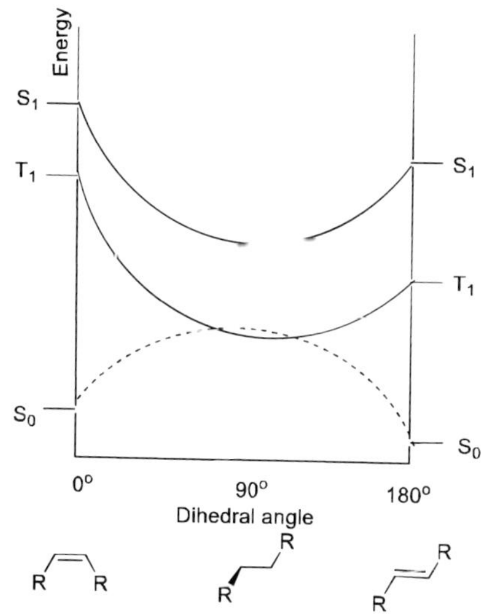

Figure 2. Energy diagram for a hypothetical alkene.

The most ubiquitous photoreaction of alkenes is *E-Z* photoisomerization, which can take place via both singlet and triplet excited state manifold [16]. Upon excitation of a simple alkene, promotion of an electron from the HOMO orbital to the LUMO results in a decrease of electron density between the C atoms of the double bond. Due to a repulsion between the substituents on the double bond, a rotation around the C-C bond takes place resulting in the formation of a perpendicular geometry with the dihedral angle of 90°. This perpendicular geometry, often named as a P-state, corresponds to an energy minimum on the S_1 or T_1 potential energy surface (PES), which coincides with the maximum on the PES in S_0 (Figure 2). From the singlet excited state, the reactions usually proceed via a conical intersection to the PES of the ground state, and from the maximum on S_0, the reaction proceeds to either of isomers. The triplet excited state has a biradical character, particularly in the P-state. For the reaction to proceed from the minimum in T_1 to S_0, an ISC is required [12, 17].

For the *E-Z* photoisomerization taking place from the triplet excited state, the composition of the irradiated solution after formation of the photostationary state depends on the energy of the triplet excited states of the *E*- and *Z*-isomers and the photosensitizer. The one-way-photoisomerization can be achieved by choosing a sensitizer that has energy of the triplet excited state in between energies of the isomers. In that way, it is possible to excite only isomer of lower triplet energy and transform it to the other isomer [17].

One recent example involves riboflavin used as a triplet sensitizer for the selective photoisomerization of α,β-unsaturated carbonyl compounds, styrene derivatives **1**, whereupon they were selectively transferred from *E* to *Z*-isomers (Scheme 2) [18]. Riboflavin was also incorporated on silica nanoparticles and used for the sensitization and selective transformation of *E*-styrenes to the corresponding *Z*-isomers [19].

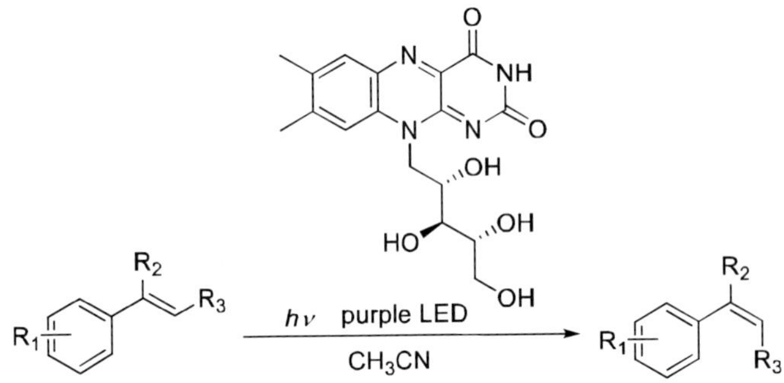

***E*-1** ***Z*-1**

R_2 = alkyl, R_3 = COOR, COOH, $CONR_2$, COR, CN, CHO

Scheme 2. Photosensitized isomerization of styrene derivatives.

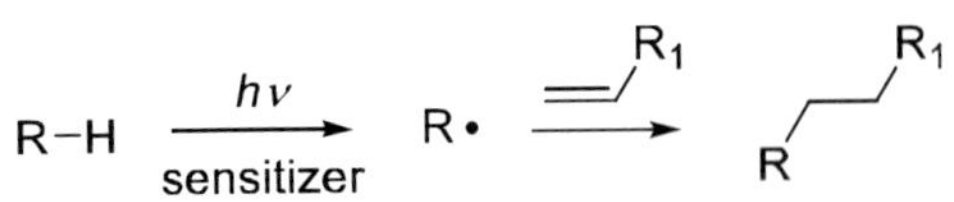

Scheme 3. Schematic photosensitized C-H activation and addition to alkenes.

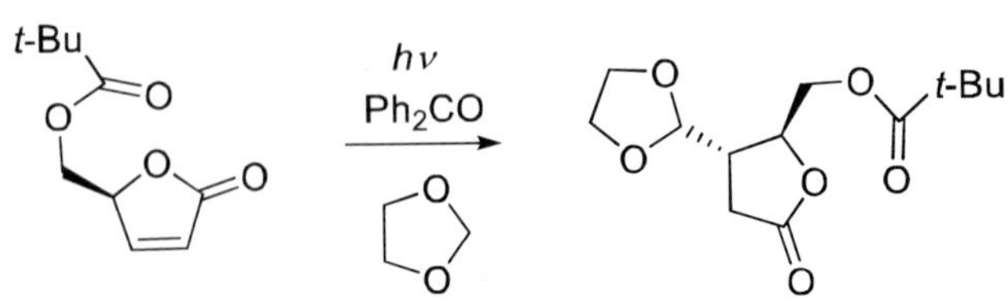

2 **3** (93%, anti/syn = 97:3)

Scheme 4. Photosensitized addition of ethers to alkene.

Photosensitizers which initiate formation of radicals can be used in radical additions to alkenes. The most interesting examples are the use of sensitizers which activate C-H bonds and allow for selective functionalizations (Scheme 3) [20]. Such approach has been used in a number of reactions, including addition of ethers on furanone **2** (Scheme 4) [21]. The reaction proceeds in high isolable yields giving **3** diastereoselectively.

Sensitized photochemical formation of radicals can also be initiated in a photodecarboxylation reaction of phthalimide derivatives. The reactions can be achieved upon direct excitation, but the yields are higher if acetone is used as a solvent and sensitizer. Upon population of phthalimide triplet excited state in **4**, PET takes place and initiates decarboxylation to give an adamantyl radical, which adds to electron deficient alkenes yielding adducts **6** (Scheme 5) [22].

Photosensitized C-H activation protocol has been used for the addition of amines to alkenes (Scheme 6). The yields on **8** and **9** are high, and the reaction proceeds with high *anti*-diastereoselectivity [23]. However, the mechanism of reaction involves PET without energy transfer from the tertiary amine to the triplet sensitizer. PET induced additions of different reagents to alkenes are numberless and the readers are referred to review articles on photocatalysis [4-10].

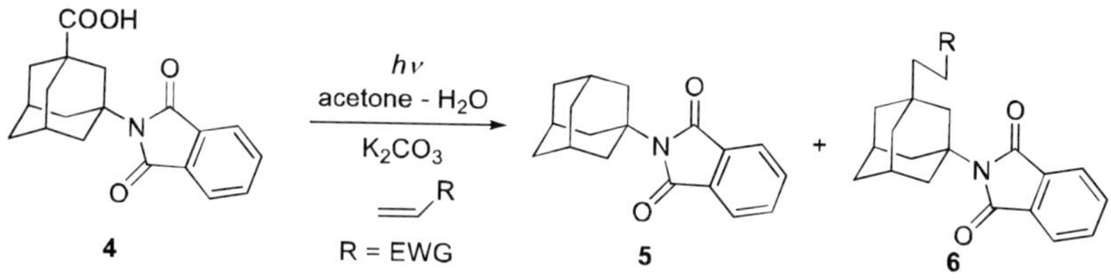

Scheme 5. Photosensitized decarboxylation and radical addition to alkenes.

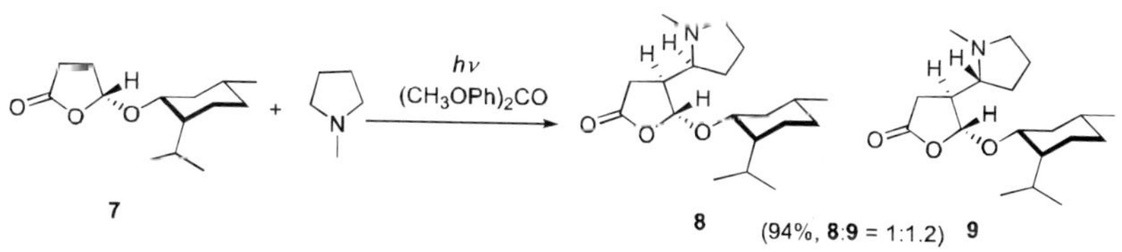

Scheme 6. Photosensitized addition of amine to alkene involving PET and not energy transfer.

Formation of Three-Membered Rings

A synthetically useful method for the formation of cyclopropane derivatives is the di-π-methane rearrangement. The reaction proceeds via triplet biradicals, and the regioselectivity of the reaction can be explained by relative stability of the radical intermediates. In the first step, 1,4-biradical is formed, which rearranges to a 1,3-biradical and vinylcyclopropane photoproduct. Barrelene (**10**), benzobarrelene (**12**) and dibenzobarrelene undergo sensitized reaction to give semibulvalene derivatives **11** and **13**, generally in excellent yields (Schemes 7 and 8) [24-26]. However, in the absence of sensitizers the reaction proceeds via singlet manifold giving products such as **14**, formed in the 6π-pericyclic ring opening [27].

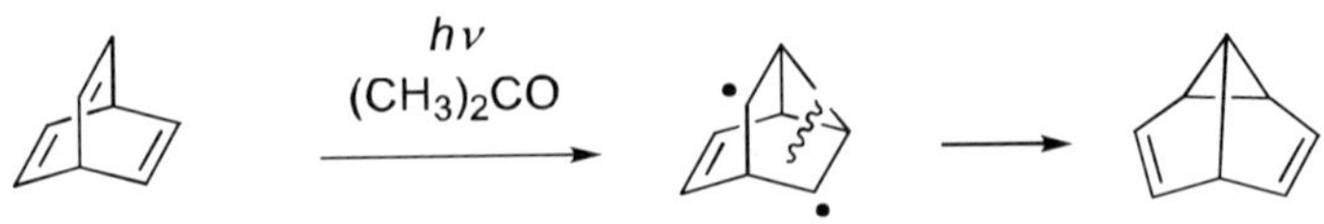

10 **11** (59%)

Scheme 7. Photosensitized di-π-methane rearrangement of barrelene (**10**).

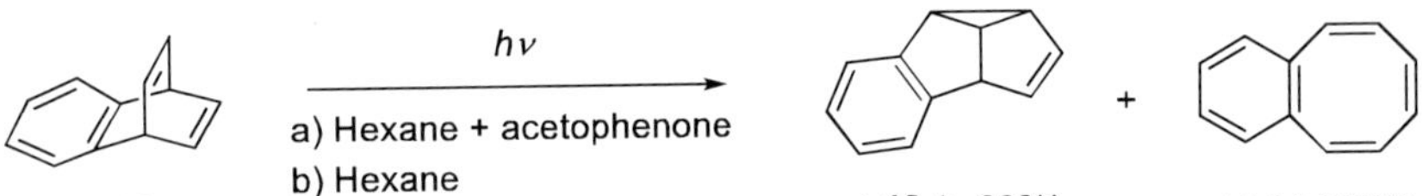

12 a) **13** (> 99%) b) **14** (97%)

Scheme 8. Photoreaction of benzobarrelene (**12**) in the presence and absence of triplet photosensitizer.

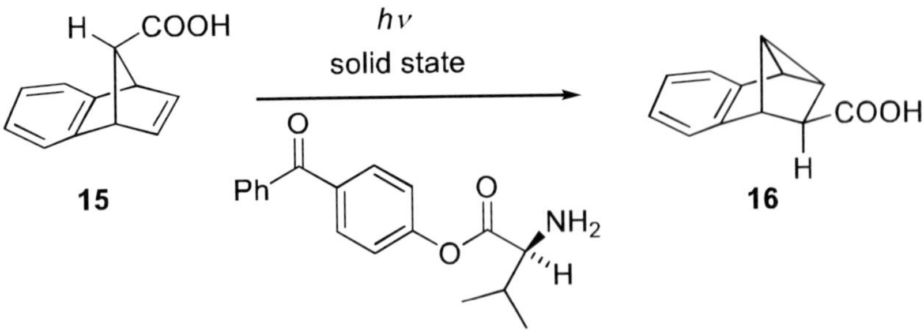

15 **16**

Scheme 9. Enantioselective di-π-methane rearrangement in the solid state in the presence of chiral sensitizer.

The di-π-methane rearrangement has also been conducted in the solid state in the presence of chiral photosensitizer, affording product **16** with 91% ee (Scheme 9) [28].

In some acyclic compounds, depending on whether the reaction is conducted in the presence of sensitizer or not, it is possible to obtain di-π-methane photoproducts with different regioselectivity, such as from compound **17** (Scheme 10) [29].

In analogy to the di-π-methane rearrangement, a similar photoreaction takes place if one double bond is replaced by a carbonyl group, called oxa-di-π-methane, or by an imine, called aza-di-π-methane rearrangement. Thus, β,γ-unsaturated carbonyl compounds in the presence of triplet sensitizers undergo oxa-di-π methane rearrangement. One example is the transformation of polycyclic molecule **20** into **21**, proceeding in high yields and with high stereoselectivity (Scheme 11) [30].

β,γ-unsaturated imine derivatives in the presence of triplet sensitizers such as acetophenone or benzophenone undergo aza-di-π-methane rearrangement [31-33]. One example is rearrangement of **22** to cyclopropane **23**, which hydrolyses on silica upon purification giving carbaldehyde **24** (Scheme 12) [34].

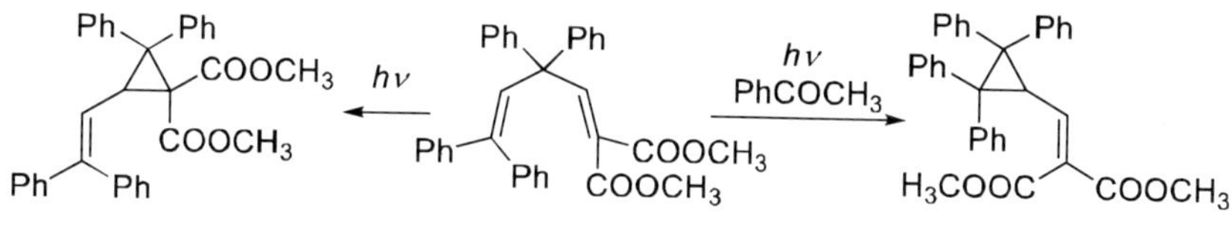

19 (95%) **17** **18** (95%)

Scheme 10. Regioselectivity of di-π-methane rearrangement depending on the spin selectivity.

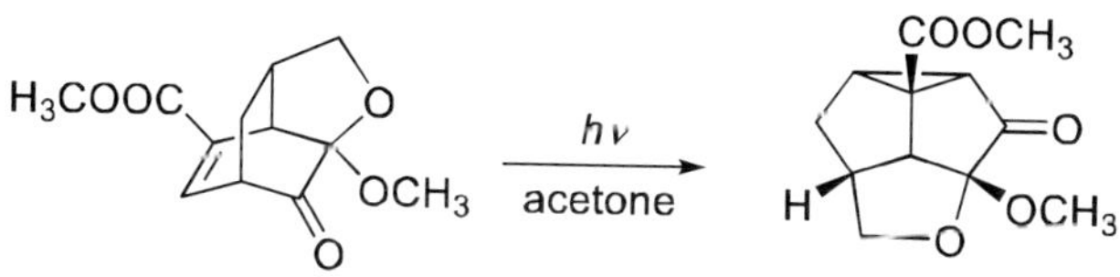

20 **21** (77%)

Scheme 11. Oxa-di-π-methane rearrangement.

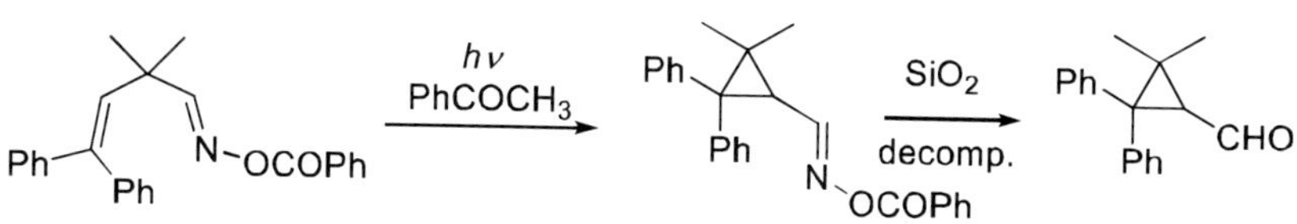

22 **23** (90%) **24** (40%)

Scheme 12. Aza-di-π-methane rearrangement.

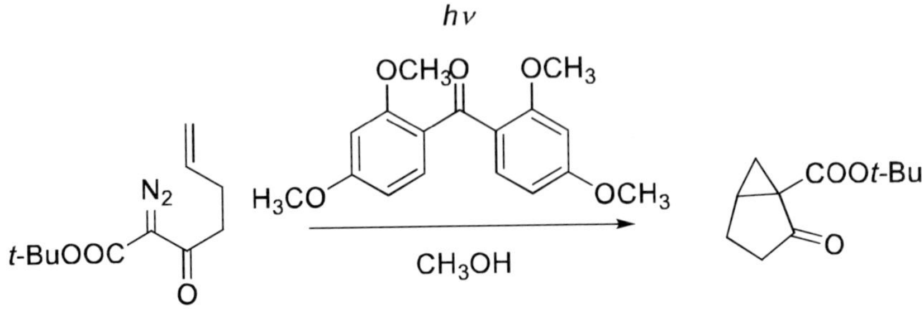

25

26 (86%)

Scheme 13. Sensitized formation of triplet carbene and addition to alkene.

A powerful method for the generation of cyclopropanes is photolysis of diazirines [35] or diazo compounds to generate carbenes [36], which add to alkenes in a [2+1] reaction. In the photochemistry of α-diazo carbonyl compounds, singlet carbenes (if formed) undergo Wolff rearrangement [37], whereas irradiations in the presence of sensitizers afford triplet carbenes that add to alkenes. The usual choice of sensitizer is benzophenone, which has to be in excess to assure that it absorbs most of the irradiation. However, better results were obtained by the use of tetramethoxybenzophenone, as shown in Scheme 13 [38].

Formation of Four-Membered Rings

Photochemical formation of cyclobutane derivatives is very useful in organic synthesis and can be achieved by [2+2] cycloaddition, electrocyclization or γ-H abstraction by a carbonyl compound followed by a cyclization (Yang cyclization) [39]. The electrocyclizations [40, 41] are only possible in the singlet excited state manifold, and for the Yang cyclizations sensitizers are not needed since the ketone derivatives can be excited by UV light and usually undergo efficient ISC to the triplet excited state [42]. Furthermore, due to a problem that alkenes without aromatic substituents do not absorb UV light, and due to their competing *E*-*Z*-photoisomerizations, alkenes are often complexed with transition metals. Upon complexation to a transition metal, olefins can be directly excited by UV light due to a metal-to-ligand charge transfer (MLCT) or via a ligand-to-metal charge transfer (LMCT) excitation. Particularly applicable are copper salts such as CuOTf, and in these examples sensitizers are not needed [43]. However, for [2+2] cycloadditions [44] sensitization is often used, particularly for the cycloadditions of alkenes which do not absorb UV light, for intramolecular reactions and formation of polycyclic cage molecules. Thus, photoreaction of diene **27** in acetone (sensitizer and solvent) affords cycloadduct **28** in a high yield (Scheme 14) [45, 46].

A remarkable dimerization of azetidine derivative **29** in acetone was also documented, giving a mixture of diastereomers **30** and **31** in a ratio 1:1 (Scheme 15) [47]. Furthermore, sensitized intramolecular cycloaddition of a dihydropyridone derivative **32** was performed, giving compound **33** in a high yield and with high ee. The enantioselectivity is possible since **32** has axial chirality. Enantiomers **32** were separated and irradiated to furnish photoproducts **33** in the enantioselective reaction (Scheme 16) [48].

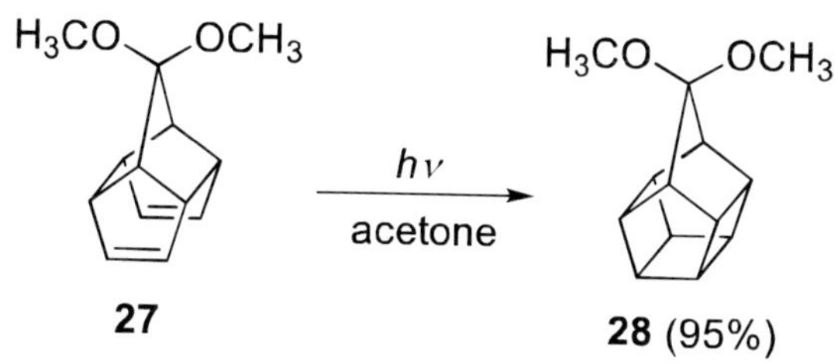

Scheme 14. Intramolecular [2+2] cycloaddition for the formation of cage molecules.

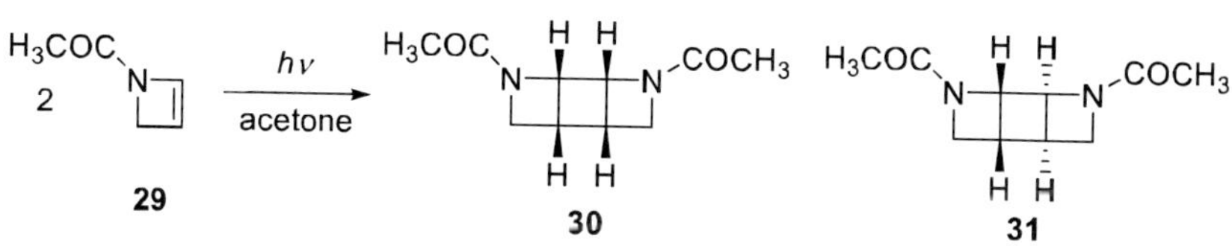

Scheme 15. Sensitized photodimerization of azetidine.

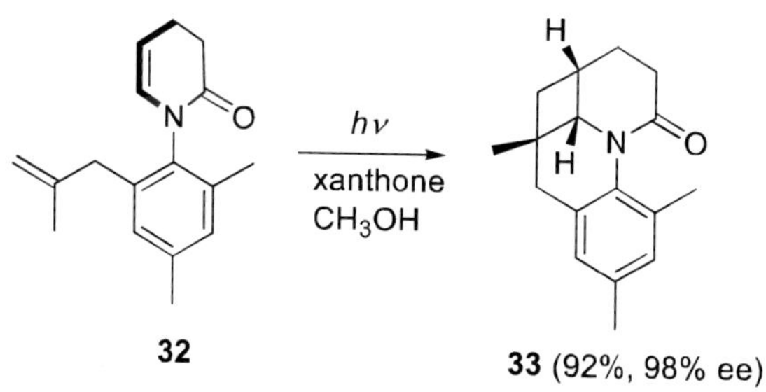

Scheme 16. Enantioselective cycloaddition of dihydropiridone derivative.

[2+2] cycloadditions of enones are also very applicable in organic synthesis as they usually proceed with high yields [44, 49]. However, triplet sensitizers are not needed in these reactions since enones undergo efficient ISC and populate triplet excited states. Further synthetically useful examples include photochemical formation of oxetanes in Paterno-Büchi reactions, which also do not need sensitizers as they involve carbonyl compounds in the reactions with alkenes [50].

Formation of Five- and Six-Membered Rings and Larger Rings

Five-membered rings can be obtained in δ-H photochemical abstraction reactions of carbonyl compounds, whereupon 1,5-diradicals are formed that cyclize to cyclopentane derivatives [51]. Due to involvement of carbonyl compounds as substrates that readily undergo ISC, the use of sensitizers is not required. The intramolecular H-abstraction by a carbonyl compound usually takes place at the γ-position due to thermodynamic reasons, since it involves formation of a stable 6-membered ring transition state. However, abstractions from δ-, or other positions, are possible if γ-H atoms are not available, or if the C-H bond becomes activated by an effect of heteroatom adjacent to it. Another important method for the photochemical formation of five-membered rings is [3+2] cycloaddition. Thus, photofragmentation of oxiranes provides carbonyl ylides, whereas aziridines provide azomethine ylides, both of which can be trapped by alkenes (or alkynes) forming tetrahydrofurans or pyrolidines, respectively [51]. These cycloadditions usually do not require sensitizers; they proceed upon direct excitation of the molecules. However, notable examples of sensitized reactions have been reported, as in the case of acetone-sensitized opening of epoxide **34** to form carbonyl ylide **35**, which was trapped in intramolecular reaction to give tetrahydrofuran **36** (Scheme 17) [52]. Further examples for the formation of five-membered rings involve photocatalytic methods in which radical cations cyclize with alkenes, and these are outside of the scope of this review [51].

Photochemical reactions for the formation of six-membered rings often rely on 6π-electrocyclizations, which however take pace in the singlet excited state. Moreover, 4π+4π cycloaddition can furnish eight membered rings, but also from singlet excited state [53].

Photochemical reactions of phthalimide derivatives provide a rich array of different synthetically useful protocols for the formation of five- or six-membered rings, as well as larger rings [54]. They proceed upon direct excitation by light, but can be sensitized, usually giving higher yields due to low ISC in some phthalimide derivatives. The photochemical reaction mechanisms include H-abstractions, PET or cycloadditions [55]. Thus, irradiation of phthalimide **37** gave polycyclic molecules **38** and **39** with the formation of five-membered ring, as well as compound **40** with a seven-membered ring. The products are formed in photoinduced H-abstraction reactions, and the process can also take place in the solid state [56]. Interestingly, seven-membered ring compound **40** can also be obtained in a photochemical reaction of phthalimide that involves photoinduced decarboxylation [57].

Formation of six- and seven-membered rings was also documented in the photoinduced H-abstraction reactions of phthalimide **41** (Scheme 19) [58], whereas a domino reaction sequence of H-abstractions gave polycyclic derivative **46** in high isolable yield (Scheme 20) [59].

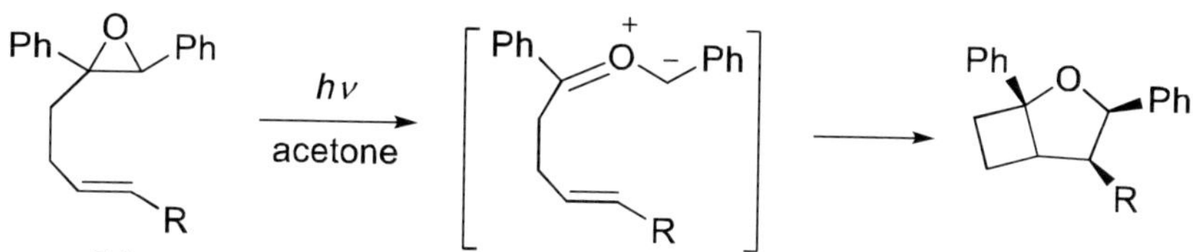

34 **35** **36** (25%)

Scheme 17. Sensitized ring-opening of epoxide and intramolecular [3+2] cycloaddition.

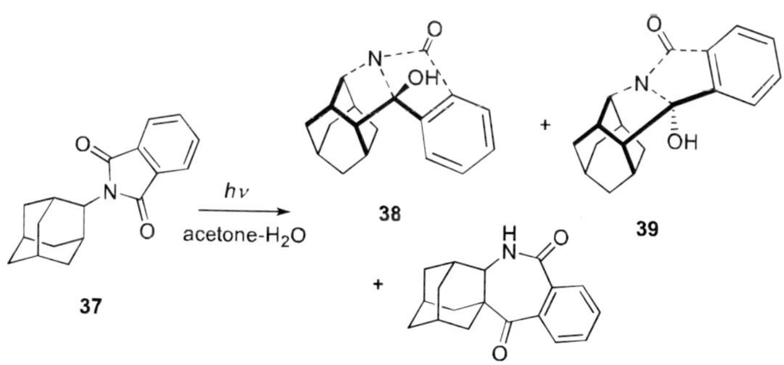

37 **38** **39** **40**

Scheme 18. Sensitized photochemical H-abstraction in phthalimide derivatives giving new five- or seven-membered rings.

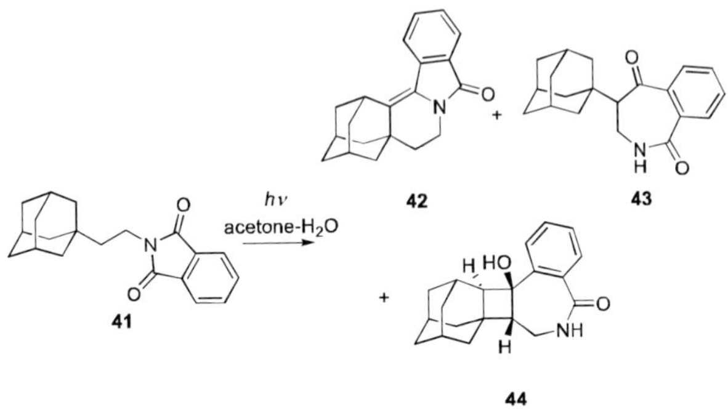

Scheme 19. Photoinduced C-H activation in phthalimide derivatives and formation of six- and seven-membered rings.

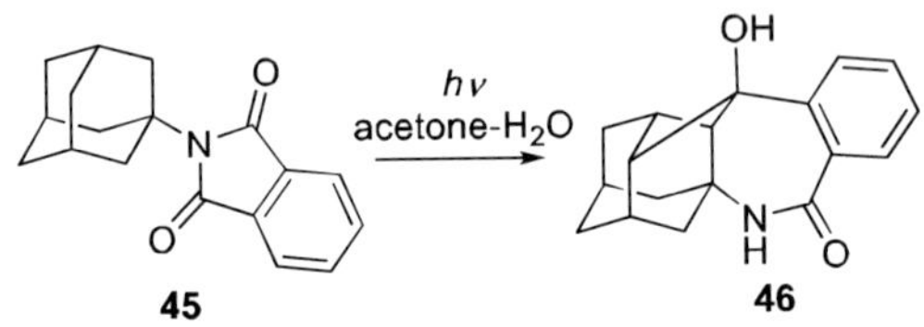

Scheme 20. Photoinduced domino reaction of phthalimide **45**.

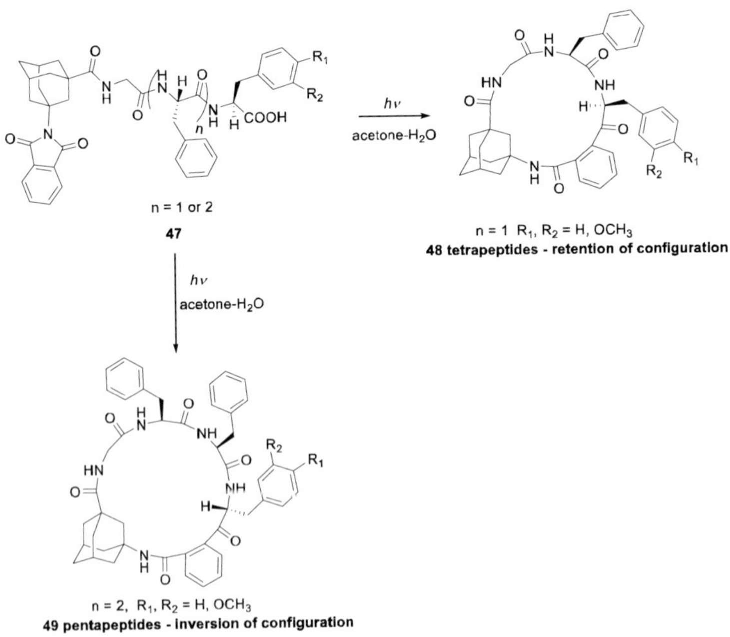

Scheme 21. Sensitized photocyclization of peptides with retention or inversion of configuration.

Photosensitized reactions of phthalimides can also be used to construct different large cycles [60], including crown ethers [61], or cyclic peptides [62, 63]. Furthermore, interesting stereoselectivity in the peptide photocyclizations was documented proceeding with the retention of configuration for tetrapeptides **48** and inversion of configuration for pentapeptides **49** (Scheme 21) [64].

Diastereoselective and Enantioselective Photosensitized Reactions

Developing new reactions that proceed with high stereoselectivity, and particularly enantioselectively, is the Holy Grail of modern synthetic organic chemistry. Consequently, similar endeavors are oriented to develop enantioselective photochemical reactions [65-68], and recently a significant effort has been directed to the use of visible light [69]. Several approaches have been used including the use of circularly polarized light, chiral auxiliaries in the diastereoselective photoreactions, supramolecular control, use of chiral templates or chiral sensitizers, which have all been reviewed [65].

In the numerous examples of diastereoselective cycloadditions of enones, or in the Paterno-Büchi reactions, the use of sensitizers is not required since the processes involve carbonyl compounds that are usually characterized by high efficiency of ISC. Furthermore, sensitizers are not needed in diastereoselective H-abstraction reactions of excited carbonyl compounds, whereas cycloadditions with aromatic compounds usually proceed via singlet excited states [70]. On the contrary, there are notable examples of diastereoselective sensitized photochemistry with singlet oxygen. Note that oxygen in the ground state is in the triplet manifold, and triplet sensitizers are used to excite it to the singlet excited state. Singlet molecular oxygen ($^1O_2$) is an electrophilic reagent that reacts with dienes and alkenes in [4+2] or [2+2] cycloadditions or ene-reactions. Thus, irradiation of sorbic acid derivative **50** having chiral oxazoline auxiliary in the presence of tetraphenylporphyrin as singlet oxygen sensitizer affords product **51** with excellent stereocontrol during the process (Scheme 22) [71].

High diastereoselectivity was also documented for [2+2] cycloaddition of $^1O_2$ to allylic alcohol **52** [72]. The irradiation was performed in the presence of tetrakis(pentafluorophenyl)porphyrin (TPFPP) as a $^1O_2$ sensitizer. Due to steric hindrance of the adamantyl and the methyl group, the reaction affords mainly product **53** (95%, Scheme 23). Addition of a protic solvent significantly reduces selectivity due to formation of H-bonds between the $^1O_2$ and the solvent.

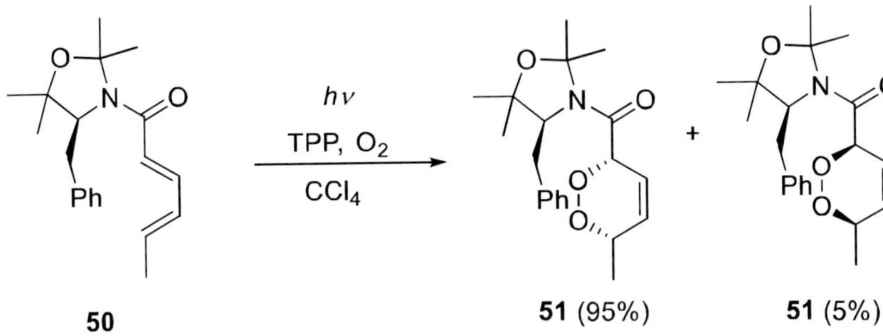

Scheme 22. Diastereoselective [4+2] cycloaddition with $^1O_2$.

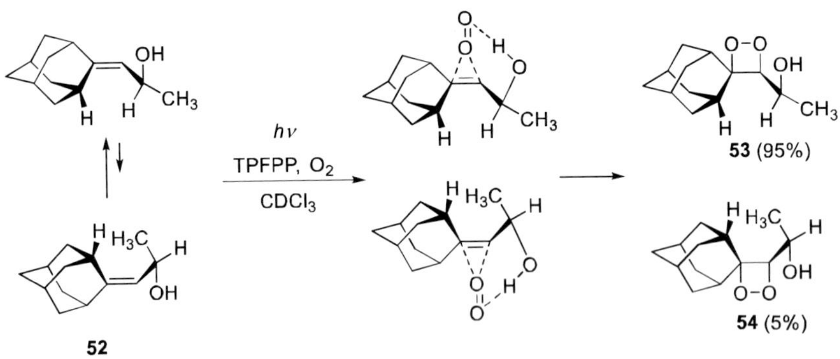

Scheme 23. Diastereoselective [2+2] cycloaddition with $^1O_2$.

One example of diastereoselective ene-reaction is shown in Scheme 24. [73] The authors used (TPFPP) as a $^1O_2$ sensitizer and oxazolidine chiral auxiliary. The site of the attack by $^1O_2$ is directed by H-bond between the amide and the $^1O_2$.

A very useful approach for the enantioselective photochemical approach is the use of chiral templates [74], or conducting photochemical reactions in mixed chiral crystals [75], but they generally do not require the use of sensitizers, although there are some sensitized examples reported (vide supra, Scheme 9). Furthermore, a recent example involves enantioselective aza Paterno-Büchi reaction of quinoxalinones **57** by use of visible light (420 nm) and triplet photosensitizers **59** embedded in the chiral template (Scheme 25) [76]. The reaction afforded high yields of azetidines such as **60**, with excellent values of ee. The same approach also afforded cycloadducts in intramolecular enantioselective reactions.

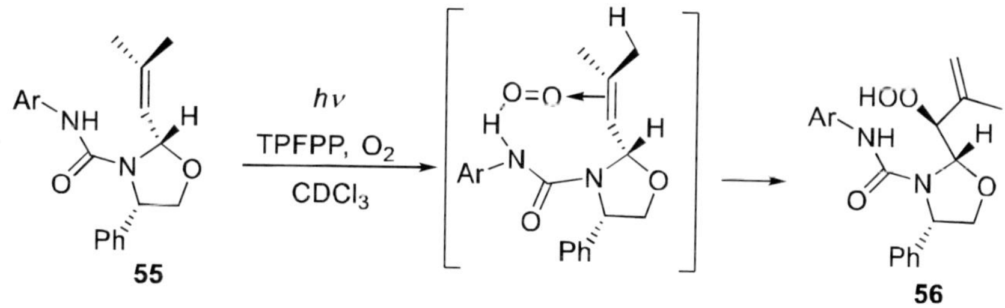

Scheme 24. Diastereoselective ene-reaction.

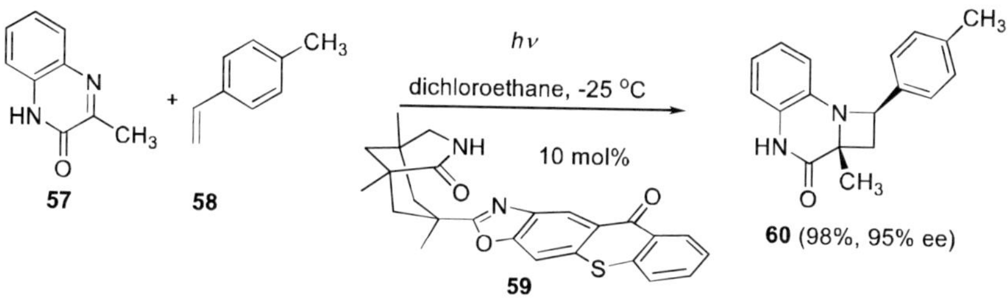

Scheme 25. Enantioselective aza Paterno–Büchi reaction in the presence of chiral template photosensitizer **59**.

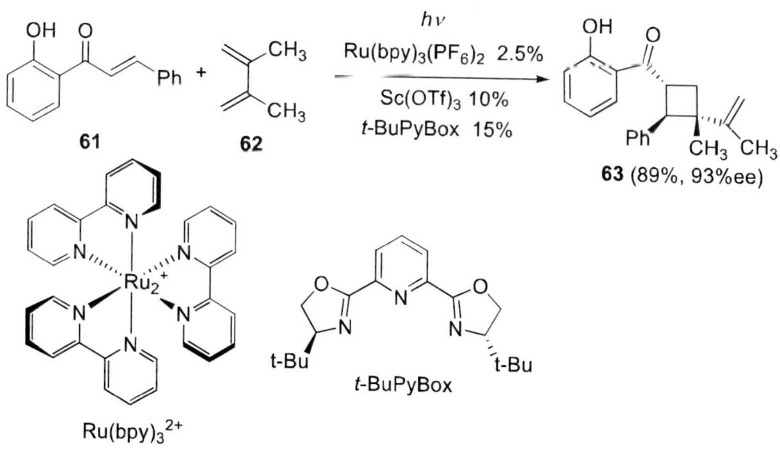

Scheme 26. Enantioselective [2+2] cycloaddition reaction in the presence of chiral Lewis acid and $Ru(bpy)_3^{2+}$ photosensitizer.

Enantiodifferentiation in photosensitized chemical reactions has also been achieved by use of chiral Lewis acids. Thus, enones such as **61** form complexes with chiral Lewis acids (*t*-BuPyBox and $Sc(OTf)_3$) and photosensitization by organomethalic compounds such as $Ru(bpy)_3^{2+}$ affords [2+2] cycloadducts **63** in high yields and with high enantioselectivity (Scheme 26) [77].

The enantiodifferentiating photosensitization is a phenomenal photochemical approach, in which chiral information of an optically active sensitizer is transferred to a prochiral or racemic substrate through noncovalent interactions in the excited state, leading to the enantioselective excitation and formation of photoproducts with high ee. The process in a way mimics the action of enzymes. The approach is different from other chiral methods in photochemistry, since only a minor amount of chiral sensitizer is required instead of the stoichiometric quantity as in the case of chiral auxiliaries, supramolecular hosts or templates. Traditionally, the method was used for unimolecular reactions, but recent advances include some bimolecular processes in which ternary complexes are formed [78].

One example of an enantiodifferentiating photoreaction was reported for oxepinone **64**. Upon photosensitization with optically active naphthalenecarboxamide **65**, *rac*-**64** first rearranges to *rac*-**66**, which undergoes di-π-methane rearrangement to give selectivity upon excitation. However, the enantioselectivity was low, the highest k_R/k_S of 1.04 was determined whereas the conversion of **64** was 50% (Scheme 27) [79].

Oxa-di-π-methane rearrangement has also been conducted in the presence of chiral sensitizer. Thus, bicyclooctenone **68** gave tricyclic compound **69** with 4.5% ee (Scheme 28) [80].

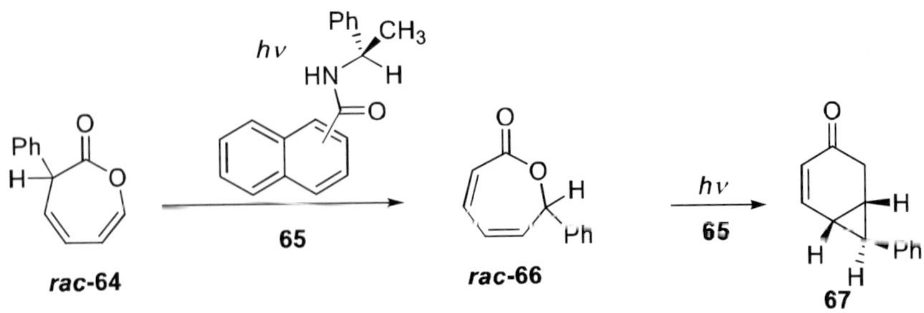

Scheme 27. Enantioselective photosensitization in the di-π-methane rearrangement.

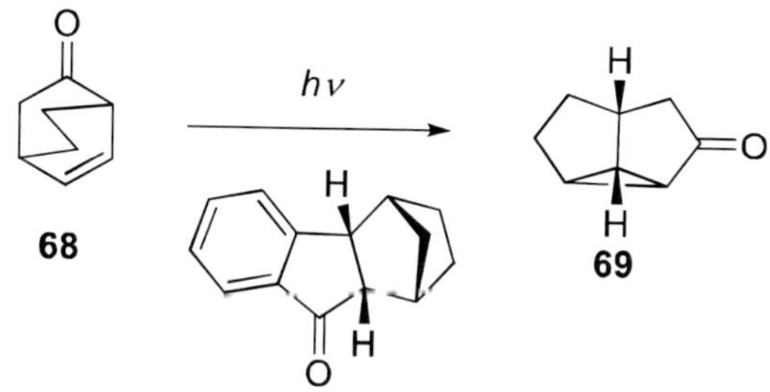

Scheme 28. Enantioselective photosensitized oxa-di-π-methane rearrangement.

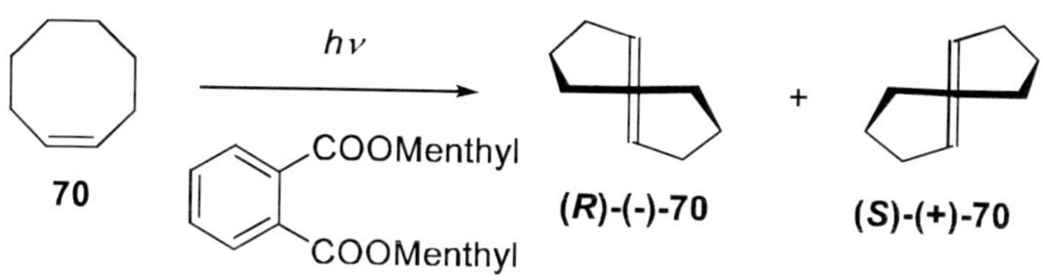

Scheme 29. Enantioselective sensitization and isomerization of *cis*-cyclooctene.

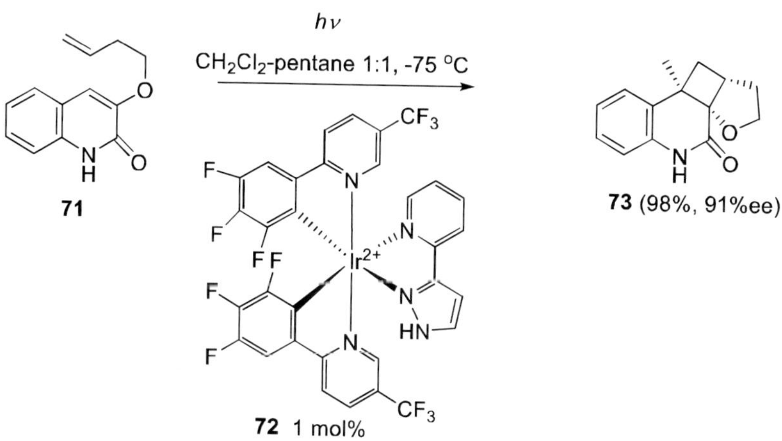

Scheme 30. Enantioselective sensitization and [2+2] cycloaddition.

Use of chiral triplet sensitizers was reported also for the isomerization of 1,2-diphenylcyclopropanes. Poor ee values (<3%) were reported with a series of chiral triplet sensitizers [81]. Furthermore, enantioselectivity in the *cis-trans* isomerization of *cis*-cyclooctene **70** was investigated. By use of chiral triplet sensitizers low values of ee were obtained. However, decrease of temperature and the use of chiral menthyl phthalates as singlet sensitizers improved ee values to 41% (Scheme 29) [82].

The use of enantioselective sensitization was reported for [2+2] cycloadditions, and excellent enantiodifferentiation has recently been achieved by use of chiral iridium complexes such as **72**, which act as photosensitizers (Scheme 30) [83]. However, in a number of cases the actual mechanism was not the energy transfer but electron transfer.

Moreover, a significant progress in the enantioselective photochemistry has been reported by use of supramolecular systems, some of which also involve photosensitization mechanisms [84-86]. For example, ee values reaching >90% were obtained by use of biomolecules such as DNA [87] or proteins [88, 89] as chiral pools.

Conclusion

Contrary to the classical chemical reactions taking place in the ground state, photochemical processes can be initiated from two electronically different species, the singlet and the triplet excited state, which often differ in reactivity. Thus, photochemical reactions, in principle, provide more opportunities to drive the process to a desired product. The use of triplet sensitizers is important as it allows for the population of triplet states, which can initiate a number of reactions. The photosensitization and photocatalysis have blossomed into a fruitful research field with increasing number of publications every year. This short review pointed to some synthetically important sensitized reactions providing different molecular structures. However, possibilities and opportunities in the use of sensitizers are endless.

Acknowledgment

The author gratefully acknowledges financing from the Croatian Science Foundation (HRZZ grants no. HRZZ-IP-2014-09 6312 and HRZZ-IP-2019-04-8008).

References

[1] *IUPAC Compendium of Chemical Terminology – Gold Book*, Version 2.3.3, 2014-02-24.

[2] Strieth-Kalthoff, F.; James, M. J.; Teders, M.; Pitzer, L.; Glorius, F. Energy transfer catalysis mediated by visible light: principles, applications, directions, *Chem. Soc. Rev.*, **2018**, *47*, 7190-7202.

[3] Michelin, C.; Hoffmann, N. Photosensitization and Photocatalysis Perspectives in Organic Synthesis, *ACS Catal.*, **2018**, *8*, 12046-12055.

[4] Rossi, R. A.; Pierini, A. B.; Peñéñory, A. B., Nucleophilic Substitution Reactions by Electron Transfer, *Chem. Rev.*, **2003**, *103*, 71-167.

[5] Xuan, J.; Xiao, W.-J. Visible-Light Photoredox Catalysis. *Angew. Chem. Int. Ed.*, **2012**, *51*, 2-13.

[6] Prier, C. K.; Rankic, D. A.; MacMillan, D. W. S. Visible Light Photoredox Catalysis with Transition Metal Complexes: Applications in Organic Synthesis, *Chem. Rev.*, **2013**, *113*, 5322-5363.

[7] Reckenthäler, M.; Griesbeck, A. G., Photoredox Catalysis for Organic Syntheses, *Adv. Synth. Catal.*, **2013**, *355*, 2727-2744.

[8] Romero N. A.; Nicewicz, D. A., Organic Photoredox Catalysis, *Chem. Rev.*, **2016**, *116*, 10075-10166.

[9] Skubi, K. L.; Blum, T. R.; Yoon, T. P., Dual Catalysis Strategies in Photochemical Synthesis, *Chem. Rev.*, **2016**, *116*, 10035-10074.

[10] Hari, D. P.; König, B., Synthetic applications of eosin Y in photoredox catalysis, *Chem. Commun.*, **2014**, *50*, 6688-6699.

[11] Albini, A.; Fagnoni, M., Eds. *Handbook of Synthetic Photochemistry*, Wiley-VCH, Weinheim, 2010.

[12] Turro, N. J.; Ramamurthy, V.; Scaiano, J. C., *Modern Molecular Photochemistry of Organic Molecules*, University Science Books, Sausalito 2010.

[13] Förster, T., Zwischenmolekulare Energiewanderung und Fluoreszenz, *Ann. Phys.*, **1948**, *437*, 55-75.

[14] Dexter, D. L., A Theory of Sensitized Luminescence in Solids, *J. Chem. Phys.*, **1953**, *21*, 836-850.

[15] Montalti, M.; Credi, A.; Prodi, L.; Gandolfi, M. T., Eds., *Handbook of Photochemistry*, third addition, CRC Taylor and Francis, Boca Raton, 2006, and references cited therein.

[16] Waldeck, D. H., Photoisomerization Dynamics of Stilbenes, *Chem. Rev.*, **1991**, *91*, 415-436.

[17] Görner, H.; Kuhn, H. J., *cis-trans* Photoisomerization of Stilbenes and Stilbene-like Molecules, in *Advances of Photochemistry*, Vol. 19, Neckers, D. C.; Volman, D. H., Eds. John Wiley and Sons, New York, 1995.

[18] Metternich, J. B.; Gilmour, G. R., A Bio-Inspired, Catalytic $E \rightarrow Z$ Isomerization of Activated Olefins, *J. Am. Chem. Soc.*, **2015**, *137*, 11254-11257.

[19] Metternich, J. B.; Sagebiel, S.; Lückener, A.; Lamping, S.; Ravoo, B. J.; Gilmour, R., Covalent Immobilization of (-)-Riboflavin on Polymer Functionalized Silica

Particles: Application in the Photocatalytic *E*→*Z* Isomerization of Polarized Alkenes, *Chem. Eur. J.*, **2018**, *24*, 4228-4233.

[20] Dichiarante, V.; Fagnoni, M., Intermolecular Addition onto C-C Multiple Bonds, in *Handbook of Synthetic Photochemistry*, Eds. Albini, A.; Fagnoni, M., Wiley-VCH, Weinheim, 2010.

[21] Ghosh, A. K.; Leshchenko, S.; Noetzel, M., Stereoselective Photochemical 1,3-Dioxolane Addition to 5-Alkoxymethyl-2(5H)-furanone: Synthesis of Bis-tetrahydrofuranyl Ligand for HIV Protease Inhibitor UIC-94017 (TMC-114), J. *Org. Chem.*, **2004**, *69*, 7822-7829.

[22] Horvat, M.; Mlinarić-Majerski, K.; Griesbeck, A. G.; Basarić, N., Photoinduced decarboxylation of 3-(*N*-phthalimido)adamantane-1-carboxylic acid and radical addition to electron deficient alkenes, *Photochem. Photobiol. Sci.*, **2011**, *10*, 610–617.

[23] Bertrand, S.; Hoffmann, N.; Pete, J.-P., Highly Efficient and Stereoselective Radical Addition of Tertiary Amines to Electron-Deficient Alkenes - Application to the Enantioselective Synthesis of Necine Bases, *Eur. J. Org. Chem.*, **2000**, 2227-2246.

[24] Zimmerman, H. E.; Armesto, D., Synthetic Aspects of the Di-π-methane Rearrangement, *Chem. Rev.*, **1996**, *96*, 3065-3112.

[25] Zimmerman, H. E., Zimmerman, H. E. Five decades of mechanistic and exploratory organic photochemistry, *Pure Appl. Chem.*, **2006**, *78*, 2193-2203.

[26] Ramaiah, D.; Sajimon, M. C.; Josepha, J.; George M. V., Photoisomerisation of dibenzobarrelenes - a facile route to polycyclic synthons, *Chem. Soc. Rev.*, **2005**, *34*, 48-57.

[27] Pitchumani, K.; Warrier, M.; Kaanumalle, L. S.; Ramamurthy, V., Triplet photochemistry within zeolites through heavy atom effect, sensitization and light atom effect, *Tetrahedron*, **2003**, *59*, 5763-5772.

[28] Janz, K. M.; Scheffer, J. R., The use of ionic chiral sensitizers in the crystalline state: application to the di-π-methane photorearrangement of a benzonorbornadiene derivative, *Tetrahedron Lett.*, **1999**, *40*, 8725-8728.

[29] Zimmerman, H. E.; Factor, R. E., Di-π-methane hypersurfaces and reactivity; multiplicity and regioselectivity; relationship between the di-π-methane and bicycle rearrangements: Mechanistic and exploratory organic photochemistry, *Tetrahedron*, **1981**, *37*, 125-141.

[30] Bach, T.; Krüger, C.; Harms, K., The Stereoselective Synthesis of 2-Substituted 3-Azabicyclo[3.2.0] heptanes by Intramolecular [2+2]-Photocycloaddition Reactions, *Synthesis*, **2000**, 305-320.

[31] Armesto, D.; Horspool, W. H.; Langa, F.; Ramos, A., Extension of the aza-di-π-methane reaction to stable derivatives. Photochemical cyclization of β,γ-unsaturated oxime acetates, *J. Chem. Soc. Perkin Trans 1*, **1991**, 223-228.

[32] Armesto, D.; Horspool, W. H.; Langa, F., The Aza-di-π-methane rearrangement of O-acetyl 2,2-dimethyl-4,4-diphenylbut-3-enal oxime, *J. Chem. Soc. Chem. Commun.*, **1987**, 1874-1875.

[33] Armesto, D.; Horspool, W. M.; Gallego, M. G.; Agarrabeitia, A. R., Steric and electronic effects on the photochemical reactivity of oxime acetates of β,γ-unsaturated aldehydes, *J. Chem. Soc. Perkin Trans 1*, **1992**, 163-169.

[34] Armesto, D.; Horspool, W. M.; Mancheño, M. J.; Ortiz, M. J., The aza-di-π-methane rearrangement of stable derivatives of 2,2-dimethyl-4,4-diphenylbut-3-enal, *J. Chem. Soc. Perkin Trans 1*, **1990**, 2348-2349.

[35] Celius, T. C.; Toscano, J. P., The Photochemistry of Diazirines, in *CRC Handbook of Photochemistry and Photobiology 2nd Ed.*, Horspool, W.; Lenci, F., Eds., CRC Press, Boca Raton, 2004.

[36] Abdel-Wahab, A.-M. A.; Ahmed, S. A.; Dürr, H., Carbene Formation by Extrusion of Nitrogen, in *CRC Handbook of Photochemistry and Photobiology 2nd Ed.*, Horspool, W.; Lenci, F., Eds., CRC Press, Boca Raton, 2004.

[37] Jones, M. Jr.; Moss, R. A., Singlet Carbenes, in *Reactive Intermediate Chemistry*, Eds.: Moss, R. A.; Platz, M. S; Jones, Jr. M., Wiley, Hoboken, 2004, and references cited therein.

[38] Pastor-Pérez, L.; Wiebe, C.; Pérez-Prieto, J.; Stiriba, S.-E., A Tetramethoxybenzophenone as Efficient Triplet Photocatalyst for the Transformation of Diazo Compounds, *J. Org. Chem.*, **2007**, *72*, 1541-1544.

[39] Hoffmann, N., Formation of a Four-Membered Ring, in *Handbook of Synthetic Photochemistry*, Eds. Albini, A.; Fagnoni, M., Wiley-VCH, Weinheim, 2010.

[40] Bian, M.; Li, L.; Ding, H., Recent Advances on the Application of Electrocyclic Reactions in Complex Natural Product Synthesis, *Synthesis* **2017**, *49*, 4383-4413.

[41] Thompson, S.; Coyne, A. G.; Knipe, P. C., Smith, M. D., Asymmetric electrocyclic reactions, *Chem. Soc. Rev.*, **2011**, *40*, 4217-4231.

[42] Chen, C., The past, present, and future of the Yang reaction, *Org. Biomol. Chem.*, **2016**, *14*, 8641-8647.

[43] Ghosh, S. Copper(I)-Catalyzed Inter- and Intramolecular [2+2]-Photocycloaddition Reactions of Alkenes, in *CRC Handbook of Photochemistry and Photobiology*, 2nd ed., Horspool, W. M., Lenci, F., Eds., CRC Press, Boca Raton, 2004.

[44] Poplata, S.; Tröster, A.; Zou, Y.-Q.; Bach, T., Recent Advances in the Synthesis of Cyclobutanes by Olefin [2+2] Photocycloaddition Reactions, *Chem. Rev.*, **2016**, *116*, 9748-9815.

[45] Eaton, P. E.; Or, Y. S.; Branca, S. J., Pentaprismane, *J. Am. Chem. Soc.*, **1981**, *103*, 2134-2136.

[46] Eaton, P. E.; Or, Y. S.; Branca, S. J., Shankar, B. K. R., Synthesis of pentaprismane, *Tetrahedron.*, **1986**, *42*, 1621-1631.

[47] Dave, P. R.; Duddu, R.; Li, J.; Surapaneni, R.; Gilardi, R., Photodimerization of *N*-acetyl-2-azetine: Synthesis of syn-Diazatricyclooctane and anti-Diazatricyclooctane (Diaza-3-ladderane), *Tetrahedron Lett.*, **1998**, *39*, 5481-5484.

[48] Kumarasamy, E.; Sivaguru, J. Light-Induced Stereospecific Intramolecular [2 + 2]-Cycloaddition of Atropisomeric 3,4-Dihydro-2-pyridones, *Chem. Commun.*, **2013**, *49*, 4346-4348.

[49] Schuster, D. I.; Lem, G.; Kaprinidis, N. A., New Insights into an Old Mechanism: [2 + 2] Photocycloaddition of Enones to Alkenes, *Chem. Rev.* **1993**, *93*, 3-22.

[50] D'Auria, M., The Paternò–Büchi reaction - a comprehensive review, *Photochem. Photobiol. Sci.*, **2019**, *18*, 2297-2367.

[51] Pandey, G.; Gadre, S. R., Formation of a Five-Membered Ring, in *Handbook of Synthetic Photochemistry*, Eds. Albini, A.; Fagnoni, M., Wiley-VCH, Weinheim, 2010.

[52] Brokatzky-Geiger, J.; Eberbach, W., Thermal and photochemical studies with stilbene oxides intramolecular trapping of carbonyl ylide intermediates, *Tetrahedron Lett.*, **1984**, 25, 1137-1140.

[53] Pérez-Prieto, J.; Miranda, M. A., Formation of Six-Memebred (and Larger) Rings, in *Handbook of Synthetic Photochemistry*, Eds. Albini, A.; Fagnoni, M., Wiley-VCH, Weinheim, 2010.

[54] Oelgemöller, M.; Griesbeck, A. G., Photoinduced electron transfer chemistry of phthalimdes: an efficient tool for C-C-bond formation *J. Photochem. Photobiol. C*, **2002**, *3*, 109-127.

[55] Horvat, M.; Mlinarić-Majerski, K.; Basarić, N., Photochemistry of *N*-alkyl and *N*-aryl Substituted Phthalimides: H-Abstractions, Single Elelectron Transfer and Cycloadditions, *Croat. Chem. Acta*, **2010**, *83*, 179–188.

[56] Basarić, N.; Horvat, M.; Franković, O.; Mlinarić-Majerski, K.; Neudörfl, J.; Griesbeck, A. G., Photoinduced hydrogen atom abstraction in *N*-(adamantyl)phthalimides: structure-reactivity study in the solid state, *Tetrahedron*, **2009**, *65*, 1438-1443.

[57] Mandić, L.; Mlinarić-Majerski, K.; Griesbeck, A. G.; Basarić, N., Photodecarboxylation of Adamantane Amino Acids Activated by Phthalimide, *Eur. J. Org. Chem.*, **2016**, 4404-4414.

[58] Horvat, M.; Görner, H., Warzecha, K.-D.; Neudörfl, J.; Griesbeck, A. G., Mlinarić-Majerski, K.; Basarić, N., Photoinitiated Domino Reactions: *N*-(Adamantyl) phthalimides and *N*-(Adamantylalkyl)phthalimides, *J. Org. Chem.*, **2009**, *74*, 8219-8231.

[59] Basarić, N.; Horvat, M.; Mlinarić-Majerski, K.; Zimmermann, E.; Neudörfl, J.; Griesbeck, A. G., Novel 2,4-Methanoadamantane-Benzazepine by Domino Photochemistry of *N*-(1-adamantyl)phthalimide, *Org. Lett.*, **2008**, *10*, 3965-3968.

[60] Griesbeck, A. G.; Henz, A.; Kramer, W.; Lex, J.; Nerowski, F.; Oelgemöller, M.; Peters, K.; Peters, E.-M., Synthesis of Medium- and Large-Ring Compounds Initiated by Photochemical Decarboxylation of ω-Phthalimidoalkanoates, *Helvetica Chim. Acta*, **1997**, *80*, 912-933.

[61] Cho, D. W.; Choi, J. H.; Oh, S. W.; Quan, C., Yoon, U. C.; Wang, R.; Yang, S.; Mariano, P. S., Single Electron Transfer-Promoted Photocyclization Reactions of Linked Acceptor-Polydonor Systems: Effects of Chain Length and Type on the Efficiencies of Macrocyclic Ring-Forming Photoreactions of Tethered r-Silyl Ether Phthalimide Substrates, *J. Am. Chem. Soc.*, **2008**, *130*, 2276-2284.

[62] Yoon, U. C.; Jin, Y. X.; Oh, S. W.; Park, C. H.; Park, J. H.; Champana, C. F.; Cai, X.; Duesler, E. N.; Mariano, P. S., A Synthetic Strategy for the Preparation of Cyclic Peptide Mimetics Based on SET-Promoted Photocyclization Processes, *J. Am. Chem. Soc.*, **2003**, *125*, 10664-10671.

[63] Griesbeck, A. G.; Heinrich, T.; Oelgemöller, M.; Lex, J.; Molis, A., A Photochemical Route for Efficient Cyclopeptide Formation with a Minimum of Protection and Activation Chemistry, *J. Am. Chem. Soc.*, **2002**, *124*, 10972-10973.

[64] Sohora, M.; Vazdar, M.; Sović, I.; Mlinarić-Majerski, K.; Basarić, N., Photocyclization of tetra- and pentapeptides containing adamantylphthalimide and phenylalanines: reaction efficiency and diastereoselectivity, *J. Org. Chem.*, **2018**, *83*, 14905-14922.

[65] Inoue, Y.; Ramamurthy, V., Eds., *Chiral Photochemistry*, Marcel Dekker, New York, 2004.

[66] Inoue Y., Asymmetric photochemical reactions in solution, *Chem. Rev.*, **1992**, *92*, 741-770.

[67] Griesbeck, A. G.; Meierhenrich, U. J., Asymmetric photochemistry and photochirogenesis, *Angew. Chem. Int. Ed.*, **2002**, *41*, 3147-3153.

[68] Brimioulle, R.; Lenhart, D.; Maturi, M. M.; Bach, T., Enantioselective Catalysis of Photochemical Reactions, *Angew. Chem. Int. Ed.*, **2015**, *54*, 3872-3890.

[69] Stephenson, C. R. J.; Yoon, T. P.; MacMillan D. W. C., Eds., *Visible Light Photocatalysis in Organic Chemistry*, Wiley-VCH, Weinheim, 2018.

[70] Hoffmann, N.; Pete, J.-P., Diastereodifferentiating Photoreactions, in *Chiral Photochemistry*, Inoue, Y.; Ramamurthy, V., Eds., Marcel Dekker, New York, 2004.

[71] Adam, W.; Güthlein, M.; Peters, E.M.; Peters. K.; Wirth, T., Chiral-Auxiliary-Induced Diastereoselectivity in the [4+2] Cycloadditions of Optically Active 2,2-Dimethyloxazolidine Derivatives of Sorbic Acid: A Model Study with Singlet Oxygen as the Smallest Dienophile, *J. Am. Chem. Soc.*, **1998**, *120*, 4091-4093.

[72] Adam, W.; Saha-Möller, C.R.; Schambony, S.B., A Highly Diastereoselective Dioxetane Formation by the Hydroxy-Directed [2+2] Cycloaddition of Singlet Oxygen to a Chiral Allylic Alcohol, *J. Am. Chem. Soc.*, **1999**, *121*, 1834-1838.

[73] Adam, W.; Peters, K.; Peters, E.M.; Schambony, S.B., Diastereoselective and Regioselective Singlet-Oxygen Ene Reaction of Oxazolidine-Substituted Alkenes: Control through Hydrogen Bonding Mediated by the Urea Functionality of Chiral Auxiliaries, *J. Am. Chem. Soc.*, **2000**, *122*, 7610-7611.

[74] Grosch, B.; Bach, T., Template-Induced Enantioselective Photochemical Reactions in Solution, in *Chiral Photochemistry*, Inoue, Y.; Ramamurthy, V., Eds., Marcel Dekker, New York, 2004.

[75] Scheffer, J. R., The Solid-State Ionic Chiral Auxiliary Approach to Asymmetric Induction in Photochemical Reactions, in *Chiral Photochemistry*, Inoue, Y.; Ramamurthy, V., Eds., Marcel Dekker, New York, 2004.

[76] Li, X.; Großkopf, J.; Jandl, C.; Bach, T., Enantioselective, Visible Light Mediated Aza Paterno–Büchi Reactions of Quinoxalinones, *Angew. Chem. Int. Ed.*, **2021**, *60*, 2684-2688.

[77] Blum, T. R.; Miller, Z. D.; Bates, D. M.; Guzei, I. A.; Yoon, T. P., Enantioselective photochemistry through Lewis acid-catalyzed triplet energy transfer, *Science*, **2016**, *354*, 1391-1395.

[78] Inoue, Y., Enantiodifferentiating Photosensitized Reactions, in *Chiral Photochemistry*, Inoue, Y.; Ramamurthy, V., Eds., Marcel Dekker, New York, 2004.

[79] Hoshi, N.; Furukawa, Y.; Hagiwara, H.; Uda, H.; Sato, K., Enantiomer differentiation in the photoinduced 1,5-phenyl shift of 3-methyl-3-phenyl-2(3H)-oxepinone utilizing chiral sensitizers, *Chem. Lett.*, **1980**, 47-50.

[80] Demuth, M.; Raghavan, P.R.; Carter, C.; Nakano, K.; Schaffner, K., Photochemical High-yield Preparation of Tricyclo [3.3.0.0^{2,8}]octan-3-ones. Potential Synthons for Polycyclopentanoid Terpenes and Prostacyclin Analogs. Preliminary Communication, *Helv Chim Acta*, **1980**, *63*, 2434-2439.
[81] Ouannes, C.; Beugelmans, R.; Roussi, G., Asymmetric induction during transfer of triplet energy, *J. Am. Chem. Soc.*, **1973**, *95*, 8472-8474.
[82] Inoue, Y.; Yokoyama, T.; Yamasaki, N.; Tai, A., Temperature switching of product chirality upon photosensitized enantiodifferentiating *cis-trans* isomerization of cyclooctene, *J. Am. Chem. Soc.*, **1989**, *111*, 6480-6482.
[83] Skubi, K. L.; Kidd, J. B.; Jung, H.; Guzei, I. A.; Baik, M.-H.; Yoon, T. P., Enantioselective Excited-State Photoreactions Controlled by a Chiral Hydrogen-Bonding Iridium Sensitizer, *J. Am. Chem. Soc.*, **2017**, *139*, 17186-17192.
[84] Yang, C.; Inoue, Y., Supramolecular photochirogenesis, *Chem. Soc. Rev.*, **2014**, *43*, 4123-4143.
[85] Vallavoju, N.; Sivaguru, J., Supramolecular photocatalysis: combining confinement and non-covalent interactions to control light initiated reactions, *Chem. Soc. Rev.*, **2014**, *43*, 4084-4101.
[86] Ramamurthy, V.; Sivaguru, J., Supramolecular Photochemistry as a Potential Synthetic Tool: Photocycloaddition, *Chem. Rev.*, **2016**, *116*, 9914-9993.
[87] Pace, T.; Müller, V.; Li, S.; Lincoln, P.; AndrØasson, J., Enantioselective Cyclization of Photochromic Dithienylethenes Bound to DNA, *Angew. Chem. Int. Ed.*, **2013**, *52*, 4393-4396.
[88] Nishijima, M.; Wada, T.; Mori, T.; Pace, T. C. S.; Bohne, C.; Inoue, Y., Highly Enantiomeric Supramolecular [4+4] Photocyclodimerization of 2-Anthracenecarboxylate Mediated by Human Serum Albumin, *J. Am. Chem. Soc.*, **2007**, *129*, 3478-3479.
[89] Fuentealba, D.; Kato, H.; Nishijima, M.; Fukuhara, G.; Mori, T.; Inoue, Y.; Bohne, C., Explaining the Highly Enantiomeric Photocyclodimerization of 2-Anthracenecarboxylate Bound to Human Serum Albumin Using Time-Resolved Anisotropy Studies, *J. Am. Chem. Soc.*, **2013**, *135*, 203-209.

Chapter 2

Photodynamic Therapy: Fundamentals, Mechanisms and Its Applications against Osteosarcoma

Júlia M. Veronese[1] and Lucas D. Dias[2,*]

[1]Faculty of Medicine, University of Gurupi (UnirG), Gurupi, Tocantins, Brazil
[2]São Carlos Institute of Physics, University of São Paulo, São Carlos, São Paulo, Brazil

Abstract

Photodynamic therapy (PDT) has been presented as a novel and adjuvant option to treat cancer. Osteosarcoma is a primary bone malignancy (mostly occuring in childhood and adolescence). This disease has a poor prognosis (survival rate 50%–80%) and current therapeutic approaches are based on neoadjuvant and adjuvant chemotherapy, surgery, and radiotherapy. PDT is an alternative therapeutic tool which is characterized by application of a photosensitizing chemical entity, oxygen (O_2), and light (of appropriate wavelength) producing highly reactive oxygen oxidative species able to selectively destroy cancer cells. Herein, we describe the fundamental aspects of photodynamic therapy through activation of a photosensitizing molecule, highlighting the mechanism. Moreover, some recent advances (2010-2020) on use of PDT against osteosarcoma are reported and discussed. In this chapter, we do not give a systematic review; instead, we highlight several fundamental concepts and the possibilities and limitations of application of photodynamic therapy for treatment of osteosarcoma.

Keywords: cancer, osteosarcoma, treatment, photodynamic therapy

* Corresponding Author's Email: lucasdanillodias@gmail.com.

In: Photosensitizers and Their Applications
Editors: Davor Margetic and Renjith Thomas
ISBN: 978-1-68507-880-5

Introduction

Cancer is still one of the challenging health problems of society [1, 2]. Alarmingly, the World Health Organization (WHO) expects an increase (ca. 70%) in the number of new cancer cases in the next 20 years [3]. Among them, osteosarcoma has an international incidence of 3.4 million people/year [4]. This sarcoma type is a primary malignancy of bone defined by a predisposition for developing in quickly growing bone [5]. Regarding its prognosis, it shows with relapse-free survival rates up to 50%-80% (young patients) [6] and it predominantly affects adolescents (preferentially during puberty) and young adults [7].

Nowadays, conventional treatment approaches for osteosarcoma are based on a combination of neoadjuvant and adjuvant chemotherapy, and sometimes radiotherapy [8]. Overall, the use of adjuvant chemotherapy for osteosarcoma increased relapse-free survival rates to approximately 50%-80% for young patients [6, 9], but there is a real need to develop alternative therapeutics that show less multiple side effects and high efficiency/specificity.

In this regard, photodynamic therapy (PDT) is a disease site-specific treatment and has been recognized as an attractive tool for osteosarcoma treatment [10]. This light-based therapy approach is characterized by administration of a photosensitizing molecule (PS) followed by light activation of the targeted region that in the presence of oxygen (O_2) is able to generate cytotoxic oxygen species, inducing death of the targeted lesion [11, 12].

Herein, this chapter will outline the urgent need on development of therapeutics for osteosarcoma as well as the fundamentals and evaluation of PDT as an alternative treatment against osteosarcoma. Moreover, this chapter covers the past 10 years on use of PDT protocols (*in vitro* and *in vivo*) against osteosarcoma. The examples reported are presented in chronological order showing photodynamic parameters such as incubation time, fluency, and photosensitizer type.

General aspects of Cancer

Cancer is described as a condition (disorder) of growth and cellular behavior, where mutations of deoxyribonucleic acid (DNA) occur by genetic factors or induced by the environment, resulting in an unbridled cell proliferation [13].

Such changes and growth disorders are inherited by daughter cells in the cell cycle, thus affecting neighboring cells and adjacent tissues/organs, giving rise to tumor formation [14]. The factors that promote and advance carcinogenesis are direct acting agents, such as chemotherapeutic agents and indirect acting agents, for instance UV exposure, infections, and cigarette smoke [15, 16].

Moreover, the accumulation of mutations generate a series of properties that characterize cancer such as self-sufficiency in growth, lack of response to growth inhibitory signals, induction of apoptosis in normal cells, high replicative potential, angiogenesis, reprogramming of metabolic pathways, and ability to escape from immune response [17]. According to The Global Burden of Disease, cancer is considered as the second cause of death worldwide after cardiovascular diseases [18]. In the U.S., an estimated annual cost of cancer treatment was approximately US $ 94.4 billion in 2015, corresponding to US $ 191,900 per cancer death [19].

The development of cancer can be divided into three stages, as described in Figure 1: i) initiation: presence of initiating agents (external or internal) causing DNA damage and activation of proto-oncogenes; ii) promotion: non-mutated cells suffer the effects of oncopromotors, which are slowly and gradually transformed into a cancer cell; iii) progression: irreversible process, uncontrolled multiplication occurs, primary tumor formation, and the first clinical manifestations appear [20].

Subsequent to the development of a primary tumor, these cells acquire the ability to colonize sites far from their initial location, forming a secondary tumor, in the process called metastasis [17]. For this displacement to occur, several genetic modifications in the cell pattern are required, e.g., insertions, substitutions, exclusion of a segment of DNA. The metastasis process is divided in the following steps: i) cells of the primary tumor acquire invasive properties, which is known as invasion of the extracellular matrix; ii) vascular dissemination with the initial of angiogenesis process, mechanism that supports cancer cells; iii) insertion of tumor cells and colonization in adjacent or distant tissues/organs followed by trapping in capillaries of distant organs adhering to the cell endothermic, proliferation within the organ and completion of the metastatic process [20].

Regarding cancer classification, it is classified according to the origin of the tissue, neoplastic cells that make up its parenchyma, as well as the position in the body where it develops, and divided into seven main groups (carcinoma, sarcoma, leukemia, lymphoma, myeloma, adenocarcinoma, and blastoma) [21] (Table 1).

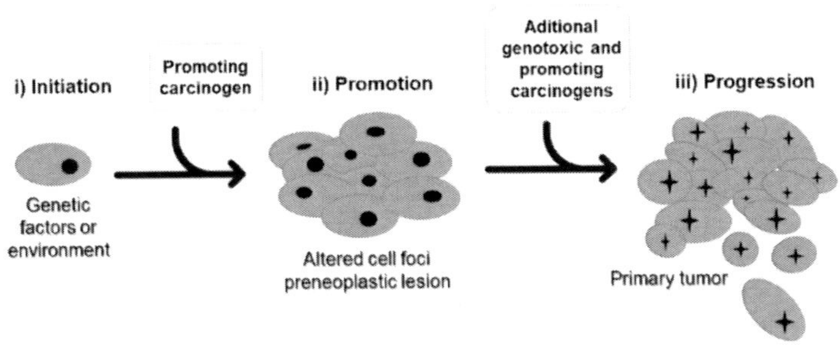

Figure 1. Carcinogenesis stages (initiation, promotion, and progression).

Table 1. Cancer classification according to to the origin of the tissue, neoplastic cells that make up its parenchyma, as well as the position in the body where it develops [22]

Type of cancer	Origin	Localizator
Carcinoma	Skin epithelial cells, GI tract, internal organs, squamous cells	Skin, lungs, pancreas
Sarcoma	Connective tissue derivatives	Muscles, adipose tissue, bones or blood vessels
Leukemia	Alter lymphocytes; do not form solid tumors	Blood
Lymphoma	Lymphoid tissue, T cells	Lymph nodes or ganglia
Myeloma	Plasmocyte (white cells)	Bone marrow
Adenocarcinoma	Squamous cells	Lung (women and non-smokers)
Blastoma	Epithelial and mesenchymal cells, blastocysts	It affects many organs, tissues and systems in children.

In this book chapter, we highlighted the sarcoma-type, specifically osteosarcoma, which is a primary malignant bone cancer, which occurs mainly in childhood and adolescence. An overview about osteosarcoma, including fundamentals, risk factors, diagnosis, treatment, followed by the use of a light-based therapy, called photodynamic therapy, and its potential application and limitations for osteosarcoma treatment was described and explained.

Osteosarcoma

Osteosarcoma is a high-grade malignant mesenchymal tumor defined by production of bone matrix and deposition of osteoid tissue [23]. It is a primary malignant bone tumor with a manifestation peak in children and young people (around 75% cases) [24, 25]. Its etiology is still unknown, but some cases are reported to be associated with mutations in tumor suppressor genes (regulation of the cell cycle) [26]. In addition, there are some factors that make children and adolescents more predisposed to developing this sarcoma-type for instance: age (10-30 years), incidence is higher among males (male:female ratio, 1.43:1), previous radiation therapy, and inherited genetic conditions e.g., Paget's, family syndromes (Li-Fraumeni, Rothmund-Thompson, retinoblastia), Bloom's syndrome, and Werner's syndrome [27].

Osteosarcoma is classified in primary and secondary types [28]. Among primary types, the WHO sub-classified according to its histological and biological parameters for example: 1) central - subdivided in: 1.1) conventional-intramedullary, most common and represents 80% of the cases, with subtypes in osteoblastic, chondroblastic, and fibroblastic, depending on the predominant cells; 1.2) telangiectatic osteosarcoma; 1.3) low-grade osteosarcoma; and 1.4) small cell osteosarcoma; and 2) surface - represents 10% of cases and are subdivided in: 2.1) parosteal; 2.2) periosteal; and 2.3) high grade surface [24, 29].

Regarding its clinical presentation, patients with osteosarcoma usually experience localized pain and swelling in the area affected by the tumor, which may worsen during sleep [30]. The findings of the physical examination reveal a palpable mass, limited movement and increased local volume (without infection or injury) [31].

Due to its poor prognosis and potential risk of metastasis, early diagnosis is needed to determine the staging and the therapeutic plan. Diagnostic imaging is also performed by radiography of the affected extremities showing destruction of the cortex, elevation of the periosteum and extra-bony extension with ossification of the adjacent soft tissues [32]. Magnetic resonance imaging (MRI) has been used for definition of the bone marrow, analysis the presence/absence and size of soft tissue, the relationship between the structures and evaluation of the intra and extramedullary extension [30]. Positron emission tomography (PET) monitors metabolic activity and identifies tumors with high-grade malignancies, as well as metastasis in lymph nodes and organs of the body, with better detection in soft tissues [33]. Computed tomography (CT) of the thorax is a sensitive non-invasive diagnostic modality used for the detection of metastatic disease within the lungs [34]. Moreover, laboratory tests that require high titers such as alkaline phosphatase and lactate dehydrogenase can indicate the presence of an advanced tumor [35].

It is important to emphasize that when a diagnosis of malignancy is suspected, tissue biopsy is performed for analysis/confirmation, and it assists in conducting the treatment, classifies the predominant tissue, and facilitates determining the stage of the disease. Age, location, metastasis and local recurrence are also taken into account to determine the prognosis; survival rates are improved after the introduction of chemotherapy and is designated by stages at the time of diagnosis [36].

Until the 1970s, osteosarcoma was treated by amputations, many patients worsened with the resurgence of the disease and lung metastases [37, 38]. The incorporation of neoadjuvant chemotherapy improved the long-term standard of care in patients with osteosarcoma who needed definitive surgical resection, bringing a good response in the histological assessment for the prognosis and opportunity for limb salvage [39, 40].

The discovery of chemotherapeutic agents helped in the inhibition of pulmonary metastases, present in at least 80% of patients at the time of diagnosis. The standard drugs widely used in the treatment of osteosarcoma are doxorubicin, high-dose methotrexate, cisplatin and ifosfamide [41-43]. Unfortunately, chemotherapy alone is not able to eradicate the tumor in high-grade localized osteosarcoma. In this case, surgical treatment has been associated with pre- and postoperative chemotherapy, since it eliminates unmet micrometastases in surgical resection [44].

Overall, a survival rate up to 60-70% (in the last 20 years) is observed. However, the severe side effects and the problem of absence of response to chemotherapy is still a challenge [45]. For these reasons, the development of non-toxic and active drugs against osteosarcoma, as well as the development of alternative techniques to combat it, is still a major challenge.

Photodynamic Therapy: Concepts and Mechanism

The application of light and a photosensitizing compounds for medical purposes has its origin in India, Egypt, and Greece [46]. As a remarkable example, the use of natural products named psoralens under UV-A irradiation was evaluated and found to be effective against psoriasis [47]. Since that, psoralens and ultraviolet A irradiation have been largely used in treatment of different dermatological diseases such as vitiligo, parapsoriasis, atopic eczema, and others [46].

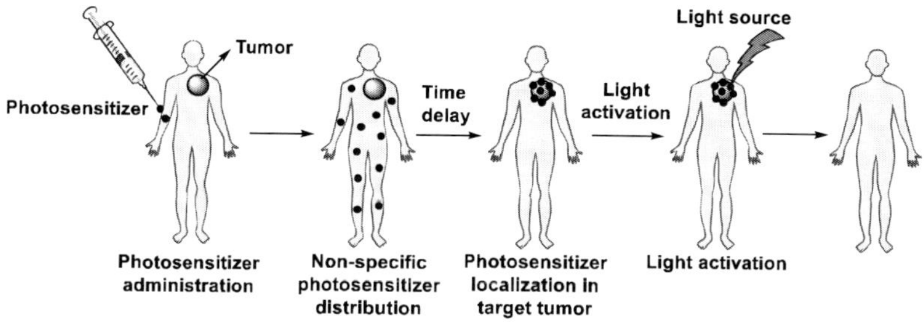

Figure 2. General clinical procedure of the photodynamic therapy.

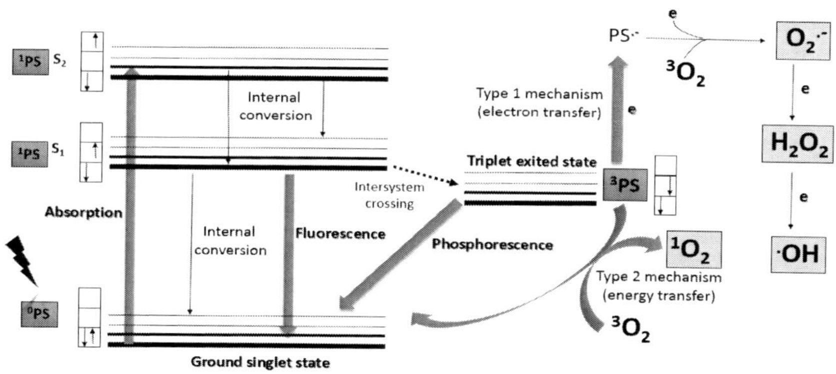

Figure 3. Simplified Jablonski´s diagram.

Along the years, several discoveries and advances on use of light and photosensitizers were made [48]. In 1904, von Tappenier and Jodlbauer reported that O_2 is required for a photosensitization process. Some years later, von Tappenier first named this process as photodynamic therapy (PDT) [48].

The general mechanism of PDT is characterized by administration and photoactivation of a photosensitizer followed by photophysical and photochemical processes yielding reactive oxidative species (ROS) i.e., superoxide radical anion ($O_2^{\bullet -}$), hydrogen peroxide (H_2O_2), hydroxyl radicals ($HO^{\bullet}$) and singlet oxygen ($^1O_2$) able to react with biomolecular targets present in cancer cells [49] (Figure 2).

The photophysical PDT mechanism is described and explained by Jablonski's diagram [50] (Figure 3). Initially, the PS in its singlet ground state (^0PS) absolves light of appropriate wavelength leading the excitation of one electron into an orbital with higher energy (S_1). Then, it can convert its energy by four pathways: i) emission of light (namely fluorescence process);

ii) internal conversion (producing heat); iii) absorbing more quantum energy into a second singlet excited state; and/or iv) intersystem crossing (IC) process, which is defined by formation of a triplet excited-state (^3PS) (more stable) showing parallel spins.

The ^3PS may decay back to the ^0PS by phosphorescent process (emitting of photon] or transfer electron to O_2 forming ROS e.g., $O_2^{\bullet -}$, H_2O_2, and $HO^{\bullet}$. On the other hand, the ^3PS can also undergo the type 2 mechanism which transfers its energy to $^3O_2$ yielding $^1O_2$, a powerful oxidative species in a photodynamic protocol. These ROS and $^1O_2$ generated can damage biomolecular targets on cancer cells *via* an oxidative process [51].

Modes of Cancer Cells Death in a Photodynamic Protocol

According to literature and the examples reported in this chapter, there is no specific pathway that leads to cancer cell death (e.g., osteosarcoma) after a PDT protocol. The application of PDT may display different cell death pathways against a tumor such as: i) direct tumor cell damage *via* apoptosis, necrosis and/or autophagy; ii) damage to the vascular system; and iii) inducing immune and inflammatory responses [52].

As described in the Figure 3, when a PS is photoactivated by an appropriate light source yielding highly reactive ROS ($^1O_2$, $O_2^{\bullet -}$, H_2O_2, $HO^{\bullet}$) able to oxidize target biomolecules e.g., peptides, lipids, proteins, and amino acids. This photodynamic approach results in signaling cascades that are simultaneously activated in cancer cells [53]. Overall, the mode of cell death depends on several factors: the type of cell, photosensitizer concentration, sub-cellular localization of photosensitizer, and the identity and light dose. Moreover, the photosensitizer's physical and chemical properties are fundamental for determination of its subcellular localization (mitochondria, lysosome, Golgi apparatus, endoplasmic reticulum (ER), and nuclear membrane) and consequently the pathway of cell death [54]. A correlation between cell death pathways generated by PDT, subcellular localization, and the key players is described in Table 2.

Apoptosis is a cell-death mechanism (multi-pathway) which is controlled by extracellular and intracellular signals. It is characterized by activation of receptors and/or mitochondrial release of cytochrome c [56] followed by stimulation of caspase cascades (executioner caspases) such as caspase-3, -6 and -7 [57]. These activated caspases can promote biochemical reactions in cellular substrates and consequently cell death.

Normally, in a photodynamic protocol, when the photosensitizer is localized on mitochondria and/or endoplasmic reticulum the cell death pathway occurs by an apoptosis process [58]. But, the initiation of an apoptotic process differs depending on the cell line evaluated, the molecular structure and physical/chemical properties of photosensitizers and subsequently, its subcellular location, the overall dose, and other photodynamic protocol conditions.

Autophagy is an intracellular degradation pathway which targets proteins, organelles, and invading microbes are degraded by enzymatic action (lysosomes) of three different types in mammal cell such as: i) micro-autophagy; ii) macroautophagy; and iii) chaperone-mediated autophagy. This programmed process possesses double functions depending on the specific circumstances, since it can promote either cell survival (recycling cells components) or cell death [59].

Necrosis is a non-programmed and external cellular injury mechanism [60]. In this cell death mechanism cell and organelle volume swelling occurs, cytoplasm vacuolization, disruption of the plasma membrane, and consequently intracellular content loss/cellular fragmentation resulting in an inflammatory reaction [61]. In a PDT protocol, necrosis pathway is a non-reversible process induced by external cellular damage via action of an overdose of oxidative cytotoxic species. Moreover, Plaetzer reported that when lower light dose is used it leads to more apoptosis, while higher light doses lead to proportionately more cell deaths *via* necrosis [62].

Table 2. Cell death mechanisms and subcellular localization in a photodynamic protocol

Mechanism	PS localization (organelles)	Characteristics
Apoptosis	Mitochrondria: Bcl-2 damage and cytochrome c release Cytoplasm: NF_kB damage	Cell damage (direct)
Autophagy	Endoplasmatic reticulum: mTOR activation and beclin 1	Cell damage (direct)
Necrosis	Cell membrane desintegration	Cell damage (direct)
Apotosis, autophagy, and necrosis	Depletion of O_2 and nutrients	Vascular shutdown

This table is cited with permission (open access) from Mroz et al. [55) published in 2011.

Overall, recent information points out that PDT can selectively kill cancer cells by the efficient induction of apoptotic or non-apoptotic cell death mechanisms. Since the cell death mechanism can be modulated by the photosensitizer sub-cellular localization, the development and design/synthesis of photosensitizers and drug delivery systems which have a specific sub-cellular localization are considered good strategies against aggressive types of cancer such as osteosarcoma.

Concerning the need for development of effective/alternative therapeutic tools against osteosarcoma, we describe some important examples (from 2010-2020) concerning the application of PDT as an alternative treatment against osteosarcoma. We will not produce a systematic review; instead we highlight the most important *in vitro*, *in vivo* studies, clinic aspects and the parameters used for photodynamic protocols.

Recent Advances of Photodynamic Therapy against Osteosarcoma (2010-2020)

In Vitro Studies (2010-2020)

In order to present the efficiency and limitations of photodynamic therapy protocols against osteosarcoma cell lines, different parameters reported for *in vitro* studies (2010-2020) such as type of photosensitizer (PS), concentration, light dose, cell line, and results are presented in Table 3.

In 2015, Tsai and co-authors [63] studied the effect of low-level light therapy on increase of cellular ATP content. It was based on the fact that some works have demonstrated that high ATP level is required for induction/execution of apoptosis pathway after a photodynamic protocol (Table 3, entry 1). For this purpose, the authors used human osteosarcoma cell line (MG-63) irradiated with near infrared (NIR) laser (at 810 nm, 1.5 J/cm^2) followed by a photodynamic procedure using mono-L-aspartyl chlorin(e_6) as photosensitizer (0-10 μM), a laser at 652 nm (1.5-6.0 J/cm^2), incubation time of 2 h. These experiments evaluated if an increased survival after PDT *via* a cytoprotective pathway was observed, or alternatively if the cells would show enhanced photosensitizer cell uptake and efficacy of PDT. The authors concluded that PDT combined with a low-level light therapy protocol led to higher cell death (up to 20% cell viability) and increased intracellular ROS when it is compared to PDT alone *via* an increase of mono-L-aspartyl chlorin(e_6) uptake.

In 2016, Aluigi and co-authors [64] described the synthesis of a chlorin e_6-conjugated keratin nanoparticles and their characterization by scanning electron microscopy (SEM), zeta potential, atomic force microscopy (AFM), UV-Vis spectroscopy, and measured ROS production (Table 3, entry 2). These chlorin e_6-conjugated keratin nanoparticles (0.5, 2.5 and 5 μg/mL of chlorin e_6) were evaluated as photosensitizers in osteosarcoma cell line (U2OS) using a white halogen lamp (at 25.5 mW/cm^2 for 10 min) and 24 h for incubation time. The authors reported that chlorin e_6-conjugated keratin nanoparticles showed a good ability to cross the membrane and accumulate in the cytoplasm. Moreover, a higher cellular uptake via analysis of intracellular signal when the photosensitizer immobilized on keratin was observed. This drug delivery system resulted in an increase of photocytotoxic effect (up to 90% cell death) in osteosarcoma cell line (U2OS).

In 2016 [65], a study was performed aiming to understand the cell death pathways (apoptosis or autophagy) when pyropheophorbide-a methyl ester (a derivative from chlorophyll) was applied as a photosensitizing molecule in MG-63 cells (Table 3, entry 3). The authors used pyropheophorbide-a methyl ester in different concentrations (0, 0.25, 0.5, 0.75, and 1.5 μmol/L), an incubation time of 20 h, and different values of light dose (1.2, 2.4, 4.8, and 9.6 J/cm^2). After optimization studies, the authors chose a photosensitizer concentration of 0.75 μmol/L and a light dose of 4.8 J/cm^2 for the cell death mechanism evaluation. Apoptosis mechanism was studied using Hoechst 33258 as a dye to specifically stain the cell nucleus and different morphological changes regarding apoptosis were observed by using a fluorescence microscope. Evaluation by western blotting also revealed an increased expression levels of cleaved caspase-3 at 3, 6, and 12 h after treatment using light and the quantification of apoptosis level was carried out by annexin V-PI staining and flow cytometry. For autophagy evaluation, the authors used MDC (autophagy marker) staining and transmission electron microscope (TEM) to detect the presence of autophagic vacuoles. In conclusion, the authors described that pyropheophorbide-a methyl ester photosensitizer induces apoptosis and autophagy mechanism in MG-63 cells.

Also in 2016, Tu and co-authors [66] investigated the efficiency and cell death mechanism of aloe-emodin as photosensitizer, an anthraquinone derivative compound that can be extracted from traditional Chinese medicine plants in MG-63 cell line (Table 3, entry 4). In this study, the authors optimized the photodynamic protocol using the following conditions: aloe-emodin (0, 5, 10, 20 and 50 μM), incubation time of 6 h and various light doses (0, 2.4, 4.8 and 6.4 J/cm^2). The optimized condition (aloe-emodin at 10 μM and a light dose of 4.8 J/cm^2) was used for cell death mechanism studies.

Table 3. *In vitro* photodynamic therapy studies against osteosarcoma (2010-2020)

Entry	PS	Concentration	Light dose	Cell line	Results	Ref
1.	mono-L-aspartyl chlorin e_6	0 - 10 μM	1.5-6.0 J/cm^2	MG-63	Up to 20% (cell viability)	[63]
2.	chlorin e_6-conjugated keratin nanoparticles	0.5 - 5 μg/mL of chlorin e_6	15.3 J/cm^2	U2OS	Up to 90% (cell death)	[64]
3.	pyropheophorbide-a methyl ester	0 - 1.5 μmol/L	1.2 - 9.6 J/cm^2	MG-63	Up to 1% (cell viability)	[65]
4.	aloe-emodin	0 - 50 μM	0 - 6.4 J/cm^2	MG-63	Up to 40% (cell viability)	[66]
5.	Aminolevulinic acid (ALA)	0 - 10 mM	0.3 - 6 J/cm^2	MG-63	Up to 70% (cell viability)	[68]
6.	PPIX in silica nanoparticles	0.061 - 0.974 mM	3.2 J/cm^2	Osteosarcoma cells	Up to 100% (cell death)	[69]
7.	Zinc phthalocyanine	0 - 80 μM	1.8 KJ/cm^2	MNNG/Hos, MG-63, U2os, Saos-2	Up to 0% (cell viability)	[70]
8.	Albumin-zinc phthalocyanine	0-12 μg/mL	1.8 KJ/cm^2	MNNG/Hos, MG-63 and K7M2	Up to 5% (cell viability)	[72]
9.	Eosin Y and porphyrin derivative	1 - 10 mM	-	MG-63	-	[73]

The authors observed that aloe-emodin as photosensitizer acts in a concentration/density-dependent manner and can induced autophagy apoptosis cell death mechanisms through activation of the ROS-JNK signaling pathway.

Li and co-authors [67] also evaluated cell death mechanism in a photodynamic protocol on the MG-63 cells when aloe-emodin is used as photosensitizer. The authors corroborated the previously reported results which apoptosis pathway is the primary mode, in concentration dependence of aloe-emodin. Moreover, subcellular localization of aloe-emodin was determined by confocal microscopy showing that mitochondria and endoplasmic reticulum serve as the subcellular targets.

The efficiency of aminolevulinic acid (precursor of protoporphyrin IX) as photosensitizer in MG-63 cells line was evaluated by White and co-authors [68] (Table 3, entry 5). Aiming to provide optimal parameters for *in vitro* ALA-PDT, the authors applied aminolevulinic acid in different concentrations (0-10 mM), incubation time of 4 or 24 h, and a range of light dose (0.3-6 J/cm^2) in MG-63 cell line and analyzed by MTT assay. The authors found that ALA-mediated PDT was most effective at 3 and 6 J/cm^2 and 24 h of incubation (maximal protoporphyrin IX formation). These data found are useful in determining translational and preclinical parameters for ALA-PDT of osteosarcoma.

In 2018, Makhadmeh and Aziz [69] promoted the encapsulation of protoporphyrin IX photosensitizer in silica nanoparticles by reverse-micellar method (Table 3, entry 6). This drug delivery system was used to avoid rapid clearance of the protoporphyrin IX photosensitizer by the reticulo-endothelial system. The authors compared the efficiency of encapsulated protoporphyrin IX and free protoporphyrin IX (0.974, 0.478, 0.244, 0.122 and 0.061 mM as final concentrations) in photodynamic protocols in osteosarcoma cells by using a light intensity of 0.89 mW/cm^2 for 60 min. In general, the authors concluded that encapsulation by silica should greatly increase the ability and the efficiency of protoporphyrin IX as photosensitizer against osteosarcoma cells (up to 100% cell death).

The development of a drug delivery system based on use of poly(ethylene glycol)-poly[2-(methylacryloyl)ethyl nicotinate] block polymer which is coordinated with a zinc phthalocyanine by π–π interactions was reported by Yu and co-authors [70] (Table 3, entry 7). This drug delivery system was applied in a photodynamic protocol in osteosarcoma (MNNG/Hos) cells using the following conditions: red light and a light dose of 1.8 KJ/cm^2. The authors described that the drug delivery prepared with the zinc phthalocyanine possesses a photocytotoxicity for osteosarcoma cells via apoptosis in a concentration-dependent manner, and its IC_{50} was 15- to 100-fold lower than free-zinc phthalocyanine and 15- to 25-fold lower than cisplatin (conventional drug used for osteosarcoma treatment). In this study, the drug delivery system releases the free zinc phthalocyanine photosensitizer which is then photoactived producing ROS inducing mitochondria and DNA damage.

Regarding the activation of immune system by PDT procedure, a research group from China (Shanghai Tenth People's Hospital) [71] observed that necrosis of osteosarcoma cells line (LM8) inhibited the function of dendritic cells. But, PDT showed the ability to restore the function of dendritic cells through upregulating heat shock protein 70.

In 2018, Yu and co-authors [72] synthetized and characterized an albumin-zinc phthalocyanine and reported its application as a photosensitizer in osteosarcoma cells (MNNG/HOS, MG-63 and K7M2) (Table 3, entry 8). Moreover, the authors evaluated cell death mechanism (apoptosis and autophagy) induced by albumin-zinc phthalocyanine and detect the effect of a combined treatment using surgery and PDT. After an optimization study, the authors reported an IC_{50} of ca. 1.04 μg/mL, 0.32 μg/mL, and 0.08 μg/mL for MNNG/HOS cells, MG-63 cells and in K7M2 cells, respectively. Also, their results show that the combined use of PDT and surgery has a promising anti-tumor effect against osteosarcoma with reduced injury of superficial tissues. Moreover, another achievement in this work is based on use of PDT with an autophagy inhibitor (3-MA); this efficiency occurred via suppression of the PD-L1 immune checkpoint, activating CD8+ T cells and inhibiting tumors with distant metastasis. This work shows a synergistic therapy on treating osteosarcoma, which may be its future application in clinical studies.

Also in 2018, Dobos and co-authors [73] evaluated two-photon excited photodynamic therapy procedure for osteosarcoma treatment using 3D osteosarcoma model (MG-63) together with adipose tissue-derived stem cells (Table 3, entry 9). This two-photon methodology promotes the excited state photosensitizer by absorbing two photons of a longer wavelength simultaneously during irradiation using two-photon photoinitiator P2CK, eosin Y, and a porphyrin derivative as photosensitizers. The authors determined the optimal working window for each photosensitizer by z-scan measurements. One of the main advantages of two-photon excited photodynamic therapy is that the cellular damage can be reduced to the highly localized irradiated region. Moreover, the authors described that contrary to the use of cisplatin, no side effects to non-irradiated regions were observed [73].

In Vivo Studies (2010-2020)

In order to present the efficiency and limitations of photodynamic therapy protocols against osteosarcoma, different parameters reported for *in vivo* studies (2010-2020) such as type of photosensitizer (PS), concentration, light dose, cell line, and results are presented in Table 4.

Sun and co-authors [74] evaluated the efficiency of a photodynamic protocol using hiporfin as photosensitizer in osteosarcoma both *in vitro* and *in vivo* (Table 4, entry 1). For *in vivo* studies, hiporfin 10 mg/kg b.w., light dose of 150 J/cm^2 at 630 nm, incubation time of 4 h were used and showed a significant decrease in tumor volume and weight (measured every 4 days) in all tumor mouse models (DLM-8 subcutaneous tumor model and subcutaneous and tibia tumor model) studied. Moreover, the authors elucidated the cell death pathways (apoptosis, necrosis, and autophagy) by western blotting after photodynamic procedure using hiporfin as photosensitizer. The authors suggested that cell death mechanism in DLM-8, 143B and HOS cells proceeds through the activation of apoptosis. Furthermore, an increased expression of RIP1, a modulator of necroptosis, after PDT in DLM-8 cells, suggestive of necroptosis is involved in the cell death of DLM-8 cells. Whereas, autophagy could be involved in a protective mechanism. Overall, these results point out that this light-based protocol is a potential future clinical tool for the treatment of osteosarcoma.

In order to evaluate effect of a bacteriochlorin derivative (cis-2, 3, 12, 13-tetracarboxymethyl-5, 10, 15, 20-tetraphenyl bacteriochlorin - TCTB) as photosensitizer against mice bearing osteosarcoma S180 tumors, a photodynamic therapy protocol with the following parameters was applied: 10 mg/kg b.w. bacteriochlorin derivative, light dose of 120 J/cm^2 at 730 nm, and incubation time of 5 min (Table 4, entry 2). The authors observed that tumors became necrotic, implying effective photodynamic protocol compared to control group (saline solution), dried over the course of the next 3 days [75].

In 2017, Meier and co-authors [76] reported the cytotoxic effect of 5,10,15,20-tetrakis(-meta-hydroxyphenyl) chlorine (mTHPC, Foscan®) and a third-generation photosensitizer namely liposomal mTHPC formulation (Foslip®) in the human 143B and a mouse K7M2- derived osteosaroma cell line (K7M2L2) *in vitro* (Table 3, entry 3). According to authors, the studied models allowed to evaluate the tumor- and metastasis-suppressive potential of both photosensitizers described in the absence and in the presence of a

Table 4. *In vivo* photodynamic therapy studies against osteosarcoma (2010-2020)

Entry	PS	Concentration	Light dose	Animal model	Results	Ref
1.	hiporfin	10 mg/kg b.w.	150 J/cm^2	DLM-8 (C3H mice) and tibia (BALB/C Nude mice)	tumor eradication for subcutaneous and tibia model	[74]
2.	bacteriochlorin derivative - TCTB	10 mg/kg b.w.	120 J/cm^2	S180 (Kunming mice)	tumors became necrotic	[75]
3.	Foscan® and Foslip®	0.15 mg/kg b.w.	20 J/cm^2	143B/lacZ cells (SCID and BALB/c mice)	Tumor growth was inhibited (Foslip most effective)	[76]
4.	5,15-bis(2-bromo-5-hydroxyphenyl) porphyrin	-	180 J/cm^2	NNG-HOS cells (Balb/c nude mice)	Up to complete reduction of tumor	[77]
5.	BODIPY derivative	100 mg mL^{-1} (100 μL) per mice	240 J/cm^2	143B cells (BALB/c nude mice)	anti-tumor activity without any toxicity	[78]

functional immune response. For experimental studies, Foscan or Foslip (mTHPC 0.15 mg/kg b.w.) was administered by intravenous injection (4–6 min), light dose of 20 J/cm^2 at 652 nm, and incubation time of 24 h was used. The results obtained in this study show that mTHPC-based PDT-mediated tumor suppressive effects in two clinically intratibial animal mouse models of osteosarcoma, demonstrating PDT as a potential tool to inhibit lung metastatic growth.

In 2017, de Miguel and co-authors [77] evaluated the cytotoxic effect of 5,15-bis(2-bromo-5-hydroxyphenyl) porphyrin as photosensitizer in osteosarcoma animal models (cranial and vertebral osteosarcoma) and analyzed by scintigraphy with 99mTc-MIBI and radiography (Table 4, entry 4). To evaluate the results, tumor growth, necrotic area, osteoid matrix area, and inflammatory infiltration were determined. The authors used a light dose of 180 J/cm^2 and an incubation time of 72 h and observed a decrease in tumor size in vertebral and cranial osteosarcoma models, compared with the untreated mice models. These results suggest that this protocol might represent a potential treatment for osteosarcoma and bone metastasis.

In 2019, Zhu and co-authors [78] synthetized a NIR triphenylamine grafted BODIPY derivative and its nanoparticles with PEG-2000 by nanoprecipitation method and both were characterized by TEM, dynamic light scattering (DLS), mass spectrometry, and nuclear magnetic resonance (Table 4, entry 5). These photosensitizers showed high capacity to form ROS and present high photothermal conversion efficiency. BODIPY derivative with PEG-2000 at different concentrations were applied in *in vitro* studies using different cell lines and a light source at 808 nm, 0.5 W/cm^2 for 8 minutes presenting an IC50 values of 4.38, 5.729, 5.862 and 3.996 mg/mL for HOS, U-2OS, MG-63 and 143B cells, respectively. Moreover, *in vivo* experiments indicate that BODIPY derivative with PEG-2000 also suppress tumor growth in nude mice. The authors concluded that NIR triphenylamine grafted BODIPY showed a satisfactory therapeutic effect for both *in vitro* and *in vivo* studies, indicating great potential for treatment of osteosarcoma.

Clinical Evaluation

Some clinical trials studies using photodynamic therapy against osteosarcoma have been performed in Japan. In 2005, researchers evaluated a surgical modality based on photodynamic protocol and acridine orange in 2 patients (osteosarcoma on radius and humerus, stage III and IIB, respectively) [79].

These patients also received a pre- and post-operative chemotherapy with cisplatin and Adriamycin.

In 2010, researchers from Japan performed a clinical proof-of-principle using photodynamic therapy against sarcoma arising around the forearm and hand. In this clinical study, 18 patients (including 12 men, 6 women, with a mean age of 45 years) with primary musculoskeletal sarcoma (forearm) of which 8 patients were treated by a photodynamic protocol using acridine orange and 10 patients were submitted to conventional wide resection followed by various limb reconstruction surgery. However, for the patients bearing osteosarcoma this study did not show results details and conclusions when photodynamic therapy was applied [80].

Conclusion

The treatment for osteosarcoma involves surgery (amputation or reconstruction with auto- or allograft) and use of neoadjuvant chemotherapy followed by adjuvant chemotherapy which show several side effects. In this regard, the development of alternative, efficient, and non-toxic treatments for osteosarcoma is required. PDT has been presented as an efficient therapeutic tool with low dark toxicity and good accumulation in tumor tissue in several *in vitro* and *in vivo* studies. This technique exhibited cell death mechanism by necrosis, apoptosis, and/or autophagy, depending on photosensitizer molecular structure, drug delivery system used, photosensitizer cell localization, and photodynamic protocol parameters applied. Moreover, PDT showed a synergic effect via activation of immunological system (e.g., dendritic cells) and regulation of inflammatory response [72]. We expect that PDT in combination with chemotherapy drugs could represent an alternative treatment solution for inoperable lesions as well as a post-surgery option after osteosarcoma resection to eradicate residual tumor cells. However, more studies are mandatory on the mechanisms and applications of PDT in osteosarcoma to translate these results to clinical trials.

References

[1] Evan GI, Vousden KH. Proliferation, cell cycle and apoptosis in cancer. *Nature*, 2001; 411(6835): 342-348.

[2] Visvader JE. Cells of origin in cancer. *Nature*, 2011; 469(7330): 314-322.

[3] Arem H, Loftfield E. Cancer Epidemiology: A Survey of Modifiable Risk Factors for Prevention and Survivorship. *Am J Lifestyle Med*, 2018; 12(3): 200-210.

[4] Mirabello L, Troisi RJ, Savage SA. International osteosarcoma incidence patterns in children and adolescents, middle ages and elderly persons. *Int J Cancer*, 2009; 125(1): 229-234.

[5] Sadykova LR, Ntekim AI, Muyangwa-Semenova M, Rutland CS, Jeyapalan JN, Blatt N, et al. Epidemiology and Risk Factors of Osteosarcoma. *Cancer Invest*, 2020; 38(5): 259-269.

[6] Bielack SS, Kempf-Bielack B, Delling G, Exner GU, Flege S, Helmke K, et al. Prognostic factors in high-grade osteosarcoma of the extremities or trunk: An analysis of 1,702 patients treated on neoadjuvant cooperative osteosarcoma study group protocols. *J Clin Oncol*, 2002; 20(3): 776-790.

[7] de Azevedo JWV, Fernandes T, Fernandes JV, de Azevedo JCV, Lanza DCF, Bezerra CM, et al. Biology and pathogenesis of human osteosarcoma. *Oncol Lett*, 2020; 19(2): 1099-1116.

[8] Hong AM, Millington S, Ahern V, McCowage G, Boyle R, Tattersall M, et al. Limb preservation surgery with extracorporeal irradiation in the management of malignant bone tumor: the oncological outcomes of 101 patients. *Ann Oncol*, 2013; 24(10): 2676-2680.

[9] Eilber F, Giuliano A, Eckardt J, Patterson K, Moseley S, Goodnight J. Adjuvant chemotherapy for osteosarcoma - a randomized prospective trial. *J Clin Oncol*, 1987; 5(1): 21-26.

[10] Yu W, Zhu J, Wang YT, Wang JJ, Fang WJ, Xia KS, et al. A review and outlook in the treatment of osteosarcoma and other deep tumors with photodynamic therapy: from basic to deep. *Oncotarget*, 2017; 8(24): 39833-39848.

[11] Tampa M, Sarbu MI, Matei C, Mitran CI, Mitran MI, Caruntu C, et al. Photodynamic therapy: A hot topic in dermato-oncology. *Oncol Lett*, 2019; 17(5): 4085-4093.

[12] Dias LD, Bagnato VS. An update on clinical photodynamic therapy for fighting respiratory tract infections: a promising tool against COVID-19 and its co-infections. *Laser Phys Let.*, 2020; 17(8): 1-9.

[13] Aguirre-Ghiso JA. Models, mechanisms and clinical evidence for cancer dormancy. *Nat Rev Cancer*, 2007; 7(11): 834-846.

[14] McDonald ER, El-Deiry WS. Cell cycle control as a basis for cancer drug development (review). *Int J Oncol*, 2000; 16(5): 871-886.

[15] Vineis P, Schatzkin A, Potter JD. Models of carcinogenesis: an overview. *Carcinogenesis*, 2010; 31(10): 1703-1709.

[16] Lacina L, Coma M, Dvorankova B, Kodet O, Melegova N, Gal P, et al. Evolution of Cancer Progression in the Context of Darwinism. *Anticancer Res*, 2019; 39(1): 1-16.

[17] Fidler IJ. Timeline - The pathogenesis of cancer metastasis: the 'seed and soil' hypothesis revisited. *Nat Rev Cancer*, 2003; 3(6): 453-458.

[18] Nagai H, Kim YH. Cancer prevention from the perspective of global cancer burden patterns. *J Thorac Dis*, 2017; 9(3): 448-451.

[19] Islami F, Miller KD, Siegel RL, Zheng ZY, Zhao JX, Han XS, et al. National and State Estimates of Lost Earnings From Cancer Deaths in the United States. *JAMA Oncol*, 2019; 5(9): E1-E8.

[20] Geiger TR, Peeper DS. Metastasis mechanisms. *Biochim Biophys Acta-Rev Cancer*, 2009; 1796(2): 293-308.

[21] Idikio HA. Human Cancer Classification: A Systems Biology-Based Model Integrating Morphology, Cancer Stem Cells, Proteomics, and Genomics. *J Cancer*, 2011; 2: 107-115.

[22] Kumar V, Abbas A, Aster J. *Basic pathology*: Elsevier, 2017. 1-952.

[23] Campanacci M. *Bone and Soft Tissue Tumors*: Springer-Verlag. Wien, 1999.

[24] Klein MJ, Siegal GP. Osteosarcoma - Anatomic and histologic variants. *Am J Clin Pathol*, 2006; 125(4): 555-581.

[25] Kager L, Zoubek A, Potschger U, Kastner U, Flege S, Kempf-Bielack B, et al. Primary metastatic osteosarcoma: Presentation and outcome of patients treated on neoadjuvant cooperative osteosarcoma study group protocols. *J Clin Oncol*, 2003; 21(10): 2011-2018.

[26] Hayden JB, Hoang BH. Osteosarcoma: Basic science and clinical implications. *Orthop Clin North Am*, 2006; 37(1): 1-7.

[27] Rech A, Castro Junior CG, Mattei J, Gregianin L, Di Leone L, David A, Rivero LF, Tarrago R, Abreu A, Brunetto AL. Clinical features in osteosarcoma and prognostic implications. *Jornal de Pediatria*, 2004; 80(1): 65-70.

[28] Kundu ZS. Classification, imaging, biopsy and staging of osteosarcoma. *Indian J Orthop*, 2014; 48(3): 238-246.

[29] Misaghi A, Goldin A, Awad M, Kulidjian AA. Osteosarcoma: a comprehensive review. *Sicot.*, 2018; 4: 8-12.

[30] Durfee RA, Mohammed M, Luu HH. Review of Osteosarcoma and Current Management. *Rheumatol Ther*, 2016; 3(2): 221-243.

[31] Scully SP, Ghert MA, Zurakowski D, Thompson RC, Gebhardt MC, Representation Musculoskeletal T. Pathologic fracture in osteosarcoma - Prognostic importance and treatment implications. *J Bone Joint Surg-Am*, Vol 2002; 84A(1): 49-57.

[32] Grimer RJ, Bielack S, Flege S, Cannon SR, Foleras G, Andreeff I, et al. Periosteal osteosarcoma - a European review of outcome. *Eur J Cancer*, 2005; 41(18): 2806-2811.

[33] Eary JF, O'Sullivan F, Powitan Y, Chandhury KR, Vernon C, Bruckner JD, et al. Sarcoma tumor FDG uptake measured by PET and patient outcome: a retrospective analysis. *Eur J Nucl Med Mol Imaging*, 2002; 29(9): 1149-1154.

[34] Kitajima K, Futani H, Fujiwara M, Minakawa G, Usugi Y, Tsuchitani T, et al. Usefulness of Quantitative Bone Single Photon Emission Computed Tomography/Computed Tomography for Evaluating Response to Neoadjuvant Chemotherapy in a Patient with Periosteal Osteosarcoma. *Cureus*, 2018; 10(11): 1-8.

[35] Meyers PA, Heller G, Healey J, Huvos A, Lane J, Marcove R, et al. Chemotherapy for nonmetastatic osteogenic-sarcoma - the memorial sloan-kettering experience. *J Clin Oncol*, 1992; 10(1): 5-15.

[36] Block KI, Gyllenhaal C, Lowe L, Amedei A, Amin A, Amin A, et al. Designing a broad-spectrum integrative approach for cancer prevention and treatment. *Semin Cancer Biol*, 2015; 35: S276-S304.

[37] Luetke A, Meyers PA, Lewis I, Juergens H. Osteosarcoma treatment - Where do we stand? A state of the art review. *Cancer Treat Rev*, 2014; 40(4): 523-532.

[38] Jaffe N. Historical Perspective on the Introduction and Use of Chemotherapy for the Treatment of Osteosarcoma. In: Kleinerman ES, editor. Current Advances in Osteosarcoma. *Advances in Experimental Medicine and Biology*. 804. Springer-Verlag Berlin; 2014. p. 1-30.

[39] Ferrari S, Bacci G, Picci P, Mercuri M, Briccoli A, Pinto D, et al. Long-term follow-up and post-relapse survival in patients with non-metastatic osteosarcoma of the extremity treated with neoadjuvant chemotherapy. *Ann Oncol*, 1997; 8(8): 765-771.

[40] Bishop MW, Chang YC, Krailo MD, Meyers PA, Provisor AJ, Schwartz CL, et al. Assessing the Prognostic Significance of Histologic Response in Osteosarcoma: A Comparison of Outcomes on CCG-782 and INT0133A Report From the Children's Oncology Group Bone Tumor Committee. *Pediatr Blood Cancer*, 2016; 63(10): 1737-1743.

[41] Huang XY, Zhao J, Bai JY, Shen H, Zhang BB, Deng LL, et al. Risk and clinicopathological features of osteosarcoma metastasis to the lung: A population-based study. *J Bone Oncol*, 2019; 16: 1-8.

[42] Ferrari S, Palmerini E. Adjuvant and neoadjuvant combination chemotherapy for osteogenic sarcoma. *Curr Opinion Oncol*, 2007; 19(4): 341-346.

[43] Jaffe N. Osteosarcoma: Review of the Past, Impact on the Future. The American Experience. In: Jaffe N, Bruland OS, Bielack SS, editors. Pediatric and Adolescent Osteosarcoma. *Cancer Treatment and Research.*, 152. Springer; Dordrecht: 2009. p. 239-362.

[44] Casali PG, Blay JY, Bertuzzi A, Bielack S, Bjerkehagen B, Bonvalot S, et al. Bone sarcomas: ESMO Clinical Practice Guidelines for diagnosis, treatment and follow-up. *Ann Oncol*, 2014; 25: 113-123.

[45] Sakamoto A, Iwamoto Y. Current Status and Perspectives Regarding the Treatment of Osteosarcoma: Chemotherapy. *Reviews on Recent Clinical Trials*, 2008; 3(3): 228-231.

[46] Grossweiner LI. Photodynamic therapy. *J Laser Appl*, 1995; 7(1): 51-57.

[47] Pathak MA, Fellman JH. Activating and fluorescent wave-lengths of furocoumarins-psoralens. *Nature*, 1960; 185(4710): 382-383.

[48] Kessel D. Photodynamic Therapy: A Brief History. *J Clin Med*, 2019; 8(10): 1581-1588.

[49] Allegra A, Pioggia G, Tonacci A, Musolino C, Gangemi S. Oxidative Stress and Photodynamic Therapy of Skin Cancers: Mechanisms, Challenges and Promising Developments. *Antioxidants*, 2020; 9(5): 448-468.

[50] Zhu TC, Finlay JC. The role of photodynamic therapy (PDT) physics. *Med Phys*, 2008; 35(7): 3127-3136.

[51] Robertson CA, Evans DH, Abraharnse H. Photodynamic therapy (PDT): A short review on cellular mechanisms and cancer research applications for PDT. *J Photochem Photobiol B-Biol*, 2009; 96(1): 1-8.

[52] Donohoe C, Senge MO, Arnaut LG, Gomes-da-Silva LC. Cell death in photodynamic therapy: From oxidative stress to anti-tumor immunity. *Biochim Biophys Acta-Rev Cancer*, 2019; 1872(2): 188308-188325.

[53] Lange C, Lehmann C, Mahler M, Bednarski PJ. Comparison of Cellular Death Pathways after mTHPC-mediated Photodynamic Therapy (PDT) in Five Human Cancer Cell Lines. *Cancers*, 2019; 11(5): 702-734.

[54] Kroemer G, Galluzzi L, Vandenabeele P, Abrams J, Alnemri ES, Baehrecke EH, et al. Classification of cell death: recommendations of the Nomenclature Committee on Cell Death 2009. *Cell Death Differ*, 2009; 16(1): 3-11.

[55] Mroz P, Yaroslavsky A, Kharkwal GB, Hamblin MR. Cell death pathways in photodynamic therapy of cancer. *Cancers*, 2011; 3(2): 2516-2539.

[56] Rustin P. Mitochondria, from cell death to proliferation. *Nature Genet*, 2002; 30(4): 352-353.

[57] Perfettini JL, Kroemer G. Caspase activation is not death. *Nat Immunol*, 2003; 4(4): 308-310.

[58] Oleinick NL, Morris RL, Belichenko T. The role of apoptosis in response to photodynamic therapy: what, where, why, and how. *Photochem Photobiol Sci*, 2002; 1(1): 1-21.

[59] Yorimitsu T, Klionsky DJ. Autophagy: molecular machinery for self-eating. *Cell Death Differ*, 2005; 12: 1542-1552.

[60] Kessel D, Oleinick NL. Cell Death Pathways Associated with Photodynamic Therapy: An Update. *Photochem Photobiol*, 2018; 94(2): 213-218.

[61] Panzarini E, Inguscio V, Dini L. Overview of Cell Death Mechanisms Induced by Rose Bengal Acetate-Photodynamic Therapy. *Int J Photoenergy*, 2011: 1-11.

[62] Plaetzer K, Kiesslich T, Krammer B, Hammerl P. Characterization of the cell death modes and the associated changes in cellular energy supply in response to AlPcS4-PDT. *Photochem Photobiol Sci*, 2002; 1(3): 172-177.

[63] Tsai SR, Yin R, Huang YY, Sheu BC, Lee SC, Hamblin MR. Low-level light therapy potentiates NPe6-mediated photodynamic therapy in a human osteosarcoma cell line via increased ATP. *Photodiagnosis Photodyn Ther*, 2015; 12(1): 123-130.

[64] Aluigi A, Sotgiu G, Ferroni C, Duchi S, Lucarelli E, Martini C, et al. Chlorin e6 keratin nanoparticles for photodynamic anticancer therapy. *RSC Adv*, 2016; 6(40): 33910-33918.

[65] Huang Q, Ou YS, Tao Y, Yin H, Tu PH. Apoptosis and autophagy induced by pyropheophorbide-alpha methyl ester-mediated photodynamic therapy in human osteosarcoma MG-63 cells. *Apoptosis*, 2016; 21(6): 749-760.

[66] Tu PH, Huang Q, Ou YS, Du X, Li KT, Tao Y, et al. Aloe-emodin-mediated photodynamic therapy induces autophagy and apoptosis in human osteosarcoma cell line MG-63 through the ROS/JNK signaling pathway. *Oncol Rep*, 2016; 35(6): 3209-3215.

[67] Li KT, Chen Q, Wang DW, Duan QQ, Tian S, He JW, et al. Mitochondrial pathway and endoplasmic reticulum stress participate in the photosensitizing effectiveness of AE-PDT in MG63 cells. *Cancer Med*, 2016; 5(11): 3186-3193.

[68] White B, Rossi V, Baugher PJ. Aminolevulinic Acid-Mediated Photodynamic Therapy Causes Cell Death in MG-63 Human Osteosarcoma Cells. *Photomed Laser Surg*, 2016; 34(9): 400-405.

[69] Makhadmeh GN, Aziz AA. Photodynamic application of protoporphyrin IX as a photosensitizer encapsulated by silica nanoparticles. *Artif Cell Nanomed Biotechnol*, 2018; 46: S1043-S1046.

[70] Yu W, Ye MZ, Zhu J, Wang YT, Liang CZ, Tang JB, et al. Zinc phthalocyanine encapsulated in polymer micelles as a potent photosensitizer for the photodynamic therapy of osteosarcoma. *Nanomed-Nanotechnol Biol Med*, 2018; 14(4): 1099-1110.

[71] Zhang F, Zhu YJ, Fan GX, Hu S. Photodynamic therapy reduces the inhibitory effect of osteosarcoma cells on dendritic cells by upregulating HSP70. *Oncol Lett*, 2018; 16(4): 5034-5040.

[72] Yu W, Wang YT, Zhu J, Jin LB, Liu B, Xia KS, et al. Autophagy inhibitor enhance ZnPc/BSA nanoparticle induced photodynamic therapy by suppressing PD-L1 expression in osteosarcoma immunotherapy. *Biomaterials*, 2019; 192: 128-139.

[73] Dobos A, Steiger W, Theiner D, Gruber P, Lunzer M, Van Hoorick J, et al. Screening of two-photon activated photodynamic therapy sensitizers using a 3D osteosarcoma model. *Analyst*, 2019; 144(9): 3056-3063.

[74] Sun MX, Zhou CH, Zeng H, Puebla-Osorio N, Damiani E, Chen J, et al. Hiporfin-Mediated Photodynamic Therapy in Preclinical Treatment of Osteosarcoma. *Photochem Photobiol*., 2015; 91(3): 533-544.

[75] Zhang LJ, O'Shea D, Zhang CY, Yan YJ, Wang L, Chen ZL. Evaluation of a bacteriochlorin-based photosensitizer's anti-tumor effect *in vitro* and *in vivo*. *J Cancer Res Clin Oncol*, 2015; 141(11): 1921-1930.

[76] Meier D, Botter SM, Campanile C, Robl B, Grafe S, Pellegrini G, et al. Foscan and foslip based photodynamic therapy in osteosarcoma *in vitro* and in intratibial mouse models. *Int J Cancer*, 2017; 140(7): 1680-1692.

[77] de Miguel GC, Abrantes AM, Laranjo M, Grizotto AYK, Camporeze B, Pereira JA, et al. A new therapeutic proposal for inoperable osteosarcoma: Photodynamic therapy. *Photodiagnosis Photodyn Ther*, 2018; 21: 79-85.

[78] Zhu JW, Zou JH, Zhang ZJ, Zhang J, Sun Y, Dong XC, et al. An NIR triphenylamine grafted BODIPY derivative with high photothermal conversion efficiency and singlet oxygen generation for imaging guided phototherapy. *Mat Chem Front*, 2019; 3(8): 1523-1531.

[79] Kusuzaki K, Murata H, Matsubara T, Miyazaki S, Okamura A, Seto M, et al. Clinical trial of photodynamic therapy using acridine orange with/without low dose radiation as new limb salvage modality in musculoskeletal sarcomas. *Anticancer Res*, 2005; 25(2B): 1225-1235.

[80] Matsubara T, Kusuzaki K, Matsumine A, Murata H, Nakamura T, Uchida A, et al. Clinical Outcomes of Minimally Invasive Surgery Using Acridine Orange for Musculoskeletal Sarcomas Around the Forearm, Compared With Conventional Limb Salvage Surgery After Wide Resection. *J Surgical Oncol*, 2010; 102(3): 271-275.

Chapter 3

Tumor-Selective and Self-Activating Chemiluminescent Photosensitizers for Photodynamic Therapy

Ricardo M. S. Sendão[1], Carla M. Magalhães[1], Patricia González-Berdullas[1] and Luís Pinto da Silva[1,2,*]

[1]Chemistry Research Unit (CIQUP), Institute of Molecular Sciences (IMS), Faculty of Sciences of University of Porto (FCUP), Porto, Portugal

[2]LACOMEPHI, GreenUPorto, Department of Geosciences, Environment and Territorial Planning, Faculty of Sciences of University of Porto (FCUP), Porto, Portugal

Abstract

Photodynamic therapy (PDT) is a minimally-invasive anti-tumor therapeutic modality, known for its selectivity and limited side-effects. PDT has three fundamental requirements: a nontoxic photosensitizer (PS), light of a specific wavelength, and molecular oxygen. The anti-tumor therapeutic effect is triggered by the photo-activation of the PS via irradiation of the target tumor site with light of a specific wavelength. This leads to a photochemical reaction in which the PS generates cytotoxic reactive oxygen species (ROS), which lead to tumor-selective cell destruction. Despite significant advantages, the necessity of an external light source impairs the use of PDT for most tumors, due to the poor penetration of light into deeply localized tissues. Given this, researchers have focused on developing strategies to improve the efficiency of this therapy and expand it for deeper tumors and metastatic

* Corresponding Author's Email: luis.silva@fc.up.pt.

In: Photosensitizers and Their Applications
Editors: Davor Margetic and Renjith Thomas
ISBN: 978-1-68507-880-5

cancer. One strategy is to use either chemi- (CL) or bioluminescent (BL) reactions as intracellular excitation sources of PSs for PDT. Namely, CL/BL are processes in which light is generated due to a chemical/biochemical reaction, without any light source, and therefore, can be used to excite the PS intracellularly. More importantly, an extra step can be taken and instead of using CL reactions as excitation sources, they can be used as the basis for tumor-selective and self-activating single-molecule drugs that are both the excitation source and the PS itself. In this chapter, we will provide a critical review about the evolving use of CL/BL reactions in PDT toward solving its light-related problems, focusing on the most recent and exciting development: tumor-selective and self-activating single-molecule PSs.

Keywords: photodynamic therapy; chemiluminescence; bioluminescence, photosensitizers; cancer

Introduction to Photodynamic Therapy

Cancer is one of the biggest scourges of our time, responsible for nearly 10 million deaths in 2020 [1]. Moreover, the incidence of cancer in humans has increased significantly as a result of the increasing life expectancy worldwide, with 19.3 million cancer cases being diagnosed in 2020 [1]. Considering the prominence of cancer in human society, it is not surprising that several therapeutics have been developed to treat the different cancer types known to science. Therapeutic modalities such as surgical removal, [2, 3] radiotherapy, [4, 5] chemotherapy, [6-8] immunotherapy, [9, 10] and hormonal therapy [11, 12] have been extensively studied by the scientific community, albeit not always with the intended efficiency. One of the most promising strategies to increase efficiency is combining different anticancer drugs or therapy modalities to treat a single disease [8, 13]. This can increase the efficacy, counter drug resistance, and diminish the required dosage for the treatment to be effective [14].

Although the aforementioned strategies have been commonly employed to treat cancer patients and have led to significant increases in the survivability of cancer patients, they also have some drawbacks: they can be invasive, they can lack selectivity towards cancer cells and be toxic to healthy tissues, they present several side effects for the patient, and the tumor might develop resistance towards the therapy. As such, there is an increasing demand for the development of more efficient anticancer therapies.

Photodynamic therapy (PDT) is a new and minimally invasive therapeutic modality to treat various diseases (e.g., bacterial and fungal infections, skin diseases) as well as some types of cancer (e.g., esophageal cancer, non-small cell lung cancer, skin cancer), which is already in clinical use [15-18]. Over the past decade, the advances in the fields of nanotechnology and materials science and the improvements in the different components of PDT systems have promoted the rapid development of this treatment [19-23]. When compared to more traditional approaches, PDT offers significant advantages: it is minimally invasive, results in fewer side effects, has a high spatiotemporal precision (it can only be activated after irradiation with light of a specific wavelength), and allows for a fast healing of the healthy tissues [15-17, 24-26].

The working mechanism of PDT requires three fundamental parts: a non-toxic photosensitizer (PS), light of a specific wavelength, and the presence of molecular oxygen [15-17]. PDT is based on the production of reactive oxygen species (ROS), mainly singlet oxygen. This is achieved by photo-activation of a PS via irradiation of the targeted tissue with light [27]. Upon photoexcitation, the PS is transiently excited from its singlet ground state to a singlet excited state. Subsequently, the population of the singlet excited state can cross to comparatively long-lived triplet states, resulting in the production of ROS through one of two different pathways (Figure 1) [26, 28]. In type I, the PS, while in the triplet excited state, can transfer an electron to molecular oxygen (or biological substrates), prompting the formation of radical species such as the superoxide anion, which can then originate the highly reactive hydroxyl radical, starting a cascade of cytotoxic free radicals. In an alternative type I pathway, we have the transference of a hydrogen atom to the excited state PS. This originates free radicals that may react with molecular oxygen and create a mixture of ROS, causing oxidative damage. In the second pathway, type II, which is thought to be the most important for PDT, the triplet excited state of the PS can transfer its energy to molecular oxygen, resulting in highly reactive singlet oxygen.

Despite its advantages when compared to more conventional therapeutics, PDT has some limitations and drawbacks. PSs typically absorb strongly at around 400 nm and weakly between 600 and 800 nm. Therefore, PSs are generally excitable using UV or visible light, which leads to problems related to light penetration into tissues [15-17, 26]. Light of wavelengths below 580 nm is usually unsuitable for biological applications since it is absorbed and scattered by tissues, resulting in a penetration depth of less than 1 cm. [15-17, 26]. This limits PDT application to tumors either on or just under the skin, or in the outer lining of internal organs/cavities [29, 30]. Additionally, given that this therapy only works at the irradiated zones, the method is inefficient against metastatic tumors, which are a frequent cause of death in oncologic patients [15, 16].

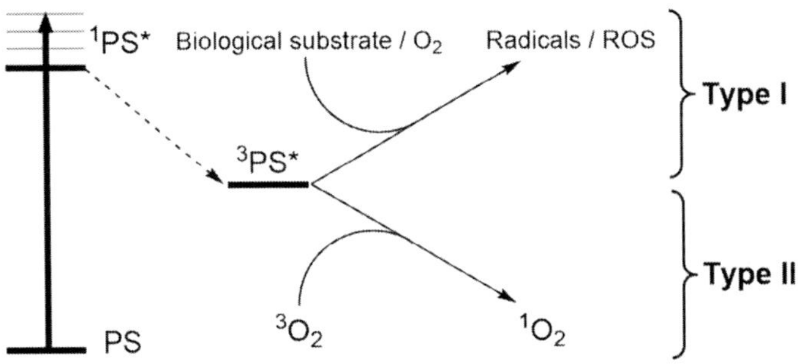

Figure 1. Schematic representation of PDT and its respective mechanisms of action.

To solve the limitation caused by the lack of penetration depth, several researchers proposed alternative strategies for PS activation, such as the use of near-infrared, [31, 32] two-photon, [33, 34] and X-ray irradiation for excitation.[35] Near-infrared light also presents light penetration limitations, low absorption efficiency, and may cause heat damage to tissues. On the other hand, X-ray excitation is unable to directly activate the PS and requires nanoparticles, which present low photo-conversion capacities, to convert high-energy radiation into photons of suitable energy [29, 35].

Thus, developing a PS that can be activated intracellularly and selectively without the need for an external light source is a relevant and challenging topic worth studying that could increase the role of PDT in routine cancer therapy. Internal light sources have become attractive alternatives to external light sources since they can solve the issue of light penetration in conventional PDT systems [27, 29, 36]. Recently, the exploit of chemiluminescence (CL) resonance energy transfer (CRET) and bioluminescence (BL) resonance energy transfer (BRET) has been studied to create self-illuminating PDT systems [15, 37-42]. In summary, the processes are based on a non-radiative energy transfer from donors (by either CL or BL) to suitable acceptor molecules. These studies aim to use CL/BL donors to intracellularly activate the PS without the need for an external light source [15, 37-42].

Introduction to Chemiluminescent and Bioluminescent Systems

CL is the conversion of thermal energy into excitation energy through a chemical reaction, usually in the presence of a catalyst and/or co-factor [43-45]. BL is a sub-type of CL in which the emission of light is catalysed by an enzyme in a biological system [43, 46]. This is a widespread phenomenon that can be observed in, among others, bacteria, dinoflagellates, worms, and insects. Several CL and BL systems have already been studied [43, 45-49]. enabling the development of new techniques that rely on a variety of intrinsic advantages: sensitivity, specificity, fast reaction rates, possibility of colour modulation, and high quantum yields. As such, BL and CL systems have gained several applications in the fields of sensing, pharmacy, biomedicine, and bioanalytics [50-54]. They are specially used for microbe detection, biosensing, bioimaging, determination of metabolites of interest, and as reporter gene systems.

Chemiluminescence

Nearly all CL reactions are characterized by the presence of a peroxide bond (O-O). This provides a way for a thermally activated ground-state reaction that results in excited-state products [43, 46, 55]. The peroxide-containing compounds can be divided into three main sub-categories, depicted in Figure 2: R_1, R_2, R_3, R_4-dioxetane-based compounds, [43, 46] R_1, R_2-dioxetanone-based compounds, [43, 46] and dioxetanedione [56].

The ability to satisfy the energetic requirement is the most important factor when considering the suitability of a compound as a CL substrate – it determines whether the molecule will thermally decompose into an electronically excited-state product. Thermochemical calculations have shown that, for typical dioxetanes and dioxetanones, the heats of activation for their thermal decompositions vary between 70 and 90 kcal mol^{-1} and their activation energies range between 20 and 30 kcal mol^{-1} [55]. Therefore, the peroxide rings can provide enough energy for the chemical excitation of a fragment during thermal decomposition.

Two different reaction mechanisms have been proposed to explain the ground-state decomposition of peroxide-containing molecules: a stepwise biradical mechanism and a mechanism that defends a partially/fully concerted decomposition in which the compounds are cleaved through a transition state [57-61]. The stepwise biradical mechanism (Scheme 1) is best supported by the literature than its counterpart. In this mechanism, the reaction starts with the homolytic cleavage of the peroxide bond, which results in a biradical transition state, and the C-C bond only breaks after this step. The participation of a biradical intermediate in the decomposition of simpler dioxetanones has been recently discussed [61-63].

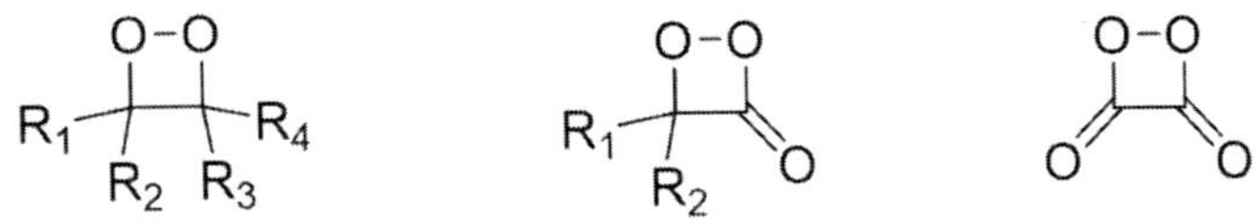

R_1,R_2,R_3,R_4-Dioxietane **R_1,R_2-Dioxietanone** **Dioxietanedione**

Figure 2. CL-capable cyclic intermediates containing peroxide bonds.

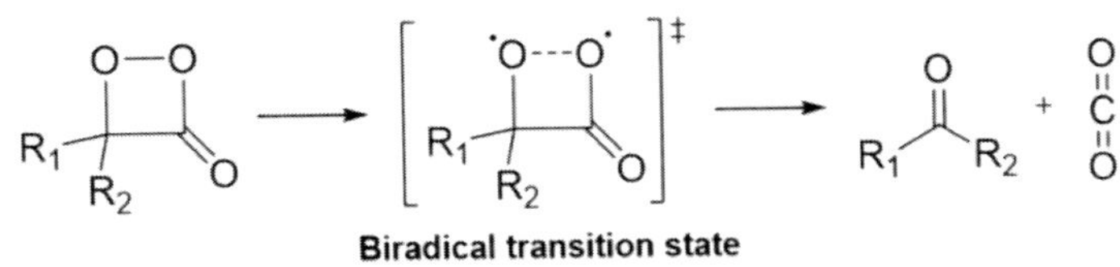

Scheme 1. Stepwise biradical mechanism for the thermal decomposition of a CL-capable cyclic peroxide, here represented by dioxetanone.

The effective generation of singlet excited states by these peroxides was originally explained by the chemically induced electron-exchange luminescence (CIEEL) [64, 65]. In CIEEL, a radical ion pair is originated by an electron transfer (ET) from an oxidizable electron-rich moiety (an activator, AM), to the peroxide. The ion pair then undergoes back electron transfer (BET) from the carbonyl radical anion to the radical cation, generating singlet excited states through charge annihilation. Scheme 2 represents the intermolecular CIEEL for dioxetanones' CL. When the electron-rich moiety is part of the organic peroxide the mechanism can be termed intramolecular CIEEL, otherwise being termed intermolecular CIEEL.

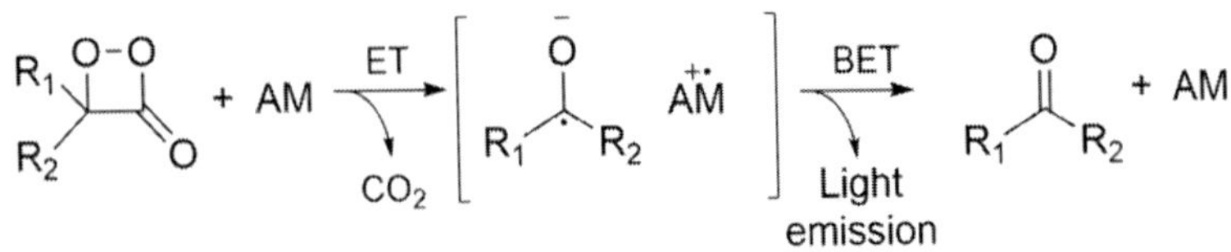

Scheme 2. Representation of the intermolecular CIEEL for dioxetanones' CL.

Koo and Schuster's mechanistic theory started to be questioned when the CL of dimethyldioxetanone and diphenoyl peroxide, important examples of CIEEL, was re-examined [66, 67]. Studies revealed that these peroxides displayed low CL quantum yields, even though they had a supposedly efficient CIEEL decay, which disproved the basis of the theory. The presence of a catalytic activator (an oxidizable electron-rich moiety) only resulted in inefficient CL, unlike the CIEEL mechanism. Another study regarding polyacene endoperoxide and two different dioxetanones also supported these conclusions [67].

Considering the failure of CIEEL, other mechanisms, in which neither the formation of a radical ion pair or a full electron transfer are involved, have been proposed to explain efficient CL reactions [66, 68, 69]. Instead of explaining the generation of S_1 states through full electron transfer, Charge Transfer-Initiated Luminescence (CTIL) mechanism is explained by gradual charge transfer and back charge transfer (BCT) between an ionized electron-rich moiety and the cyclic peroxide [68-72]. Nonetheless, even these revised forms of CIEEL are not yet fully in line with the available data, as theoretical results have suggested that charge-transfer (CT) processes should also be observed for CL systems with no electron-rich moieties [58, 62]. Additionally, the presence of electron-rich moieties is insufficient for the efficient formation of singlet excited states, both in intra- and intermolecular CL reactions [55, 66, 73].

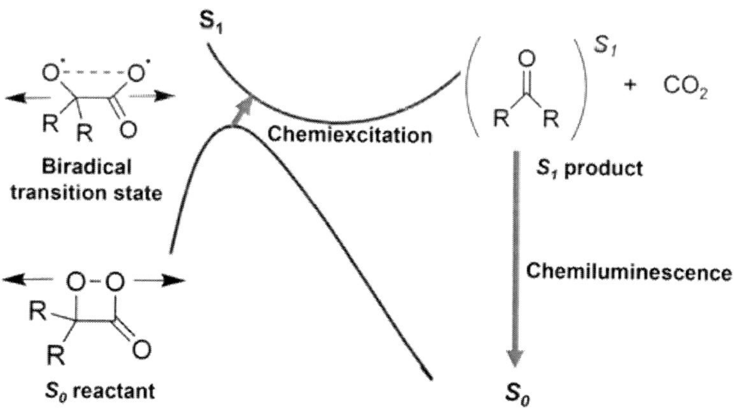

Scheme 3. Schematic representation of the thermolysis mechanism for dioxetanone molecules and subsequent chemiexcitation.

CIEEL and CTIL were also not in conformity with the report by Pinto da Silva et al. in which the luminescent reaction of *Cypridina hilgendorfii* ("sea firefly") was studied [74]. The study of the key step of chemiexcitation allowed the team to obtain results indicating that neutral dioxetanone is responsible for chemiexcitation. Its thermolysis provides access to a long potential energy surface (PES) region of S_0-S_1 degeneracy (Scheme 3) [74]. However, given that there is no clear correlation between ET/CT (between dioxetanone and the electron-rich moiety) and chemiexcitation, neither CIEEL nor CTIL can explain this imidazopyrazinone-based BL [74]. In the study, the researchers concluded that attractive electrostatic interactions between oxyluciferin moieties and CO_2 allow neutral dioxetanone to spend time in the PES region of degeneracy between singlet ground and excited states, thus explaining efficient chemiexcitation. On the other hand, anionic dioxetanone quickly detaches from CO_2 due to repulsive interactions, preventing access to the region of near-degeneracy [74].

Following this, Magalhães et al. studied the chemiexcitation step responsible for the light-emission associated with Coelenterazine (Clz), a BL substrate that is part of a prototypical system for marine CL/BL [75]. The team found evidence that supports the identification of a neutral dioxetanone intermediate as responsible for efficient chemiexcitation. Again, the researchers proposed an explanation in which the dioxetanone spends time in a PES region of degeneracy between S_0-S_1, whereas for the anionic dioxetanone there is a quicker release of CO_2 due to the repulsion between Coelenteramide and CO_2 that prevents access to the region of degeneracy [75]. No evidence of a relationship between ET/CT and efficient chemiexcitation was found and, therefore, neither CIEEL nor CTIL could be used to explain the luminescence [75].

Finally, a theoretical study focusing on dioxetanones supported the aforementioned reports and provided evidence suggesting that efficient chemiexcitation is the result of the reacting molecules reaching a PES region, since in inefficient chemiexcitation the molecules are unable to reach it [76]. The authors were able to hypothesize why this region only occurs in the thermolysis of certain dioxetanones: the main finding was that the access to the region of degeneracy between S_0-S_1 is a result of the increased interaction between the moieties of the dioxetanones during the thermolysis (CO_2 and keto moieties), which appear to extend the biradical region [76]. The increased attractive interaction seems to be caused mainly electrostatic interactions between the moieties of dioxetanone originated by charge separation. However, this electrostatic interaction does not appear to be so relevant for dioxetanes, and further study is required. Considering their results, the authors hypothesized that efficient chemiexcitation results from the interaction between the keto and CO_2 moieties (which control access to the PES region between S_0-S_1), instead of being just a result of ET/CT and charge annihilation, as defended by CIEEL and CTIL [76]. It is worth noting that, in energetic terms, the same factors leading to efficient singlet chemiexcitation can also lead to efficient triplet chemiexcitation, which is consistent with previous studies [76].

In summary, despite decades of research, the mechanism responsible for the efficient chemiexcitation of cyclic peroxides in luminescent reactions is not fully understood. Long-standing theories, such as CIEEL and CTIL are not able to explain this phenomenon in all cases. More recent results indicate that efficient chemiexcitation is a result of reacting molecules having access to a region of PES where S_0 and S_1 are degenerate [74-76]. Access to this region of degeneracy seems to be the result of the increased interaction between the two moieties that compose the peroxide ring. Overall, these reports support the hypothesis that efficient chemiexcitation is not just a result of ET/CT and subsequent charge annihilation, but is instead based on the degree of interaction between the CO_2/keto and keto moieties (responsible for the access to the S_0-S_1 region of degeneracy) [74-76].

Bioluminescence

Light emission catalysed in a biological system by an enzyme is known as BL [43, 46].

The study of several CL and BL systems, which ranged from simple dioxetanes/dioxetanones to complex BL substrates, allowed for the identification of three common key structural moieties: a peroxide bond, an electron-rich moiety, and an ionizable group [46, 55].

The first key structural moiety is the peroxide bond. When cleaved, it allows thermally activated ground-to-excited state chemiexcitation [43, 46, 55]. This type of feature is ubiquitous among BL systems, even though some differences can be found among them. In particular, the bacterial and dinoflagellate luciferins eject different fragments through a single bond-breaking step [46]. The second key structural moiety is an electron-rich group, responsible for tuning the activation energy of the peroxide ring decomposition (by charge/electron transfer) [43, 46, 55, 58, 66, 69]. This can be found for nearly all BL cases, except for *latia* luciferin [46] for which it is possible that the electron-rich moiety is provided by the enzyme, in the form of an aromatic amino acid. The electron-rich moiety can consist of different types of aromatic systems, as exemplified by firefly and coelenterazine dioxetanones [68, 77, 78]. It can be concluded that this electron-donating moiety is more relevant than the specific chemical composition. The third and last key structural moiety is an ionizable group, which is responsible for triggering the charge/electron transfer and sharply tune the activation energy for the thermal reaction [58, 66, 68, 69, 77]. These oxidizable groups are crucial for modulating the colour of the light emitted by the BL product [79-82]. More recently, the importance of these moieties has been debated: evidence suggests that some of them (ionizable groups) might not be required, whereas the role and function of others (electron-rich moieties) is not yet fully understood [58, 83].

In the same way, luminol is the most studied CL substrate,[84-86] the firefly luciferin–luciferase reaction (Scheme 4) is, undoubtedly, the most studied and well-known BL system. Firefly luciferase (Fluc) catalyses a two-step reaction: it starts with an AMPylation reaction between D-luciferin and adenosine triphosphate (ATP) that leads to an adenylyl intermediate. In the second step, the adenylyl intermediate is oxidized by molecular oxygen, adenosine monophosphate is released, and firefly dioxetanone is formed [43, 46, 87-89]. Subsequently, the thermal decomposition of the peroxide results in the chemiexcitation of firefly oxyluciferin (the light emitter) [43, 68, 90] The Fluc/luciferin system is a pH-sensitive BL system. While the emitted light has a peak at 560 nm for basic pH, the maximum shifts to around 620 nm in acidic pH [43, 46, 89]. It is worth noting that the identity of the light emitter is still a matter of debate: while some researchers defend the participation of anionic keto species [47, 68] others support the presence of the oxyluciferin enolate [48, 91].

Another example of BL is Clz, a luciferin that can be found among marine living organisms [45]. Clz, which has an imidazopyrazinone core, is essential to the BL of several marine species and is involved, as a substrate, in reactions catalysed by several luciferases, such as *Gaussia princeps* (Gluc) and *Renilla reniformis* (Rluc) luciferases [92]. Besides being a functional BL substrate, Clz can also be used for CL reactions involving molecular oxygen or ROS (such as superoxide anion), [92, 93] and is even routinely used as a probe for ROS [94, 95]. The reactions involving Clz (Scheme 5) start with an electron transfer from Clz to an oxidizing agent, leading to the formation of a radical intermediate [92, 93]. A fast radical coupling between the intermediate and a superoxide anion follows, forming the Clz dioxetanone. The thermal decomposition of this dioxetanone generates a singlet excited state of coelenteramide, with CO_2 loss [92, 93]. Coelenteramide, which is the light-emitting form, produces blue-green light with a spectral peak at 480 nm [92, 96].

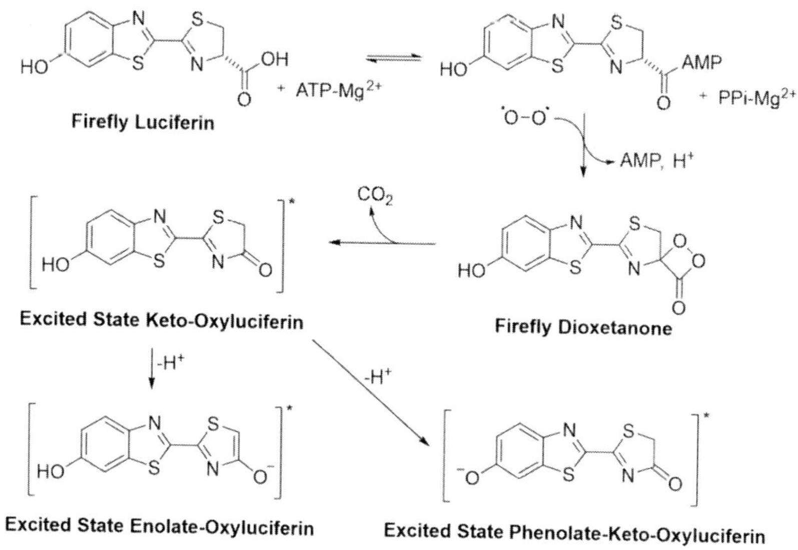

Scheme 4. Reaction mechanism of firefly luciferin-luciferase BL.

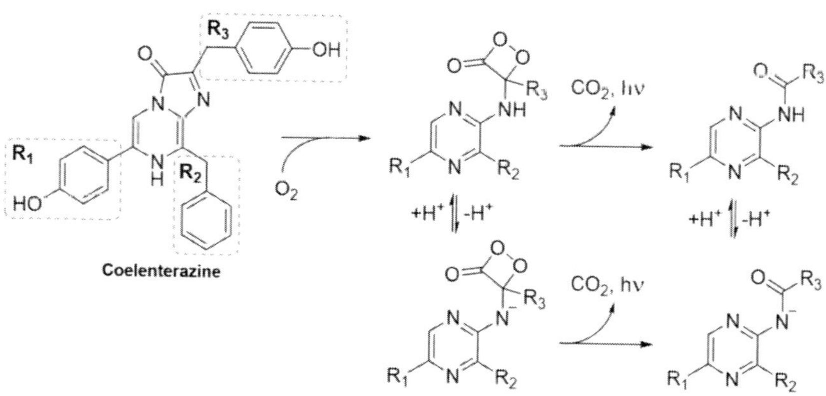

Scheme 5. Reaction mechanism of Clz BL.

As in the case of firefly BL, there is still no full consensus between the true ionization state of coelenteramide undergoing BL emission [81, 92].

Coelenteramide is thought to occur in one of the chemical forms present in the equilibrium between the neutral, protonated, and deprotonated amino and hydroxyl groups, and the phenolate-amide tautomerism. However, two other forms can occur considering the nature of the emitter: the phenolate-NH_2^+ and the diphenolate. A study by Min et al. predicted the behaviour of the fluorescent and CL properties of coelenteramide in several conditions using time-dependent density functional theory.[97] The team determined the excited state equilibrium constants to estimate the relative stabilities of the different species of coelenteramide in the fluorescent and CL states and obtained the minimum-energy structures for each state for all possible conformations [97]. They concluded that CL states are "dark" states with low oscillator strengths that must evolve into bright fluorescent states and that the photoacidity of the phenol group in the molecules is considerably higher in the fluorescent state when compared to the CL state [97]. Finally, they concluded that the higher photoacidity permits the formation of phenolate coelenteramide in the fluorescent state instead of the CL state, enabling its role as the luminescent emitter [97].

Applications for Chemiluminescence and Bioluminescence in PDT

Designing PS that can be intracellularly activated, without the need for an external light source, is a challenging but rewarding research topic. Considering the limited depth penetration of external light sources, the range of applications for PDT could increase significantly when they are not dependent on said sources. CL and BL systems have been studied as intracellular excitation sources for PDT, with luminol (for CL-based systems), and firefly luciferase–luciferin and *Renilla* luciferase–Clz (for BL-based systems) being the best examples.

CL-Based PDT Systems

Light emission is the result of a large number of chemical reactions, and one of the most known and efficient examples is the oxidation of luminol (Scheme 6) [15, 84-86]. Luminol is a well-studied system that has been thoroughly tested as a substitute for external laser activation in PDT [15]. Luminol undergoes oxidation in basic solution in the presence of hydrogen peroxide and catalysts (e.g., Fe^{2+}, Cu^{2+}, Co^{2+}, periodate ions, or hydrogen peroxidase). This CL reaction originates from the reaction of luminol with hydroxide anions, from which a dianion is formed [15, 84-86]. This dianionic intermediate reacts with oxygen to yield the corresponding cyclic peroxide. After releasing N_2, this peroxide evolves into the excited state of the aminophthalate ion, which emits blue CL with a maximum wavelength of 425 nm [15, 84-86].

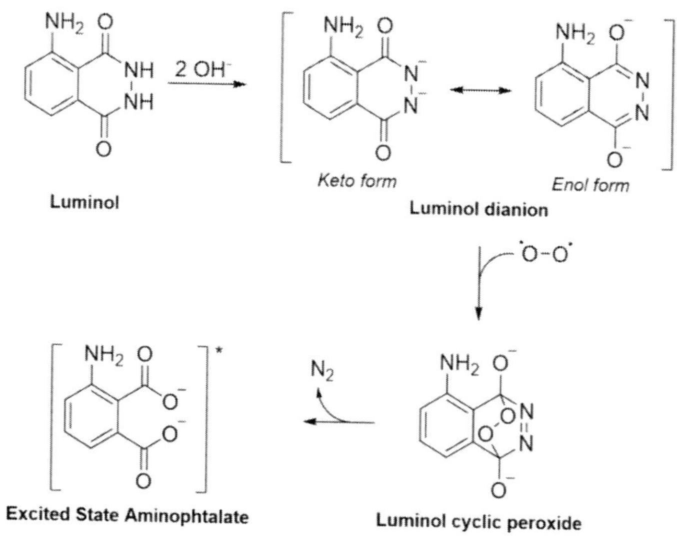

Scheme 6. Reaction mechanism for the CL of luminol.

Laptev et al. who first studied the luminol CL system, added hydrogen peroxide and ferrous sulphate to cultures of hybridoma cells to assess the efficiency of luminol in activating the PS hematoporphyrin (Hp) [15]. They found that, when reacting, luminol could be turned into the excited state aminophthalate ion, which in turn could activate a conjugate involving Hp, resulting in nearly 100% of cytotoxicity, out of which around 15% was directly induced by luminol (indicating some biocompatibility issues) [15]. Additionally, while the luminol system was able to lead to high cytotoxicity rates in the presence of the PS, the amount of Hp conjugate needed to attain LD_{MAX} with the CL-based system was 6.7 times higher than with an external source of radiation [15]. Thus, the system proposed by Laptev et al. required further optimization.

Yuan et al. also developed a new CL-based PDT [38] in which they used a luminol/hydrogen peroxide/horseradish peroxidase CL system to excite the PS oligo(*p*-phenylene vinylene) (OPV). Considering that luminol has a maximum at 425 nm and that OPV exhibits broad absorption ranging from 350 to 550 nm, the spectral overlap required for CRET is met [98, 99]. Additionally, CRET should be favoured by electrostatic interactions between the cationic OPV and the dianionic CL emitter [38]. To confirm that a CRET process takes place, they measured the luminescence of luminol in the presence of several concentrations of OPV [38]. The addition of OPV resulted in a large decrease of the luminescence intensity at 425 nm (corresponding to luminol) and, at the same time, an increase of the luminescence intensity at 550 nm (corresponding to OPV), which confirmed that CRET occurs in this system [38].

The capacity of the luminol system to activate OPV was first evaluated *in vitro* in cervical cancer (HeLa) cells [38]. The use of the luminol CL system with increasing OPV concentrations resulted in a cellular viability lower than 10%. However, the luminol CL system was itself responsible for killing about 30% of the HeLa cells and was also cytotoxic to healthy human epithelial cells [38]. These results show that the luminol CL system has some biocompatibility limitations that might impede its clinical use, as observed in some studies [15, 38]. It is worth noting that these results, besides demonstrating that the CL system by itself is cytotoxic, also show that it has no specificity or selectivity for tumor cells.

The efficacy of this CL system was also evaluated by Yuan et al. as an *in vivo* excitation source in nude mice with HeLa cells-derived tumors [38]. They demonstrated that the system has, in fact, some potential in PDT, as around 30% of tumor inhibition was achieved with the CL-based PDT when compared to the CL system itself. Moreover, the addition of OPV alone did not cause any tumor inhibition. In this way, it was shown that CL-mediated PDT can result in *in vivo* tumor growth inhibition [38]. Remarkably, the CL-mediated PDT did not show any signs of side effects, as the mouse body weight growth was not inhibited in comparison with the control groups. The authors concluded that the luminol CL system had no apparent toxicity to normal healthy tissues, even though it presented a lack of specificity. However, tumor cells can be more susceptible to ROS-mediated damage than normal cells since they are characterized by their state of oxidative stress [100]. This being said, a lower amount of exogenous ROS is required to prompt cellular destruction in cancer cells. Considering the low cytotoxicity of the CL PDT system, this lack of toxicity might be the result of ROS generation in amounts that are insufficient to affect normal cells. Should the ROS output of the CL system be increased, as would occur in a trial for clinical applications, toxicity toward normal cells might also be observed. The use of a luminol signal enhancer impacted the anti-tumor activity of the CL-mediated PDT, resulting in a 55% tumor inhibition ratio, which pointed to signal enhancers as one way of increasing the effectiveness of CL PDT systems [38].

More recently, the same research group proposed an electroluminescence-based PDT system, [101] and reported the preparation of a multifunctional nanoplatform for self-illuminating phototherapy [102]. The haemoglobin-nanoparticle platform was prepared by coupling haemoglobin (Hb) to polymeric nanoparticles made of poly(styrene-co-maleic anhydride) and poly[2-methoxy-5-(2-ethylhexyloxy)-1,4-phenylenevinylene]. The polymer-conjugated Hb can function simultaneously as the oxygen carrier, as the catalyst for the CL reaction of luminol with hydrogen peroxide, and as the PS for the generation of ROS [102]. Thus, the system does not require an external light source for the PS excitation and overcomes the problem of insufficient oxygen under hypoxia. Moreover, the system can be applied to control drug release and therefore be used simultaneously as PDT and chemotherapy [102].

Another innovative report involving luminol was made by Zhang et al. The team constructed a "smart" nanoconjugate in which semiconducting polymer dots acted as a hydrophobic matrix to incorporate the PS *meta*-tetra(hydroxyphenyl)chlorin (*m*-THPC) (Scheme 7) [40]. Amphiphilic Janus dendrimers were used as surface-functionalizing agents to conjugate horseradish peroxidase and aminated folic acid onto the surface of the polymer dots [40]. When adding luminol and hydrogen peroxide, the PS can be activated either by CRET from luminol CL to *m*-THPC or by CRET to the polymer dots, followed by fluorescence resonance energy transfer to the polymer dots. The photodynamic effect was evaluated by incubating polymer dot nanoconjugates with different types of cells (C6 glioma, MCF-7 breast cancer, and NIH3T3 fibroblast cells) *in vitro*, in the presence of CL substrates [40]. For a nanoconjugate concentration of 10 $\mu g\ mL^{-1}$, viabilities of 32%, 17%, and 72% were observed for the C6, MCF-7, and NIH3T3 cells, respectively. However, when the CL substrate is absent, a cell viability of over 95% is observed for all three cancer cell types [40]. These values showed that luminol can indeed activate the PS, inducing a photodynamic effect. Nevertheless, from a clinical use perspective, it could also be said that the cell viability was too low for healthy non-cancerous cells, while too high for tumor cells [40]. Moreover, given the overexpression of folate receptors by tumor cells, it is worth noting that the difference in cell viabilities between the cancerous receptors (C6 and MCF-7) and non-cancerous (NIH3T3) could be ascribed to the aminated folic acid present on the surface of the polymer dots.

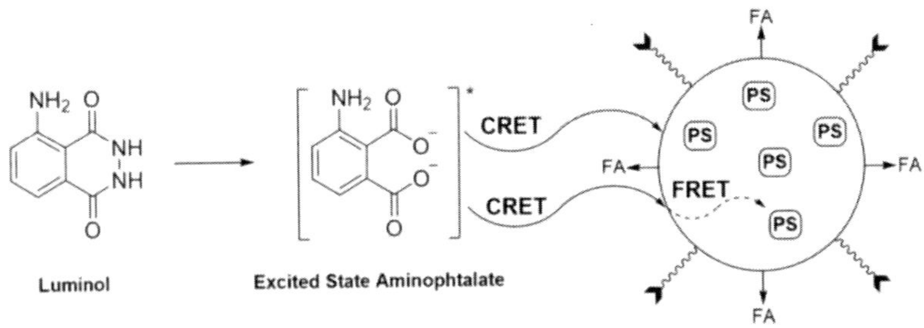

Scheme 7. Representation of the luminol CL-based system in conjugation with polymer dots. FA – folic acid.

Combining different forms of PSs with luminol CL resulted in a variety of *in situ* self-illuminating PDT systems including, among others, 5-aminolevulinic acid, [103] chlorin e6 conjugated with carbon dots possessing yellow emission, [104] or porphyrinic metal-organic frameworks (MOFs) [105]. The covalent attachment of luminol to the PS is another strategy that enables the construction of a CL-PDT system through direct energy transfer from luminol to PS. Yesilgul et al. reported the development of the modular, unimolecular, erythrosine-luminol, which is capable of producing singlet oxygen in the presence of Cu^{2+} and hydrogen peroxide [106]. Xu et al. synthesized a chlorin e6-luminol-PEG conjugate (CLP), capable of self-assembling into core-shell nanoparticles, which can be exploited for imaging as well as specifically kill cancer cells [107].

Another known CL reaction is that of peroxyoxalate, which rapidly transforms into the high-energy compound 1,2-dioxetanedione (Scheme 8) [108-110]. The reaction ensues between active derivatives of oxalic acid and hydrogen peroxide. First, the nucleophilic attack by a peroxide results in the formation of the 1,2-dioxetanedione intermediate [108-110]. In certain conditions, while decomposing into two CO_2 molecules, this intermediate is capable of transferring energy to fluorescent energy acceptors, thus generating light. Philip and Maximuke, who were among the first to employ the CL reaction of peroxyoxalate as an internal light source, used it in conjugation with Photofrin II (a Haematoporphyrin derivative) to treat mammary adenocarcinomas in mice [111] and observed some effectiveness in certain studied mice groups. They used exogenous hydrogen peroxide to trigger the CL reaction, however, large amounts of endogenous hydrogen peroxide are often found in malignant cells and inflamed tissues, [112] enabling CL even without the addition of exogenous compounds. These results suggested that a peroxyoxalate-based CL system could be a viable alternative to laser activation in PDT [108-110].

Romaniuk et al. studied polymeric oxalate as the substrate for CL reactions with tetramethylhaematoporphyrin (TMHP), acting both as PS and activator for the CL reaction [113]. Polymeric oxalate and TMHP were dispersed and stabilized with a surfactant. The dispersions could efficiently produce singlet oxygen and exert relevant cytotoxicity [113]. Mao et al. designed a nanoplatform for image-guided PDT by co-encapsulating bis [2,4,5-trichloro-6-(pentyloxycarbonyl)phenyl]oxalate (CPPO) with a specially designed PS (TBD) into pluronic F-127 and soybean oil [114]. The smartly designed TBD displayed bright NIR emission and efficient singlet oxygen generation. This system allows the surveillance of the tumor site via CL imaging and inhibits tumor growth through CL-based PDT singlet oxygen generation, thus accomplishing both tumor imaging and treatment [114]. The CL reaction of tumoral hydrogen peroxide with CPPO produces singlet oxygen, which results in the photo-excitation of the PS TBD via CRET [114]. A system with a similar basis was developed by Yu et al., with the tumoral hydrogen peroxide initiating the CL reaction of CPPO, which in turn activates the PS chlorin e6 via CRET [115].

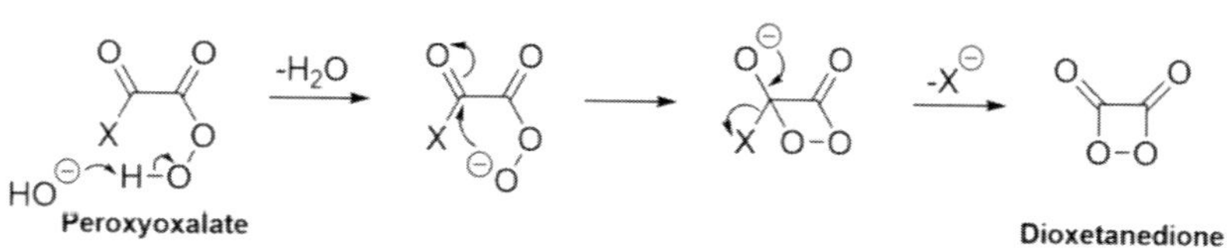

Scheme 8. Peroxyoxalate transformation into dioxetanedione. X = RO or Cl.

BL-Based PDT Systems

The first study aiming to assess the potential of BL for PDT was performed by Theodossiou et al. who, in 2003, used the Fluc-luciferin BL as an intracellular excitation source for Rose Bengal [42]. The PS is a water-soluble dye with a high singlet oxygen quantum yield (around 0.75) [42]. The study was performed *in vitro* using NIH3T3 murine fibroblasts as a model. The cells were transfected with a modified Fluc gene (Luc+, with cytosolic expression), and both luciferin and the PS were later added to the cell cultures. These results supported the use of firefly BL as an excitation source for PDT: the combination of the firefly BL system with the PS led to a 90% apoptosis rate, as opposed to the control groups, which exhibited a 100% survival rate [42]. Moreover, by using a singlet oxygen quencher, the authors concluded that the cytotoxic effect results mostly from the production of this particular ROS species.

One interesting aspect of the work by Theodossiou et al. is that it demonstrated that the addition of ATP to the cells is not required to induce the Fluc-mediated PDT, as the concentration of ATP in the fibroblasts is itself sufficient to trigger BL [42]. Additionally, it was shown that, upon uptake by the cells, both luciferin and Rose Bengal adopt a diffuse cytosolic distribution [42]. Knowing this is crucial, as BRET should only elicit sufficient PDT response if both the BL donor and the PS are co-localized within the cell. It is worth remembering that the Fluc enzyme was not added to the cell culture and instead, its gene was transferred into the cell, which means no information about the possible uptake of this enzyme by the model cells was obtained [42].

This study was followed by the work of Schipper et al. who, in 2006, also tested the firefly BL system for application in PDT [39]. This team claimed that, despite the results of Theodissiou et al., [42] there were still doubts regarding the photon output of the Fluc-transfected cells and questioned whether it was enough to induce a photodynamic effect when compared to doses typically associated with clinical PDT (around 50 mW cm^{-2} and above 1 J cm^{-2}) [39]. The authors demonstrated that firefly BL is not able to generate enough *in vitro* photons to result in a cytotoxic effect from two different PS (Rose Bengal and hypericin) [39]. The measured photon output only reached 2.3×10^3 photons per second per cell, which is approximately 1.2×10^{-9} mW cm^{-2}, orders of magnitude lower than the 50 mW cm^{-2} used in clinical trials [39]. Even assuming that maximal levels of photons were emitted during 24 h, the total amount of energy delivered would only be 1.03×10^{-7} mJ cm^{-2}, considerably lower than 1 J cm^{-2}, and higher than what is observed in clinical trials involving laser irradiation.

These results [39] contrast starkly with those presented by Theodossiou et al. [42] and the differences between them have yet to be explained. Schipper et al. [39] suggested that the cell line used by Theodossiou and his team could have an unknown mutation, causing an increase in their sensitivity to PDT treatment. However, since no follow-up study was made regarding this, no data is available to support this suggestion. A simpler justification could come from the difference in luciferin concentration. While Theodossiou et al. [42] used concentrations of 500 μM, Schipper et al. [39] used a maximum luciferin concentration of 20 μM. Knowing that BL intensity is also dependent on luciferin concentration, this could explain how Theodossiou et al. achieved higher light production than Schipper and his team.

In a recent work, Yang et al. reported a novel firefly BL-mediated PDT system based on a polymer nanoparticle made of biodegradable poly(lactic-co-glycolic acid) [116]. The particle was loaded with the PS (Rose Bengal) and conjugated with Fluc. Spectroscopic characterization suggested that BRET effectively activated the PS to generate ROS, inducing oxidative stress [116]. *In vitro* photodynamic studies demonstrated that this BL-mediated PDT results in significant cytotoxicity towards cancer cells [116]. The growth of subcutaneous tumors can be considerably inhibited while healthy, normal organs, such as the heart, liver, spleen, lung, and kidney, remain remarkably undamaged [116]. Another work on the use of firefly BL for PDT was performed by Yang et al. who conjugated carbon dots (with an excitation-independent emission) with protoporphyrin IX (PIX) to construct a PDT system [117]. In this case, carbon dots solved the limitations of PIX and served as an intermediate to bridge the excitation from the firefly BL to the PIX. This BL-mediated PDT resulted in the production of singlet oxygen in SMMC-7721 hepatocarcinoma cells through a fluorescence resonance energy transfer (FRET) process [117].

Besides the firefly BL, the Rluc-Clz BL system (using the *Renilla reniformis* luciferase) has also been tested as an alternative intracellular excitation source. The first study on such a system was performed by Hsu et al. in 2013 [41] who conjugated the Rluc enzyme with carboxylate-containing quantum dots (QDs), which absorb the photons resulting from the BL reaction via BRET in the presence of Clz [41, 118]. In this system, instead of the BL reaction, the QDs are responsible for PS activation by emitting light at 655 nm [41]. The rationale behind this is that the emission of the QDs is easily tuneable when compared to that of the BL reaction, greatly increasing their versatility for PDT applications and allowing the system to be used with a larger number of PS.

The bioluminescent conjugate was used to intracellularly activate the PS *m*-THPC (already in clinical use) in mice transfected with human lung adenocarcinoma epithelial A549 cells [41]. The results were promising, as the mean relative volumes of PDT-treated tumors were 4.5 to 6 times lower (in comparison with the initial tumor volumes) than those from untreated animals on day 20. In the mice treated with the Rluc-QDs conjugate, tumor growth was significantly inhibited [41]. Additionally, the isolated tumor sizes of the group treated with PDT (Rluc/QDs/PS/Clz) were considerably lower than those of the PS/Clz (growth inhibition of 4.2%) and Rluc/Clz/QDs (growth inhibition of 23.3%) [41]. Nevertheless, these results also showed that the Rluc-QDs conjugate has considerable cytotoxicity itself. This can be problematic, given that unwanted cytotoxicity to healthy cells can cause side effects, unacceptable in clinical applications.

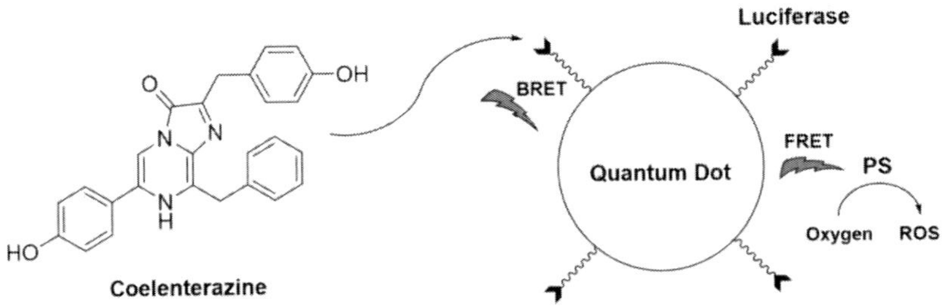

Figure 3. Schematic representation of the use of the Rluc-Clz BL system in conjugation with carboxylate-containing quantum dots to activate a PS.

The results of Hsu et al. also showed that tumors treated with BL-mediated PDT present considerably lower cell proliferation when compared to control groups (27.1 ± 2.6% *versus* 87.4 ± 2.2% / 92.2 ± 2.6% /75.2 ± 2%, respectively) [41]. The Rluc-QDs conjugate also led to a decreased degree of vascularization of the tumors, suppressing their growth. [41] This study also evaluated the effectiveness of BL-mediated PDT when compared to conventional irradiation by an external light source [41]. The team found that, for equal PS concentrations, the BL-mediated PDT reaction yielded an irradiation comparable to that resulting from externally applying 0.6-0.8 J cm^{-2} [41]. While 0.6-0.8 J cm^{-2} can be considered low, even more when compared to the light doses used in clinical PDT (higher than 1 J cm^{-2}), the efficiency of this BL-mediated PDT is enough to induce a photodynamic effect *in vivo*, as can be seen in the work of Hsu et al. [41] It is worth noting that the efficiency obtained by this complex, 0.6-0.8 J cm^{-2}, largely surpasses that obtained for firefly BL-mediated PDT (1.03 x 10^{-4} mJ cm^{-2}) [39, 41].

In summary, the study by Hsu et al. showcased the potential of the application of Rluc-QDs conjugates as an alternative and internal excitation source for PDT [41]. Despite that, the system could be further optimized, as the obtained photodynamic effect might not suffice for clinical application. The decreased efficiency of the system might originate from the fact that it is based on two energy transfer steps. First, we observe a BRET step, in which the energy is transferred from the BL reaction to the QDs. Subsequently, we observe a FRET process, in which the QDs transfer energy to the PS. A higher number of energy-transfer steps result in an increased number of points at which the overall efficiency of the therapy decreases.

Rluc-Clz BL-mediated PDT was also the focus of a more recent study by Kim and co-workers [119]. Their objective was to determine if BL, with its lower energy output, can actually be used in PDT as an alternative excitation source. They created self-illuminating Rluc-QDs, conjugated to intracellularly activate the PS (chlorin e6). The effects of the PDT on tumor growth in mice were studied *in vivo* using three different cancer cell lines: colorectal cancer cells (CT26), lung cancer cells (LLC), and melanoma cells (B16F10) [119]. The authors calculated the efficiency of BRET (from the BL reaction to the QDs), which was between 60-65%. These values support the hypothesis that the low efficiency of the QDs-based PDT results from a loss of efficiency starting at the initial BRET step. In addition, the authors observed that the conjugates did not enter the cell, but accumulated at their external surface in high concentrations instead [119]. Interestingly, this study demonstrated that whereas chlorin e6 molecules (at a concentration of 100 μM) are activated 4 x 10^7 times per minute (by a flux from a 660 nm, 2.2 mW cm^{-2} laser), the Rluc-QDs complex increases activation to 3 x 10^8 times per minute [119]. These results support the conclusion that BRET energy (in the order of 100 mJ) could generate a stronger photochemical activation in the cellular membrane than energy from laser irradiation (of the same order, 100 mJ). Cell imaging and cytotoxicity tests demonstrated that this BL-mediated PDT system can lead to significant intracellular ROS, causing membrane damage and cellular death [119].

Gluc (*Gaussia princeps* luciferase) was also used in a PDT system. Al-Ani et al. reported the preparation of a Gluc-LiDps (*Listeria innocua* DNA binding protein from starved cells) fusion protein, with chemical conjugation of Zn^{2+}-protoporphyrin IX (ZnPP) to lysine residues [120]. The Gluc-LiDps-ZnPP conjugate could generate ROS via BRET between Gluc (470-490 nm) and ZnPP. An *in vitro* study demonstrated that the conjugate was efficiently taken up by tumor cells (SKBR3 and MDA-MB231 breast cancer cells) [120]. In the presence of Clz, the conjugate inhibits the proliferation of SKBR3 cells due to high ROS levels, with the exposure to the conjugate resulting in significant suppression of the migration of the remaining cells [120] which showcases the potential of the Gluc-LiDps-ZnPP conjugate as a nanoplatform for the development of an anticancer PDT.

Alternatives to Conventional CL/BL-Mediated PDT

CL and BL have been extensively studied as alternative and intracellular light sources for PDT. However, these kinds of systems present limitations when considered for clinical applications.

The CL/BL substrates and the catalyst/co-factor must be present in the same microenvironment to generate the required light emission. However, it is not easy to ensure that all the components required for the therapy reach the same cellular space without reacting with one another during delivery. Moreover, given that the efficiency of energy transfer processes (such as CRET and BRET) is inversely proportional to the distance between the donor and the acceptor, the PS must be present in the same localization as the CL emitter to guarantee sufficient excitation. It is noteworthy that even if these conditions are met, the overall efficiency of the process still depends on several sequential steps: 1) CL or BL; 2) energy transfer from the CL/BL donor to the PS acceptor; 3) subsequent intersystem crossing of the PS; 4) ROS production [38, 40, 41, 114]. The existence of several steps will likely decrease the efficiency and reproducibility of PDT. Thus, before CL/BL-mediated PDT is available for clinical use, each individual step should be optimized to minimize the overall yield losses of the therapy.

In addition, one of the major advantages of PDT is its selectivity in terms of the targeted area, allowing the selective targeting of tumoral tissue [16, 17]. This comes from the fact that only the tissue that is irradiated by light is affected by the photodynamic effect. The substitution of laser light by CL/BL-mediated intracellular excitation as an activation source means this important advantage is lost, which can be observed in some of the studies mentioned earlier in this chapter (e.g., CL/BL systems lacking features that permitted tumor-selective activation) [38, 41]. Considering that some of the components involved in CL/BL present significant cytotoxicity towards both tumoral and healthy cells (e.g., luminol), this is of particular concern.

Clz and its analogues have been studied as self-activating agents for PDT [121-123]. Recently, Clz-based alternatives overcoming the limitations imposed by conventional CL/BL-mediated PDT have been reported. One example of this is the work published by Pinto da Silva et al. in 2019 [121]. The team developed a single-molecule Clz-based PS, which is capable of intracellular self-activation and is also tumor-selective since it is based on a CL reaction involving only a tumoral marker: superoxide anion (Scheme 9) [121]. The PS is an analogue of Clz, which is known for its use as a superoxide anion probe [94, 95]. When Clz reacts with this radical in the absence of a catalyst/co-factor, it can be oxidized and lead to the formation of an unstable dioxetanone intermediate, which readily decomposes into CO_2 and triplet excited state coelenteramide. Coelenteramide promptly transfers its energy to molecular oxygen, inducing the generation of singlet oxygen [121]. The direct chemiexcitation into triplet states is allowed due to the heavy atom effect, achieved by the addition of a halogen atom (bromine), which increases the rate of intersystem crossing [121].

Since tumor cells are under oxidative stress, mostly caused by the overexpression of the superoxide anion, this ROS species can be considered as a cancer marker [100]. Considering that Clz leads to the formation of excited-state products when in the presence of this same cancer marker, it is possible to explore this to obtain a CL reaction that is only triggered by the presence of the superoxide anion, making the reaction intrinsically tumor-selective. Additionally, given the fact that the CL reaction can be used as a self-excitation mechanism, only the PS would need to be administered to the patients, eliminating the problem of delivering several reaction components to the same localization within the tumor [121]. Interestingly, the system developed by Pinto da Silva et al. is not about the production of ROS but rather about the exchange of superoxide anion into singlet oxygen. This provides another advantage, as superoxide is a major target of the cell antioxidant machinery while singlet oxygen is not. Thus, the resulting oxidative stress is able to bypass the defenses of the cell [124].

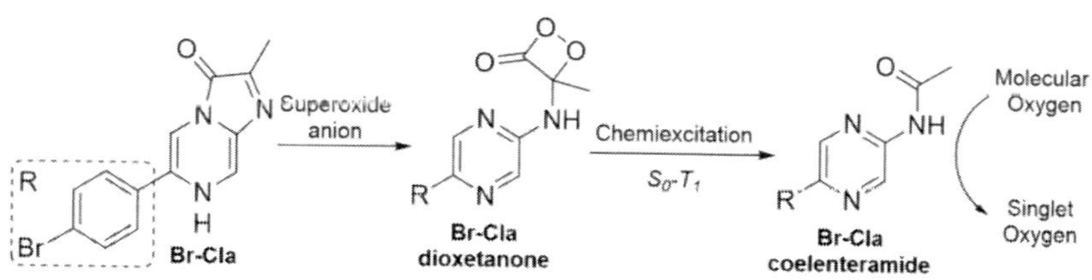

Scheme 9. Schematic representation of the BL-based PDT using the Clz derivative, Br-Cla [6-(4-bromophenyl)-2-methylimidazo(1,2-a)pyrazin-3(7H)-one].

The system proposed by Pinto da Silva et al. was based on the direct generation of triplet states via the Clz CL reaction [121] which is known for the efficient production of singlet excited states [45, 125]. To promote the formation of triplet states, they increased the efficiency of intersystem crossing during the chemiexcitation step, which can be achieved by using heavy atoms [126]. The team prepared a Clz-based analogue in which the phenol group was substituted by a bromobenzene group (Br-Cla) [121]. The *p*-cresol and benzyl moieties of Clz were substituted by a methyl group and a hydrogen atom, respectively, to simplify the structure [121].

The *in silico* modelling of Br-Cla predicted that it should be present in a neutral form in acidic tumor microenvironments (pH 4.5-5 *versus* pH 7.4 for normal tissues) [121, 127, 128]. Moreover, the theoretical calculations indicate that all the involved reactional steps should be exothermic, and so the S_0 CL reaction should be feasible. However, in normal tissues, when the Br-Cla core is ionized, the initial oxidation step is endothermic, meaning that the CL reaction is not favoured at the pH associated with healthy tissues [121, 127, 128]. Thus, the pH difference between tumoral and healthy tissues is another factor that can provide selectivity for this system [121].

Another advantage of Br-Cla is its conformity with Lipinski's rule of five [129, 130] according to theoretical modelling. Lipinski observed that new drugs tend to have higher success rates in clinical trials when they have: [131] molecular masses lower than 500 Da; less than 5 hydrogen bond donors (sum of NH and OH groups) and 10 acceptors (sum of N and O atoms); and an octanol-water partition coefficient lower than 5 (lipophilicity) [121].

The ability of the superoxide anion to trigger the CL of Br-Cla reaction was tested measuring the CL output of the PS while adding different amounts of potassium superoxide (superoxide anion source). Light-emission was detected, which demonstrated the reactivity of Br-Cla towards superoxide [121]. Furthermore, to confirm the involvement of triplet states, the CL output was measured both under normal and inert atmosphere (in which O_2, a known triplet state quencher, was removed by bubbling the solution with N_2). The results showed a 26% higher CL output under inert atmosphere, indicating the involvement of triplet excited states [121]. The same test was conducted for normal Clz, however, N_2-bubbling did not have any effect on the CL output [121]. These results support the hypothesis that the addition of the heavy atom provided Br-Cla with triplet excited states, as opposed to commercial Clz. The production of singlet oxygen as a result of the CL reaction of Br-Cla was confirmed in fluorescent assays using 9,10-anthracenediyl-bis(methylene) dimalonic acid (ABDA) as a singlet oxygen sensor [132]. ABDA is a known singlet oxygen quencher and, given that Br-Cla induces its photobleaching, it allows the monitoring of singlet oxygen generation by fluorescent analysis [132]. CL reactions with increased amounts of Br-Cla significantly quenched ABDA's fluorescence, proving the Br-Cla-mediated generation of singlet oxygen [121]. Moreover, to confirm that these results were due to the CL reaction of Br-Cla and a non-specific interaction with the PS, ABDA's fluorescence with Br-Cla was measured both in the presence and absence of potassium superoxide. Br-Cla only induces quenching in ABDA's fluorescence when potassium superoxide is present, suggesting that the quenching is a direct result of Br-Cla's CL reaction [121]. The CL reaction of commercial Clz does not affect the fluorescence of ABDA, both in the presence or absence of potassium superoxide [121]. However, it should be noted that Clz itself induces some quenching of ABDA's fluorescence, which the authors attribute to non-specific interactions [121].

To confirm the suitability of Br-Cla as a tumor-selective self-activating PS for PDT, the authors conducted *in vitro* cytotoxicity assays for both cancer (MCF-7 breast and PC-3 prostate cancer cells) and healthy (MCF-10A breast cells) cellular lines [121]. In aprotic solvents, Clz and its analogues can undergo CL triggered only by molecular oxygen, [75, 125, 133] whereas in aqueous solutions and biological media this only occurs in the presence of the superoxide anion, [92, 133] which is why, considering this radical is overexpressed tumor cells, [100] the reaction with superoxide anion should provide intrinsic tumor-selectivity to Br-Cla.

Br-Cla induced considerable cytotoxicity for MCF-7 and PC-3 tumor cell lines. Toxicity of more than 50% was observed with a Br-Cla concentration of 75 μM in PC-3 cells, whereas for MCF-7 toxicity was observed for concentrations as low as 0.1 μM [121]. Remarkably, Br-Cla showed better results than reference drugs such as Tamoxifen (breast cancer) [134] and Metformin (prostate cancer), [135] which highlighted its potential as a PS for self-activating PS [121]. While cytotoxicity towards tumor cells is required, selectivity is also needed: when incubated with MCF-10A cells, Br-Cla showed no toxicity in concentrations between 0.1 and 100 μM, indicating a high selectivity for superoxide-overexpressing tumor cells [121]. However, *in vivo* studies are required given that overexpression of superoxide anion might also result from causes other than cancer, such as senescence and inflammation. Commercial Clz, did not induce any toxicity towards MCF-7 cells, and only induced slight cytotoxicity, at higher concentrations for PC-3 cells (but at levels largely below the 50% toxicity resulting from Br-Cla) [121].

In summary, the work by Pinto da Silva et al. [121] describes a single-molecule self-activating PS for PDT with tumor-selective anti-tumoral effects. The PS is directly chemiexcited into a triplet excited state via a CL reaction induced by the superoxide anion, a known cancer marker, culminating in the generation of singlet oxygen without the need for other excitation sources. Another factor, the fact that the process is favoured by the acidic pH of tumoral microenvironments, further improves the tumor-selectivity. Additionally, Br-

Cla does not require additional catalysts/co-factors, facilitating its delivery process. The fact that it also showed cytotoxicity to several cancer cell lines while being safe towards normal cellular lines, turns Br-Cla into a promising anti-tumoral agent, even more effective than the reference drugs. This single-molecule PS provides a pathway to avoid the limitations associated with the need for an external light source (both in terms of tumor size and localization), providing an alternative PDT that is effective even in deep tissues or non-localized tumors.

Further building up on their work, Pinto da Silva et al. reported the syntheses of three new self-activating PSs that, like Br-Cla, worked through the generation of singlet oxygen via a CL reaction activated by the superoxide anion [122]. The performance of these new coelenterazine analogues as antitumor agents was also assessed *in vitro*: they performed better in co-treatment with reference drugs (Tamoxifen and Metformin) for breast and prostate cancers [122].

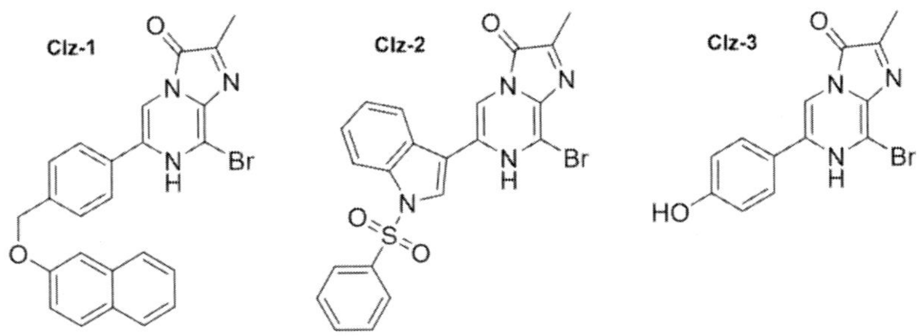

Figure 4. Clz derivatives Clz-1, Clz-2 and Clz-3.

The structures of these new single-molecule self-activating PSs (Clz-1, Clz-2, and Clz-3), also based on Clz, are depicted in Figure [122]. Similarly to Clz, the light-free activation of the new PSs is based on the production of triplet states via a CL reaction. Additionally, all three PSs were designed to incorporate bromine heavy atoms, in order to increase the efficiency of intersystem crossing (ISC) and generate more triplet states [122].

The ability of the superoxide anion for triggering the reaction in the PSs was demonstrated when all the PSs emitted light in the presence of the cancer marker. Among these three PSs, Clz-3 had the best CL, and Clz-1 showed appreciable luminescence. In comparison, Clz-2 had an almost negligible CL [122]. It is worth noting that the intensity of the emission by the PSs was dependent on the amount of potassium superoxide used. Emission by Clz-2 increased with higher potassium superoxide concentrations, whereas emission of Clz-1 and Clz-3 decreased slightly, suggesting they might undergo excessive oxidation for superoxide anion concentrations above a certain threshold [122]. The production of singlet oxygen by the PSs was assessed through fluorescent assays with ABDA, revealing that all the PSs were able to quench the fluorescence of the singlet oxygen sensor. Clz-2 and Clz-3 showed moderate results, quenching around 20-25% of ABDA's fluorescence (for a PS concentration of 50 μM), while Clz-1 achieved quenching values considerably higher than 25% [122].

The PSs' cytotoxicity towards breast and prostate cancer was evaluated *in vitro*, incubating them with MCF-7 and PC-3 cell lines, respectively [122]. Regarding breast cancer, the PSs require longer incubation times, as the best results were obtained with an incubation period of 72 hours. This agreed with the IC_{50} values obtained by the team (Table 1), which decreased considerably as the incubation progressed. For an incubation of 72 hours, the IC_{50} values of Clz-1 were similar to those presented by Tamoxifen, while those obtained for Clz-2 were up to three times lower than those of the reference drug [122]. Regarding prostate cancer, the PSs performed better than Metformin after 24 hours of incubation, suggesting that the new compounds were very active for a short time period for PC-3 cells [122]. The results showed that the new PSs were cytotoxic towards MCF-7 and PC-3 cancer cell lines, performing even better than the reference drug in some cases. The cytotoxicity towards MCF-10A (non-tumorigenic epithelial cell line) was assessed, and no cytotoxicity was observed, which supported the selective nature of the treatment [122]. Hence, the Clz-based derivatives possess potential to be safely used as PSs in PDT without the need for external light sources.

Table 1. IC_{50} values obtained for Clz-1, Clz-2, Clz-3, and the drugs Metformin and Tamoxifen in MCF-7 and PC-3 tumor cell lines (for treatments of 24 and 72 hours)

	MCF-7 IC_{50}		***PC-3 IC_{50}***	
	24 h	**72 h**	**24 h**	**72 h**
Metformin	N.D.	N.D.	1.270 ± 0.261	0.813 ± 0.261
Tamoxifen	2.219 ± 0.194	11.07 ± 0.02	N.D.	N.D.
Clz-1	>100	12.18 ± 0.06	0.048 ± 0.426	12.11 ± 0.15
Clz-2	47.31*	3.00 ± 0.08	0.388 ± 0.459	1.647 ± 0.366
Clz-3	>100	49.59*	0.530 ± 0.525	3.949 ± 0.362

IC_{50} values (μM) represent a half-maximal inhibitory concentration.
* - Approximate estimate due to lack of data points.

Having observed the selective cytotoxic activity of the compounds towards breast and prostate cancer cell lines, the team continued studying the effects of co-treating these cell lines with the Clz-based PSs and the reference drugs (Tamoxifen and Metformin,) [122]. Different concentrations of each PS, along with the reported IC_{50} dose for Tamoxifen (2.22 μM) and Metformin (1.27 μM), were used to co-treat the cells for 24 and 72 hours.

Co-treatment for the breast cancer cell line with an incubation period of 24 hours was the most efficient of all the studied combinations [122]. The most prominent effects were observed with higher concentrations of PSs (more than 10 μM). These results are quite promising, as the co-treatment with PS/drug was more effective than with the PSs themselves, or even than with the reference drug Tamoxifen [122]. Moreover, increasing the incubation time from 24 to 72 hours led to better results, which is in agreement with the IC_{50} for the individual compounds. After 72 hours of incubation, the cytotoxicity for all combinations was higher than that of the individual PSs and reference drugs, with a higher combination efficiency [122].

Opposite to the co-treatment of breast cancer, co-treatment for prostate tumor cell lines was not very effective for an incubation period of 24 hours when compared to the PSs and Metformin alone [122]. At the very best, the co-treatment could be somewhat more efficient when compared to just Metformin. However, it is worth nothing that the PSs themselves already resulted in good individual responses. As for the incubation time, increasing it to 72 hours did enhance the co-treatment with Clz-3 and Metformin. When referring to higher concentrations of the PS, the combination yielded better results than the individual compounds [122].

In summary, Pinto da Silva et al. [122] synthesized three Clz-derivatives that presented selective cytotoxicity toward breast and prostate tumor cell lines. The potential of these PSs as co-treatments with the reference drugs was evaluated. For breast cancer, co-treatment yielded better results (for both 24- and 72-hours incubation periods) than the individual components. For prostate cancer (24-hour incubation period) co-treatment did not yield better results than PSs or Metformin alone. Only when the incubation time was increased did the combination of one of the Clz-based PSs with Metformin become better than the individual components.

As a final remark regarding the work, while the use of light-emitting reactions can be considered an alternative to external light sources and solves the issue of light's depth of penetration, the fact that several components have to be delivered to the tumor and co-localized complicates reaching the target. Also, it means that conventional PDT loses its main advantage, its selectivity towards tumor cells (as only irradiated areas undergo photodynamic effect). Last, considering that several energy transfer steps are involved in the process, CL/BL-based multi-component systems tend to present a somewhat low efficiency and reproducibility, which is not acceptable in PDT. The use of a single-molecule self-activating PSs, suggested by Pinto da Silva et al. [121, 122] solves all the issues presented by normal CL/BL-mediated PDT. The use of a single molecule means that there is only one component that needs to reach the target destination, facilitating treatment delivery. Even though the system does not require external light sources, it still retains its selectivity as the luminescent reaction of the PS is triggered by the superoxide anion, a known cancer marker overexpressed in tumor cells. The cytotoxicity presented by the Clz-based systems was acceptable for its use in PDT and proved that, when activated, these PSs caused tumor cell death while not damaging normal cells. Furthermore, it was demonstrated that the PSs, which are effective by themselves, could be combined with reference drugs to potentiate the effect of the treatment, thus increasing the efficiency of the therapy.

Conclusion

Cancer is a global concern, and the cause of millions of deaths each year worldwide. Several types of therapeutics are employed to battle this condition, such as chemotherapy, radiotherapy, or immunotherapy, among others. One of such therapies is PDT, a modality known for its minimally invasive nature, selectivity towards tumoral cells, and limited side effects. However, the low depth of penetration provided by conventional light sources makes PDT effective only when considering more superficial tumors.

To solve this limitation, researchers have focused on creating strategies to improve the efficacy of PDT and allow it to be used in deeper tumors and even metastatic cancer. Chemiluminescence (CL)/bioluminescence (BL) reactions were exploited as alternative internal excitation sources for PDT. In these systems, the CL/BL reaction was used to excite the PS, eliminating the need for an external light source. However, the use of CL/BL reactions to excite a PS also has its limitations: these systems require different components to be present in the same cellular location; the efficiency of the energy transfer is limited given the participation of several components (which limits the efficacy of the photodynamic effect). The biggest advantage of PDT, its selectivity, is lost as most CL/BL reactions are not intrinsically tumor-selective. Even obtaining considerable cytotoxicity towards cancer cells, the residual toxicity towards normal cells (in some cases), the lack of selectivity and reproducibility, and the low efficiency in the energy transfer to the PS, ultimately mean that multi-component CL/BL-based PDT is not yet suitable for clinical application.

Because of the aforementioned limitations for CL/BL-based multi-component systems, new systems consisting of a tumor-selective and self-activating single-molecule PSs are being developed. These systems are based on the CL reaction of marine Coelenterazine (Clz). Without the need for a light source, the new PSs are chemiexcited directly into triplet states due to a CL reaction triggered solely by a cancer marker, which subsequently sensitizes the highly cytotoxic singlet oxygen. The use of a single-molecule system means that only one component needs to be delivered to the target area, facilitating treatment delivery. By being only activated by a cancer marker, these systems are intrinsically tumor-selective, meaning that even though the system is not activated by localized light irradiation it is still selective and thus cytotoxic only towards cancer cells. The cytotoxicity towards tumor cells is enough for the PS to be considered useful for clinical applications, while there is virtually no damage done to normal cells. The effectiveness of the system is comparable to that of reference drugs, and co-treatment with both potentiates the effects, leading to more effective treatment for the patient. Thus, these novel single-molecule PSs show great potential for eliminating the light-related restrictions regarding tumor size and location that PDT currently presents, while maintaining its most relevant advantages (selectivity, minimally invasive nature, and broad scope).

Acknowledgments

This work was made within the framework of projects PTDC/QUI-QFI/2870/2020, UIDB/00081/2020 (CIQUP), and UIDB/05748/2020 (GreenUPorto), funded by "Fundação para a Ciência e Tecnologia" (FCT, Portugal). Luís Pinto da Silva acknowledges funding from FCT under the Scientific Employment Stimulus (CEECIND701425/2017 | 2021.00768. CEECIND). FCT is also acknowledged for funding the Associated Laboratory IMS (LA/P/0056/2020). Patricia González-Berdullas acknowledges FCT-funded project PTDC/QUI-QFI/2870/2020, for funding her Post-Doc position. Carla Magalhães acknowledges FCT for her Ph.D. grant (SFRH/BD/143211/2019). Ricardo Sendão acknowledges FCT for his PhD grant (2021.06149.BD).

References

[1] Sung, H.; Ferlay, J.; Siegel, R. L.; Laversanne, M.; Soerjomataram, I.; Jemal, A.; Bray, F., Global cancer statistics 2020: GLOBOCAN estimates of incidence and mortality worldwide for 36 cancers in 185 countries. *CA Cancer J. Clin.* **2021**, 71 (3), 209-249.

[2] Andrade, D.; Vinck, E. E.; Torres, L. N.; Citarella, D., Surgical removal of intravascular leiomyomas with intracardiac extension: A two-case report. *J. Card. Surg.* **2020,** *35* (5), 1094-1097.

[3] Kotsopoulos, J.; Narod, S. A., Prophylactic salpingectomy for the prevention of ovarian cancer: Who should we target? *Int. J. Cancer* **2020,** *147* (5), 1245-1251.

[4] Chua, B.; Jackson, J. E.; Lin, C.; Veness, M. J., Radiotherapy for early non-melanoma skin cancer. *Oral Oncol.* **2019,** *98*, 96-101.

[5] Westerveld, H.; Nesvacil, N.; Fokdal, L.; Chargari, C.; Schmid, M. P.; Milosevic, M.; Mahantshetty, U. M.; Nout, R. A., Definitive radiotherapy with image-guided adaptive brachytherapy for primary vaginal cancer. *Lancet Oncol.* **2020,** *21* (3), e157-e167.

[6] Birtle, A.; Johnson, M.; Chester, J.; Jones, R.; Dolling, D.; Bryan, R. T.; Harris, C.; Winterbottom, A.; Blacker, A.; Catto, J. W. F.; Chakraborti, P.; Donovan, J. L.; Elliott, P. A.; French, A.; Jagdev, S.; Jenkins, B.; Keeley, F. X.; Kockelbergh, R.; Powles, T.; Wagstaff, J.; Wilson, C.; Todd, R.; Lewis, R.; Hall, E., Adjuvant chemotherapy in upper tract urothelial carcinoma (the POUT trial): a phase 3, open-label, randomised controlled trial. *Lancet* **2020,** *395* (10232), 1268-1277.

[7] Haddadin, M.; Taylor, J., Chemotherapy Options for Blastic Plasmacytoid Dendritic Cell Neoplasm. *Hematol. Oncol. Clin. North. Am.* **2020,** *34* (3), 539-552.

[8] Mizrak Kaya, D.; Nogueras Gonzalez, G. M.; Harada, K.; Blum Murphy, M. A.; Lee, J. H.; Bhutani, M. S.; Weston, B.; Thomas, I.; Rogers, J. E.; Das, P.; Badgwell, B. D.; Ajani, J. A., Efficacy of Three-Drug Induction Chemotherapy Followed by Preoperative Chemoradiation in Patients with Localized Gastric Adenocarcinoma. *Oncology* **2020,** *98* (8), 542-548.

[9] Klichinsky, M.; Ruella, M.; Shestova, O.; Lu, X. M.; Best, A.; Zeeman, M.; Schmierer, M.; Gabrusiewicz, K.; Anderson, N. R.; Petty, N. E.; Cummins, K. D.; Shen, F.; Shan, X.; Veliz, K.; Blouch, K.; Yashiro-Ohtani, Y.; Kenderian, S. S.; Kim, M. Y.; O'Connor, R. S.; Wallace, S. R.; Kozlowski, M. S.; Marchione, D. M.; Shestov, M.; Garcia, B. A.; June, C. H.; Gill, S., Human chimeric antigen receptor macrophages for cancer immunotherapy. *Nat. Biotechnol.* **2020,** *38* (8), 947-953.

[10] Lichtenstern, C. R.; Ngu, R. K.; Shalapour, S.; Karin, M., Immunotherapy, Inflammation and Colorectal Cancer. *Cells* **2020,** *9* (3), 618.

[11] Caffa, I.; Spagnolo, V.; Vernieri, C.; Valdemarin, F.; Becherini, P.; Wei, M.; Brandhorst, S.; Zucal, C.; Driehuis, E.; Ferrando, L.; Piacente, F.; Tagliafico, A.; Cilli, M.; Mastracci, L.; Vellone, V. G.; Piazza, S.; Cremonini, A. L.; Gradaschi, R.; Mantero, C.; Passalacqua, M.; Ballestrero, A.; Zoppoli, G.; Cea, M.; Arrighi, A.; Odetti, P.; Monacelli, F.; Salvadori, G.; Cortellino, S.; Clevers, H.; De Braud, F.; Sukkar, S. G.; Provenzani, A.; Longo, V. D.; Nencioni, A., Fasting-mimicking

diet and hormone therapy induce breast cancer regression. *Nature* **2020,** *583* (7817), 620-624.

[12] Student, S.; Hejmo, T.; Poterala-Hejmo, A.; Lesniak, A.; Buldak, R., Anti-androgen hormonal therapy for cancer and other diseases. *Eur. J. Pharmacol.* **2020,** *866*, 172783.

[13] Silva, S.; Almeida, A. J.; Vale, N., Combination of Cell-Penetrating Peptides with Nanoparticles for Therapeutic Application: A Review. *Biomolecules* **2019,** *9* (1), 22.

[14] Han, K.; Jeng, E. E.; Hess, G. T.; Morgens, D. W.; Li, A.; Bassik, M. C., Synergistic drug combinations for cancer identified in a CRISPR screen for pairwise genetic interactions. *Nat. Biotechnol.* **2017,** *35* (5), 463-474.

[15] Laptev, R.; Nisnevitch, M.; Siboni, G.; Malik, Z.; Firer, M. A., Intracellular chemiluminescence activates targeted photodynamic destruction of leukaemic cells. *Br. J. Cancer* **2006,** *95* (2), 189-196.

[16] Yano, S.; Hirohara, S.; Obata, M.; Hagiya, Y.; Ogura, S.-i.; Ikeda, A.; Kataoka, H.; Tanaka, M.; Joh, T., Current states and future views in photodynamic therapy. *J. Photochem. Photobiol. C* **2011,** *12* (1), 46-67.

[17] Ethirajan, M.; Chen, Y.; Joshi, P.; Pandey, R. K., The role of porphyrin chemistry in tumor imaging and photodynamic therapy. *Chem. Soc. Rev.* **2011,** *40* (1), 340-362.

[18] van Straten, D.; Mashayekhi, V.; de Bruijn, H. S.; Oliveira, S.; Robinson, D. J., Oncologic Photodynamic Therapy: Basic Principles, Current Clinical Status and Future Directions. *Cancers (Basel)* **2017,** *9* (2), 19.

[19] Yue, J.; Liang, L.; Shen, Y.; Guan, X.; Zhang, J.; Li, Z.; Deng, R.; Xu, S.; Liang, C.; Shi, W.; Xu, W., Investigating Dynamic Molecular Events in Melanoma Cell Nucleus During Photodynamic Therapy by SERS. *Front. Chem.* **2019,** *6*, 665.

[20] Zeng, K.; Xu, Q.; Ouyang, J.; Han, Y.; Sheng, J.; Wen, M.; Chen, W.; Liu, Y. N., Coordination Nanosheets of Phthalocyanine as Multifunctional Platform for Imaging-Guided Synergistic Therapy of Cancer. *ACS Appl. Mater. Interfaces* **2019,** *11* (7), 6840-6849.

[21] Fan, J. H.; Fan, G. L.; Yuan, P.; Deng, F. A.; Liu, L. S.; Zhou, X.; Yu, X. Y.; Cheng, H.; Li, S. Y., A Theranostic Nanoprobe for Hypoxia Imaging and Photodynamic Tumor Therapy. *Front. Chem.* **2019,** *7*, 868.

[22] Glass, S.; Trinklein, B.; Abel, B.; Schulze, A., TiO_2 as Photosensitizer and Photoinitiator for Synthesis of Photoactive TiO_2-PEGDA Hydrogel Without Organic Photoinitiator. *Front. Chem.* **2018,** *6*, 340.

[23] Ouyang, J.; Deng, Y.; Chen, W.; Xu, Q.; Wang, L.; Liu, Z.; Tang, F.; Deng, L.; Liu, Y. N., Marriage of artificial catalase and black phosphorus nanosheets for reinforced photodynamic antitumor therapy. *J. Mater. Chem. B* **2018,** *6* (14), 2057-2064.

[24] Zang, Q.; Yu, J.; Yu, W.; Qian, J.; Hu, R.; Tang, B. Z., Red-emissive azabenzanthrone derivatives for photodynamic therapy irradiated with ultralow light power density and two-photon imaging. *Chem. Sci.* **2018,** *9* (23), 5165-5171.

[25] Li, B.; Lin, L.; Lin, H.; Wilson, B. C., Photosensitized singlet oxygen generation and detection: Recent advances and future perspectives in cancer photodynamic therapy. *J. Biophotonics* **2016,** *9* (11-12), 1314-1325.

[26] Lovell, J. F.; Liu, T. W.; Chen, J.; Zheng, G., Activatable photosensitizers for imaging and therapy. *Chem. Rev.* **2010,** *110* (5), 2839-2857.

[27] Magalhaes, C. M.; Esteves da Silva, J. C.; Pinto da Silva, L., Chemiluminescence and Bioluminescence as an Excitation Source in the Photodynamic Therapy of Cancer: A Critical Review. *Chemphyschem* **2016,** *17* (15), 2286-2294.

[28] Josefsen, L. B.; Boyle, R. W., Photodynamic therapy and the development of metal-based photosensitisers. *Met. Based Drugs* **2008,** *2008*, 276109.

[29] Fan, W.; Huang, P.; Chen, X., Overcoming the Achilles' heel of photodynamic therapy. *Chem. Soc. Rev.* **2016,** *45* (23), 6488-6519.

[30] Zhou, Z.; Song, J.; Nie, L.; Chen, X., Reactive oxygen species generating systems meeting challenges of photodynamic cancer therapy. *Chem. Soc. Rev.* **2016,** *45* (23), 6597-6626.

[31] Xu, S.; Zhu, X.; Zhang, C.; Huang, W.; Zhou, Y.; Yan, D., Oxygen and Pt(II) self-generating conjugate for synergistic photo-chemo therapy of hypoxic tumor. *Nat. Commun.* **2018,** *9* (1), 2053.

[32] Lan, M.; Zhao, S.; Liu, W.; Lee, C. S.; Zhang, W.; Wang, P., Photosensitizers for Photodynamic Therapy. *Adv. Healthc. Mater.* **2019,** *8* (13), e1900132.

[33] Bolze, F.; Jenni, S.; Sour, A.; Heitz, V., Molecular photosensitisers for two-photon photodynamic therapy. *Chem. Commun.* **2017,** *53* (96), 12857-12877.

[34] Zhou, Z.; Liu, J.; Rees, T. W.; Wang, H.; Li, X.; Chao, H.; Stang, P. J., Heterometallic Ru-Pt metallacycle for two-photon photodynamic therapy. *Proc. Natl. Acad. Sci. USA.* **2018,** *115* (22), 5664-5669.

[35] Chen, H.; Wang, G. D.; Chuang, Y. J.; Zhen, Z.; Chen, X.; Biddinger, P.; Hao, Z.; Liu, F.; Shen, B.; Pan, Z.; Xie, J., Nanoscintillator-mediated X-ray inducible photodynamic therapy for in vivo cancer treatment. *Nano Lett.* **2015,** *15* (4), 2249-2256.

[36] Jiang, L.; Liu, L.; Lv, F.; Wang, S.; Ren, X., Integration of Self-Luminescence and Oxygen Self-Supply: A Potential Photodynamic Therapy Strategy for Deep Tumor Treatment. *Chempluschem* **2020,** *85* (3), 510-518.

[37] Bancirova, M.; Lasovsky, J., The photodynamic effect: the comparison of chemiexcitation by luminol and phthalhydrazide. *Luminescence* **2011,** *26* (6), 410-415.

[38] Yuan, H.; Chong, H.; Wang, B.; Zhu, C.; Liu, L.; Yang, Q.; Lv, F.; Wang, S., Chemical molecule-induced light-activated system for anticancer and antifungal activities. *J. Am. Chem. Soc.* **2012,** *134* (32), 13184-13187.

[39] Schipper, M. L.; Patel, M. R.; Gambhir, S. S., Evaluation of firefly luciferase bioluminescence mediated photodynamic toxicity in cancer cells. *Mol. Imaging. Biol.* **2006,** *8* (4), 218-225.

[40] Zhang, Y.; Pang, L.; Ma, C.; Tu, Q.; Zhang, R.; Saeed, E.; Mahmoud, A. E.; Wang, J., Small molecule-initiated light-activated semiconducting polymer dots: an integrated nanoplatform for targeted photodynamic therapy and imaging of cancer cells. *Anal. Chem.* **2014,** *86* (6), 3092-3099.

[71] Yue, L.; Liu, Y. J., Mechanism of AMPPD Chemiluminescence in a Different Voice. *J. Chem. Theory Comput.* **2013,** *9* (5), 2300-2312.

[72] Ding, B. W.; Liu, Y. J., Bioluminescence of Firefly Squid via Mechanism of Single Electron-Transfer Oxygenation and Charge-Transfer-Induced Luminescence. *J. Am. Chem. Soc.* **2017,** *139* (3), 1106-1119.

[73] McCapra, F., Chemiluminescence and bioluminescence. *J. Photochem. Photobiol. A* **1990,** *51* (1), 21-28.

[74] Pinto da Silva, L.; Pereira, R. F. J.; Magalhães, C. M.; Esteves da Silva, J. C. G., Mechanistic Insight into Cypridina Bioluminescence with a Combined Experimental and Theoretical Chemiluminescent Approach. *J. Phys. Chem. B* **2017,** *121* (33), 7862-7871.

[75] Magalhães, C. M.; Esteves da Silva, J. C. G.; Pinto da Silva, L., Study of coelenterazine luminescence: Electrostatic interactions as the controlling factor for efficient chemiexcitation. *J. Lumin.* **2018,** *199*, 339-347.

[76] Pinto da Silva, L.; Magalhães, C. M., Mechanistic insights into the efficient intramolecular chemiexcitation of dioxetanones from TD-DFT and multireference calculations. *Int. J. Quantum Chem.* **2018,** *119* (9), e25881.

[77] Ding, B. W.; Naumov, P.; Liu, Y. J., Mechanistic insight into marine bioluminescence: photochemistry of the chemiexcited Cypridina (sea firefly) lumophore. *J. Chem. Theory Comput.* **2015,** *11* (2), 591-599.

[78] Roca-Sanjuan, D.; Delcey, M. G.; Navizet, I.; Ferre, N.; Liu, Y. J.; Lindh, R., Chemiluminescence and Fluorescence States of a Small Model for Coelenteramide and Cypridina Oxyluciferin: A CASSCF/CASPT2 Study. *J. Chem. Theory Comput.* **2011,** *7* (12), 4060-4069.

[79] Pinto da Silva, L.; Esteves da Silva, J. C., Theoretical fingerprinting of the photophysical properties of four firefly bioluminophores. *Photochem. Photobiol. Sci.* **2013,** *12* (11), 2028-2035.

[80] Naumov, P.; Kochunnoonny, M., Spectral-structural effects of the keto-enol-enolate and phenol-phenolate equilibria of oxyluciferin. *J. Am. Chem. Soc.* **2010,** *132* (33), 11566-11579.

[81] Chen, S.; Navizet, I.; Lindh, R.; Liu, Y.; Ferre, N., Hybrid QM/MM simulations of the obelin bioluminescence and fluorescence reveal an unexpected light emitter. *J. Phys. Chem. B* **2014,** *118* (11), 2896-2903.

[82] Ghose, A.; Rebarz, M.; Maltsev, O. V.; Hintermann, L.; Ruckebusch, C.; Fron, E.; Hofkens, J.; Mely, Y.; Naumov, P.; Sliwa, M.; Didier, P., Emission properties of oxyluciferin and its derivatives in water: revealing the nature of the emissive species in firefly bioluminescence. *J. Phys. Chem. B* **2015,** *119* (6), 2638-2649.

[83] Pinto da Silva, L.; Magalhães, C. M.; Esteves da Silva, J. C. G., Interstate Crossing-Induced Chemiexcitation Mechanism as the Basis for Imidazopyrazinone Bioluminescence. *Chemistry Select* **2016,** *1* (12), 3343-3356.

[84] Merenyi, G.; Lind, J.; Eriksen, T. E., Luminol chemiluminescence: chemistry, excitation, emitter. *J. Biolumin. Chemilumin.* **1990,** *5* (1), 53-56.

[85] Gaikwad, A.; Silva, M.; Pérez-Bendito, D., Selective stopped-flow determination of manganese with luminol in the absence of hydrogen peroxide. *Anal. Chim. Acta* **1995,** *302* (2-3), 275-282.

[86] Rose, A. L.; Waite, T. D., Chemiluminescence of luminol in the presence of iron(II) and oxygen: oxidation mechanism and implications for its analytical use. *Anal. Chem.* **2001,** *73* (24), 5909-5920.

[87] Hosseinkhani, S., Molecular enigma of multicolor bioluminescence of firefly luciferase. *Cell Mol. Life. Sci.* **2011,** *68* (7), 1167-1182.

[88] Inouye, S., Firefly luciferase: an adenylate-forming enzyme for multicatalytic functions. *Cell Mol. Life. Sci.* **2010,** *67* (3), 387-404.

[89] Pinto da Silva, L.; Esteves da Silva, J. C., Computational Studies of the Luciferase Light-Emitting Product: Oxyluciferin. *J. Chem. Theory. Comput.* **2011,** *7* (4), 809-817.

[90] Pinto da Silva, L.; Santos, A. J.; Esteves da Silva, J. C., Efficient firefly chemi/bioluminescence: evidence for chemiexcitation resulting from the decomposition of a neutral firefly dioxetanone molecule. *J. Phys. Chem. A* **2013,** *117* (1), 94-100.

[91] Modestova, Y.; Koksharov, M. I.; Ugarova, N. N., Point mutations in firefly luciferase C-domain demonstrate its significance in green color of bioluminescence. *Biochim. Biophys. Acta* **2014,** *1844* (9), 1463-1471.

[92] Teranishi, K., Luminescence of imidazo[1,2-*a*]pyrazin-3(7*H*)-one compounds. *Bioorg. Chem.* **2007,** *35* (1), 82-111.

[93] Kondo, H.; Igarashi, T.; Maki, S.; Niwa, H.; Ikeda, H.; Hirano, T., Substituent effects on the kinetics for the chemiluminescence reaction of 6-arylimidazo[1,2-*a*]pyrazin-3(7*H*)-ones (Cypridina luciferin analogues): support for the single electron transfer (SET)–oxygenation mechanism with triplet molecular oxygen. *Tetrahedron Lett.* **2005,** *46* (45), 7701-7704.

[94] Bhattacharya, A.; Muller, F. L.; Liu, Y.; Sabia, M.; Liang, H.; Song, W.; Jang, Y. C.; Ran, Q.; Van Remmen, H., Denervation induces cytosolic phospholipase A2-mediated fatty acid hydroperoxide generation by muscle mitochondria. *J. Biol. Chem.* **2009,** *284* (1), 46-55.

[95] Zhang, P.; Hou, M.; Li, Y.; Xu, X.; Barsoum, M.; Chen, Y.; Bache, R. J., NADPH oxidase contributes to coronary endothelial dysfunction in the failing heart. *Am. J. Physiol. Heart Circ. Physiol.* **2009,** *296* (3), H840-H846.

[96] Stepanyuk, G. A.; Liu, Z. J.; Markova, S. S.; Frank, L. A.; Lee, J.; Vysotski, E. S.; Wang, B. C., Crystal structure of coelenterazine-binding protein from *Renilla muelleri* at 1.7 Å: why it is not a calcium-regulated photoprotein. *Photochem. Photobiol. Sci.* **2008,** *7* (4), 442-447.

[97] Min, C. G.; Pinto da Silva, L.; Esteves da Silva, J. C.; Yang, X. K.; Huang, S. J.; Ren, A. M.; Zhu, Y. Q., A Computational Investigation of the Equilibrium Constants for the Fluorescent and Chemiluminescent States of Coelenteramide. *Chemphyschem* **2017,** *18* (1), 117-123.

[98] Xia, Z.; Rao, J., Biosensing and imaging based on bioluminescence resonance energy transfer. *Curr. Opin. Biotechnol.* **2009,** *20* (1), 37-44.

[99] Yao, H.; Zhang, Y.; Xiao, F.; Xia, Z.; Rao, J., Quantum dot/bioluminescence resonance energy transfer based highly sensitive detection of proteases. *Angew. Chem. Int. Ed. Engl.* **2007,** *46* (23), 4346-4349.

[100] Figueira, T. R.; Barros, M. H.; Camargo, A. A.; Castilho, R. F.; Ferreira, J. C.; Kowaltowski, A. J.; Sluse, F. E.; Souza-Pinto, N. C.; Vercesi, A. E., Mitochondria as a source of reactive oxygen and nitrogen species: from molecular mechanisms to human health. *Antioxid. Redox Signal* **2013,** *18* (16), 2029-2074.

[101] Liu, S.; Yuan, H.; Bai, H.; Zhang, P.; Lv, F.; Liu, L.; Dai, Z.; Bao, J.; Wang, S., Electrochemiluminescence for Electric-Driven Antibacterial Therapeutics. *J. Am. Chem. Soc.* **2018,** *140* (6), 2284-2291.

[102] Jiang, L.; Bai, H.; Liu, L.; Lv, F.; Ren, X.; Wang, S., Luminescent, Oxygen-Supplying, Hemoglobin-Linked Conjugated Polymer Nanoparticles for Photodynamic Therapy. *Angew. Chem. Int. Ed. Engl.* **2019,** *58* (31), 10660-10665.

[103] Chen, T. C.; Huang, L.; Liu, C. C.; Chao, P. J.; Lin, F. H., Luminol as the light source for in situ photodynamic therapy. *Process Biochem.* **2012,** *47* (12), 1903-1908.

[104] Yang, K.; Wang, C.; Wei, X.; Ding, S.; Liu, C.; Tian, F.; Li, F., Self-Illuminating Photodynamic Therapy with Enhanced Therapeutic Effect by Optimization of the Chemiluminescence Resonance Energy Transfer Step to the Photosensitizer. *Bioconjug. Chem.* **2020,** *31* (3), 595-604.

[105] Fang, L.; Hu, Q.; Jiang, K.; Zhang, X.; Li, B.; Cui, Y.; Yang, Y.; Qian, G., An inner light integrated metal-organic framework photodynamic therapy system for effective elimination of deep-seated tumor cells. *J. Solid State Chem.* **2019,** *276*, 205-209.

[106] Yesilgul, N.; Uyar, T. B.; Seven, O.; Akkaya, E. U., Singlet Oxygen Generation with Chemical Excitation of an Erythrosine-Luminol Conjugate. *ACS Omega* **2017,** *2* (4), 1367-1371.

[107] Xu, X.; An, H.; Zhang, D.; Tao, H.; Dou, Y.; Li, X.; Huang, J.; Zhang, J., A self-illuminating nanoparticle for inflammation imaging and cancer therapy. *Sci. Adv.* **2019,** *5* (1), eaat2953.

[108] Rauhut, M. M.; Bollyky, L. J.; Roberts, B. G.; Loy, M.; Whitman, R. H.; Iannotta, A. V.; Semsel, A. M.; Clarke, R. A., Chemiluminescence from reactions of electronegatively substituted aryl oxalates with hydrogen peroxide and fluorescent compounds. *J. Am. Chem. Soc.* **1967,** *89* (25), 6515-6522.

[109] Schuster, G. B., Chemiluminescence of organic peroxides. Conversion of ground-state reactants to excited-state products by the chemically initiated electron-exchange luminescence mechanism. *Acc. Chem. Res.* **1979,** *12* (10), 366-373.

[110] Schuster, G. B.; Schmidt, S. P., Chemiluminescence of Organic Compounds. In *Advances in Physical Organic Chemistry*, Gold, V.; Bethell, D., Eds. Academic Press: 1982; Vol. 18, pp 187-238.

[111] Phillip, M. J.; Maximuke, P. P., Chemiluminescence and hematoporphyrin derivative: a novel therapy for mammary adenocarcinomas in mice. *Oncology* **1989,** *46* (4), 266-272.

[112] Lopez-Lazaro, M., A new view of carcinogenesis and an alternative approach to cancer therapy. *Mol. Med.* **2010,** *16* (3-4), 144-153.

[113] Romanyuk, A. V.; Grozdova, I. D.; Ezhov, A. A.; Melik-Nubarov, N. S., Peroxyoxalate Chemiluminescent Reaction as a Tool for Elimination of Tumour Cells Under Oxidative Stress. *Sci. Rep.* **2017,** *7* (1), 3410.

[114] Mao, D.; Wu, W.; Ji, S.; Chen, C.; Hu, F.; Kong, D.; Ding, D.; Liu, B., Chemiluminescence-Guided Cancer Therapy Using a Chemiexcited Photosensitizer. *Chem* **2017,** *3* (6), 991-1007.

[115] Yu, Z.; Zhou, P.; Pan, W.; Li, N.; Tang, B., A biomimetic nanoreactor for synergistic chemiexcited photodynamic therapy and starvation therapy against tumor metastasis. *Nat. Commun.* **2018,** *9* (1), 5044.

[116] Yang, Y.; Hou, W.; Liu, S.; Sun, K.; Li, M.; Wu, C., Biodegradable Polymer Nanoparticles for Photodynamic Therapy by Bioluminescence Resonance Energy Transfer. *Biomacromolecules* **2018,** *19* (1), 201-208.

[117] Yang, K.; Wang, C.; Liu, C.; Ding, S.; Tian, F.; Li, F., Bioluminescence-initiated photodynamic therapy bridged on high-luminescent carbon dots-conjugated protoporphyrin IX. *J. Mater. Sci.* **2019,** *54* (4), 3383-3391.

[118] So, M. K.; Xu, C.; Loening, A. M.; Gambhir, S. S.; Rao, J., Self-illuminating quantum dot conjugates for in vivo imaging. *Nat. Biotechnol.* **2006,** *24* (3), 339-343.

[119] Kim, Y. R.; Kim, S.; Choi, J. W.; Choi, S. Y.; Lee, S. H.; Kim, H.; Hahn, S. K.; Koh, G. Y.; Yun, S. H., Bioluminescence-activated deep-tissue photodynamic therapy of cancer. *Theranostics* **2015,** *5* (8), 805-817.

[120] Al-Ani, A. W.; Zhang, L.; Ferreira, L.; Turyanska, L.; Bradshaw, T. D.; Thomas, N. R., Listeria innocua Dps as a nanoplatform for bioluminescence based photodynamic therapy utilizing Gaussia princeps luciferase and zinc protoporphyrin IX. *Nanomedicine* **2019,** *20*, 102005.

[121] Pinto da Silva, L.; Núñez-Montenegro, A.; Magalhães, C. M.; Ferreira, P. J. O.; Duarte, D.; González-Berdullas, P.; Rodríguez-Borges, J. E.; Vale, N.; Esteves da Silva, J. C. G., Single-molecule chemiluminescent photosensitizer for a self-activating and tumor-selective photodynamic therapy of cancer. *Eur. J. Med. Chem.* **2019,** *183*, 111683.

[122] Pinto da Silva, L.; Magalhães, C. M.; Núñez-Montenegro, A.; Ferreira, P. J. O.; Duarte, D.; Rodríguez-Borges, J. E.; Vale, N.; Esteves da Silva, J. C. G., Study of the Combination of Self-Activating Photodynamic Therapy and Chemotherapy for Cancer Treatment. *Biomolecules* **2019,** *9* (8), 384.

[123] Sun, S. K.; Wu, J. C.; Wang, H.; Zhou, L.; Zhang, C.; Cheng, R.; Kan, D.; Zhang, X.; Yu, C., Turning solid into gel for high-efficient persistent luminescence-sensitized photodynamic therapy. *Biomaterials* **2019,** *218*, 119328.

[124] Pelicano, H.; Carney, D.; Huang, P., ROS stress in cancer cells and therapeutic implications. *Drug Resist. Updat.* **2004,** *7* (2), 97-110.

[125] Magalhães, C. M.; Esteves da Silva, J. C. G.; Pinto da Silva, L., Comparative study of the chemiluminescence of coelenterazine, coelenterazine-e and Cypridina luciferin with an experimental and theoretical approach. *J. Photochem. Photobiol. B* **2019,** *190*, 21-31.

[126] Zou, J.; Yin, Z.; Ding, K.; Tang, Q.; Li, J.; Si, W.; Shao, J.; Zhang, Q.; Huang, W.; Dong, X., BODIPY Derivatives for Photodynamic Therapy: Influence of Configuration versus Heavy Atom Effect. *ACS Appl. Mater. Interfaces* **2017,** *9* (38), 32475-32481.

[127] Bi, H.; Dai, Y.; Yang, P.; Xu, J.; Yang, D.; Gai, S.; He, F.; An, G.; Zhong, C.; Lin, J., Glutathione and H_2O_2 consumption promoted photodynamic and chemotherapy based on biodegradable MnO_2–Pt@Au_{25} nanosheets. *Chem. Eng. Sci.* **2019,** *356*, 543-553.

[128] Wang, Y.; Huang, X.; Tang, Y.; Zou, J.; Wang, P.; Zhang, Y.; Si, W.; Huang, W.; Dong, X., A light-induced nitric oxide controllable release nano-platform based on diketopyrrolopyrrole derivatives for pH-responsive photodynamic/photothermal synergistic cancer therapy. *Chem. Sci.* **2018,** *9* (42), 8103-8109.

[129] Lipinski, C. A.; Lombardo, F.; Dominy, B. W.; Feeney, P. J., Experimental and computational approaches to estimate solubility and permeability in drug discovery and development settings 1PII of original article: S0169-409X(96)00423-1. The article was originally published in Advanced Drug Delivery Reviews 23 (1997) 3–25. 1. *Adv. Drug Deliv. Rev.* **2001,** *46* (1-3), 3-26.

[130] Lipinski, C. A., Lead- and drug-like compounds: the rule-of-five revolution. *Drug Discov. Today Technol.* **2004,** *1* (4), 337-41.

[131] Leeson, P. D.; Springthorpe, B., The influence of drug-like concepts on decision-making in medicinal chemistry. *Nat. Rev. Drug. Discov.* **2007,** *6* (11), 881-90.

[132] Dong, C.; Liu, Z.; Wang, S.; Zheng, B.; Guo, W.; Yang, W.; Gong, X.; Wu, X.; Wang, H.; Chang, J., A Protein-Polymer Bioconjugate-Coated Upconversion Nanosystem for Simultaneous Tumor Cell Imaging, Photodynamic Therapy, and Chemotherapy. *ACS Appl. Mater. Interfaces* **2016,** *8* (48), 32688-32698.

[133] Lourenço, J. M.; Esteves da Silva, J. C. G.; Pinto da Silva, L., Combined experimental and theoretical study of Coelenterazine chemiluminescence in aqueous solution. *J. Lumin.* **2018,** *194*, 139-145.

[134] Shagufta; Ahmad, I., Tamoxifen a pioneering drug: An update on the therapeutic potential of tamoxifen derivatives. *Eur. J. Med. Chem.* **2018,** *143*, 515-531.

[135] Whitburn, J.; Edwards, C. M.; Sooriakumaran, P., Metformin and Prostate Cancer: a New Role for an Old Drug. *Curr. Urol. Rep.* **2017,** *18* (6), 46.

Chapter 4

Fluorescent Photosensitizers: A Promising Tool for Biomedicine

Athira Maria John, Anjana Baby and Sreeja Puthanveetil Balakrishanan*
Department of Chemistry, CHRIST (Deemed to be University), Bengaluru, Karnataka, India

Abstract

The growing demand for the detection for biomedical imaging has taken the interest of researchers as there is a significant increase in cancer and malignant diseases across the globe. Photosensitizers assisted with fluorescent properties can be a pioneer in this field. Photosensitizers generally damage the living cell, however, selected fluorescent photosensitizers can offer a minimal threat to the living cells. Medically relevant processes such as live cell imaging and photodynamic therapy can be monitored using this technique. Some of the commonly used fluorescent photosensitizers include porphyrins, chlorins, and bacteriochlorins. This chapter addresses the significance, limitations, and future perspectives of fluorescent photosensitizers in biological applications. Scientists can develop efficient fluorescent photosensitizers for easy detection and cure of different infectious diseases. The chapter also puts forward a deeper understanding of the principle underlying tunable fluorescent properties, and the recent efforts for developing efficient fluorescent photosensitizers.

Keywords: fluorescent photosensitizer, photodynamic therapy, live cell imaging

---

* Corresponding Author's Email; sreeja.pb@christuniversity.in.

In: Photosensitizers and Their Applications
Editors: Davor Margetic and Renjith Thomas
ISBN: 978-1-68507-880-5

Introduction

Photosensitized reactions have attracted chemists globally because of their cost-effective and simple mode of action compared to other photochemical reactions [1]. Photosensitized reactions are not restricted to a single application because of their properties. The most attractive feature is that only a small quantity of photosensitizers (PSs) is required to complete the reaction. The PS is a molecule that absorbs photons and transfers them to the neighboring molecule, and itself comes to the ground state. The chemical reaction is given by the energy accepting molecule in the excited state [2]. Medical applications such as photodynamic therapy (PDT) [3] and live-cell imaging are based on this principle [4]. PSs applications in energy field such as dye-sensitized solar cells (DSSCs) [5] and hydrogen-generating catalyst also rely on this principle [6]. For developing a PS, the presence of light is an essential condition. Altering the wavelengths can give a better PS depending on the applications and PSs can have fluorescent behavior. Fluorescent materials have higher wavelengths, and therefore can be used for biomedical applications such as PDT [7].

PSs are expected to have low dark toxicity, inhibition by the target tissue, minimum pain to the patient, and they should be appropriately administered. In addition to this, they should be water-soluble for easy transport in the body and chemically stable. These are classified into first, second, and third-generation PSs. First-generation PSs developed in the 1970s which were used in clinical trials are hematoporphyrin derivatives. Their main flaw was poor light absorption in the red spectral range, where light tissue penetration is optimal [8]. Second-generation PSs developed in the late 1980s could overcome several drawbacks of the previous one. These were based on porphyrin structure, such as benzoporphyrins, purpurins, texaphyrins, phthalocyanines, naphthalocyanines, and protoporphyrin IX. These classes of substances had less post-treatment skin photosensitization and longer absorption wavelength [9, 10, 11]. Third-generation PSs are modified versions of the previous generation with better delivery and tumor specificity [12, 13].

Photosensitized reactions can be categorized based on electron and energy transfer reactions. Electron transfer reaction involves hydrogen atom abstraction or electron transfer, resulting in a radical, which then reacts with oxygen to produce an oxidation product. In contrast, energy transfer reactions involve energy transfer between the excited sensitizer and oxygen, resulting in the formation of excited singlet oxygen, from which the oxidation product is formed. Singlet oxygen has unique physical properties that allow it to be characterized and the photosensitized process for a specific reaction to be determined [14]. Detailed mechanism is mentioned in the following sections.

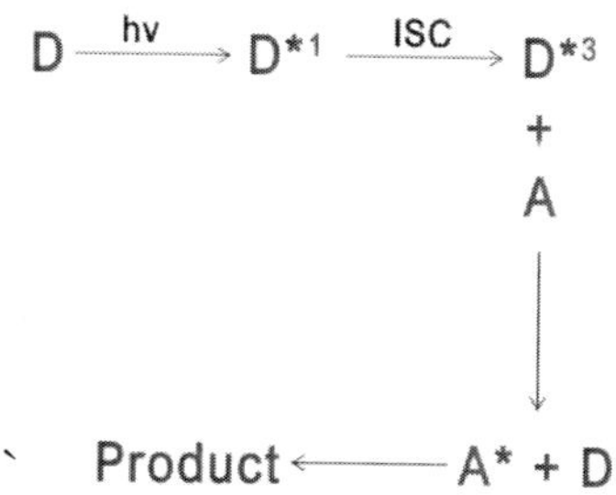

Figure 1. General mechanism of photosensitized reactions.

Figure 1 represents the general mechanism of the photosensitized reaction. The PS molecule, the donor (D), absorbs visible light first and then gets into the singlet excited state (D^{*1}). It will enter the triplet excited state (D^{*3}) through inter system crossing (ISC), where the life period of the excited species is concise, and the PS will return to the ground state (A^{*1}) by transferring energy from the triplet state to the substrate molecule of the acceptor (A). The substrate molecule will finally enter the singlet excited state due to this energy transfer mechanism, which will then cause chemical changes in the other molecules in the system. This yields the product [15].

Thus, a PS having fluorescent properties can be used as an efficient tool in biomedical applications. Photodynamic therapy (PDT) and cell imaging could work effectively when a PS is assisted with fluorescent properties. PDT is a technique in which the patient is exposed to light to treat a disease. Psoriasis, vitiligo, and skin cancer were treated by exposure to sunlight in the ancient Indian, Chinese, and Egyptian civilizations [16]. This treatment technique which involves light can be termed as phototherapy. There are two kinds of phototherapy: photodynamic therapy (PDT) and photothermal therapy (PTT). However, the former is preferred due to the low power modality [17]. PDT is thus a promising treatment for oncological, cardiovascular, ophthalmologic, and dermatological diseases [18]. The importance of cell imaging technique is that it can distinguish between the normal cells and the tumor cells [19]. This technique can monitor cellular integrity [20], enzyme activity [21], endocytosis [22], exocytosis [23], signal transduction [24], and protein trafficking [25]. Furthermore, it is possible to monitor molecular location, interaction and response within the cell [26]. This chapter discusses the medical applications, particularly the PDT and live-cell imaging using fluorescent PSs.

Fluorescent Photosensitizers in Photodynamic Therapy

Photodynamic therapy is a treatment method that uses a photosensitizing agent to kill cancerous cells. The photosensitizing agents are administered into the bloodstream through a vein or directly to the skin. Depending on the material, over time, these can be absorbed by the cancerous cells. When light is illuminated on this cell, highly reactive singlet oxygen is formed, killing the cancerous cells [27]. The reactive singlet oxygen can cause direct cellular damage and even vascular shutdown and activate the targeted cells' immune responses. PDT was used in 1963 by ophthalmologists with lasers for the photocoagulation of detached retinas [28]. However, lasers were least preferred later, as they caused high heating and cell damage [29]. Also PDT is used to treat bacterial and fungal infections [30]. PDT can selectively destroy the target tissue, so this is a boon to the therapeutic modalities such as chemotherapy, radiotherapy or immunotherapy [31].

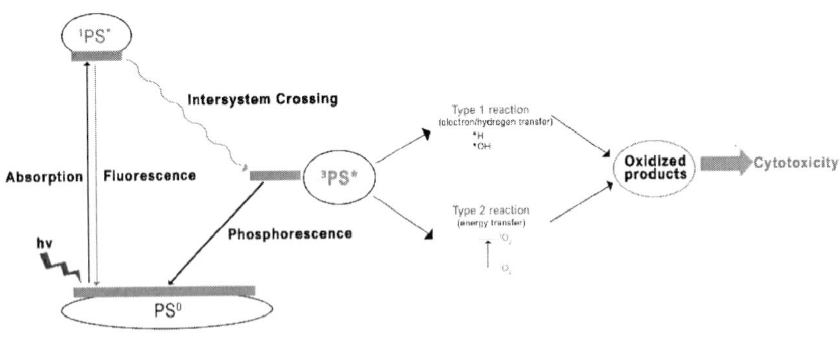

Figure 2. Mechanism of action of photodynamic therapy.

Generally, for a PDT to generate singlet oxygen, a PS, appropriate light, and oxygen are required. When exposed to light, a PS in the ground state gets promoted to the excited singlet state by absorbing photons. This energy can be dissipated by thermal decay or emission of fluorescence or even by intersystem crossing (triplet state). The reactive singlet oxygen can be generated through two mechanisms: type I and type II processes. In the type I mechanism, the PS transfers an electron to form radicals and radical ions. While in the type II mechanism, the PS in the excited triplet state directly interacts with the molecular form of oxygen to produce the reactive cytotoxic singlet oxygen [32-35]. The mechanism of action of photodynamic therapy is demonstrated in Figure 2. The latter's lifetime is of the order of 10^{-6} s, so this can lead to irreversible cell damage [36]. Thus, it is clear that PDT effects are oxygen-dependent. So, PDT will depend on the type of PS, the energy of the light, and the depth of penetration of the PS to the cancerous cell.

Figure 3 demonstrates the procedures of performing PDT [17]. The black spot indicates the cancerous cell. The PS (green circle) is injected into the patient. The PS now acts as a PDT agent and distributed around the body. The accumulation of the PDT agents to the targeted cell is achieved through active or passive targeting strategies and optional molecular activation. Then, the light of a specific wavelength is applied to the target cell, finally resulting in the ablation of cancerous cells.

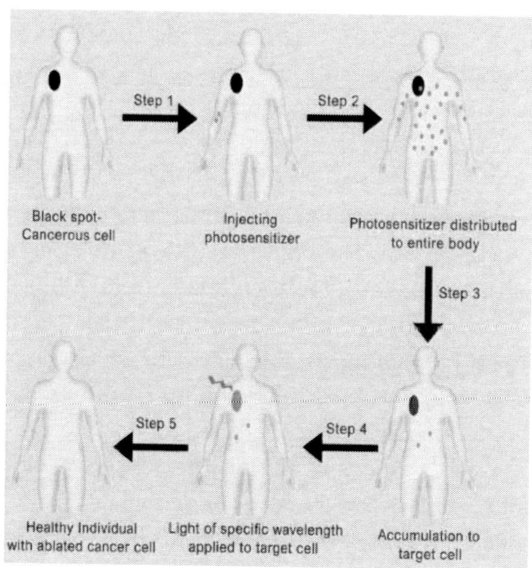

Figure 3. Procedure of photodynamic therapy using a PS.

Several PSs have been useful clinically and pre-clinically, including porphyrin, chlorin, or phthalocyanine derivatives [37, 38]. Methylene blue, rose bengal, and hypericin have been used as photodynamic agents in several clinical trials [39]. Boron dipyrromethane, cyanine dyes, and fullerenes have photodynamic properties for oncological applications [40, 41]. Some of photosensitizing agents are approved by the US Food and Drug Administration (FDA) to treat precancers and certain cancers. Two of the widely used agents are Porfimer sodium [42] and aminolevulinic acid. Porfimer sodium is activated by the red light from a laser source and is widely used to treat certain lung cancers. Aminolevulinic acid is activated by blue light and is widely used to treat actinic keratosis.

Esophageal cancer accounted for more than half a million deaths in 2018 [43]. Surgery of esophageal tumor has a huge risk for patients after the operation. Therefore, PDT is widely used to cure this type of tumor [44]. Lung cancers account for 18.4% of all cancer-related deaths and PDT effectively treats lung cancer with Porfimer sodium [45]. Talaporfin, approved in Japan in 2004, is used to treat early-stage endobronchial cancer [46]. Annually thousands of deaths are reported on head and neck squamous cell carcinoma (HNSCC) by the treatment methods [47]. European Medical Agency (EMA) has approved Temoporfin for the advanced treatment of HNSCC. This is advised to the patient when no other treatment is effective [48]. Clinically PDT has been tested for breast cancer [49], pancreatic cancer [50], cholangiocarcinoma [51] and even gynecological cancer [52]. Researchers nowadays are interested in finding combination treatments using PDT agents to leave the normal cells unaffected and for the deeper penetration of light treatment.

In the PDT approach, common side effects are skin redness or burning sensation to the exposed area. The PS is in the always-on mode, so the patients must keep away from sunlight for a long time until the PS is eliminated from the body. It was reported that the non-malignant tissue can be phototoxic if exposed to sunlight, resulting in blisters and swollen conditions [53]. However, several developments are made in this field to design an activatable PSs which will work only when activated [54, 55]. So ideally, a PS should act as a switch. The main drawback of PDT is the nonspecific binding of the PS to both the healthy and cancerous cells. For the therapeutic activation of photosensitizer, the useful range of wavelength is from 400 to 800 nm to avoid the interference of the endogenous chromophore. Furthermore, a PS that has fluorescent properties can be used for the photodynamic diagnosis [56, 57]. Hence in addition to the role of therapeutic agent in PDT, PS also serves as an imaging agent that fluoresces in the visible region. This mode of action is termed as Photodynamic Diagnosis (PDD) in the literature [58]. Celli *et al.* had suggested this terminology to be rephrased as Photosensitizer Fluorescence Detection (PFD). As shown in Figure 4, Protoporphyrin IX (PpIX) had maximum absorbance at 409 nm and several weaker bands, at 500-

635 nm. PDT treatment prefers a longer wavelength as a deeper light penetration into the tissues is required and this is achieved by irradiation at 635 nm [34].

In addition, endoscopy procedures for imaging gastrointestinal tract can adopt this methodology to deliver fluorescence excitation and detect emission. Fluorescence quenching is an important phenomenon that should be observed in designing an efficient fluorescent PS. The absorbance intensity, which was close to 1.4, was quenched to 0.9 in fluorescence intensity for PpIX in methanol as seen in Figure 4. Fluorescence intensity was reduced in both the PSs, talaporfin sodium and sodium-pheophorbide. However, the quenching rate was much faster in cancerous cells than in normal cells [59]. Hence, a PS that can be excited at 400 nm range and emission at 600 nm or excitation towards the near-infrared region can be modeled as a fluorescent PS in PDT.

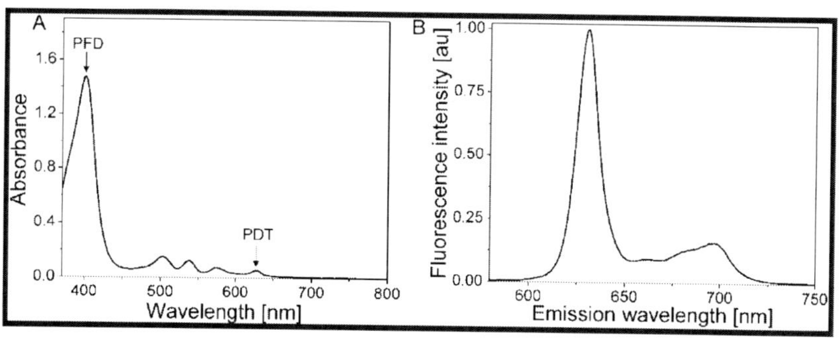

Figure 4. (A) Absorbance spectra; (B) Emission spectra of PpIX in methanol. Reprinted with permission from Ref. [34]. Copyright 2010 American Chemical Society.

In order to measure the PDT dosage and for detection, several techniques such as photobleaching have been introduced in the literature [60,61]. Another photosensitizer basic metric is the PS reaccumulation [62]. With the variation in the cellular localization of PS in the body, the PDT activity will decrease, but fluorescence emission is not affected [63,64]. Non-singlet oxygen ($^1O_2$) mediated mechanism occurs in photobleaching [65]. Such a specific biophysical pathway is preferred as these can influence the PDT outcomes and cytotoxic species production [66,67]. Hence, in general, a fluorescence-based PS can be easily used to monitor the amount of PDT dosage [68].

Fluorescence Photosensitizers in Live Cell Imaging

Live-cell imaging is an imaging technique applied to living cells and tissues. This involves maintaining healthy cells with high-quality imaging [69]. This technique can be used to observe internal structures and cellular processes in real-time. The results from live-cell imaging are more reliable than fixed cell microscopy as these are less prone to experimental artifacts [70]. PSs can act as an ideal imaging probe which can distinguish target cells from normal cells. PSs that are not phototoxic outside the infected cell can be used, so this type of PS can be termed activatable photosensitizer (aPS). These have a similar activation mechanism as activatable fluorophores [71]. As discussed, for an effective design of PDT, which requires deeper penetration of light into the cells and tissues, excitation in Near Infrared (NIR) region will be much effective [72]. Using the NIR system, Mahmood and co-workers pioneered *in vivo* imaging in mice using smart probes activated by the tumor proteases [73].

Some of the techniques used for cell imaging include: (a) Phase-contrast microscopy, which can detect intact cells [74], (b) Fluorescent microscopy is prominent in detecting single molecules or organelles [75], (c) Brightfields microscopy was used to visualize whole cells [76], (d) Confocal microscopy can give high-quality 3D images [77], (e) Time-lapse microscopy gives the 4D live-cell image which can be combined with confocal microscopy [78], and (f) Automated cell imaging has software and hardware features giving a sophisticated analysis of images [79]. Fluorescent microscopy is the most common imaging tool in studying cellular biology [80]. This technique is identified as most efficient as it can detect all kinds of proteins when compared to standard microscopy or phase-contrast microscopy. Fluorescence imaging is a non-ionizing and non-invasive technique that requires only nanomoles of fluorophores [81]. Many PSs can behave as a fluorophore and can emit in the NIR region and this property can be used for *in vivo* imaging. Techniques such as fluorescent tagging [82], with Fluorescent Proteins (FPs) can be employed in live-cell imaging. The most widely used tool for imaging is FPs [83]. PSs can be used as FPs effectively [84]. Accordingly, a PDT aided with cell imaging will be a boon in distinguishing the cells. FPs can range from blue to far-red, which can be visualized using an imaging technique. High extinction coefficients are generally observed for FPs as these will have high quantum yields up to 0.8 making them bright [85]. In the *in vivo* studies of deep tissue imaging, two-photon absorption is important [86] and spectral distinctions can be made using FPs [87].

The most commonly used FP is the green fluorescent protein (GFP). This is a protein of the jellyfish (*Aequorea Victoria*); it has green fluorescence when exposed to light. It was first isolated in 1960 by Shimomura [88]. Jellyfish releases calcium ions [89] that bind to aequorin, a protein that releases blue light upon calcium-binding. GFP absorbs this blue light and is re-emitted as green fluorescence [90].

The GFP is barrel in shape and the fluorescent portion is the chromophore which is made up of three amino acids. Upon absorbing the blue light by the chromophore, it emits green fluorescence [91]. GFPs can be used as a reference for environmental toxicity as the fluorescence intensity is decreased with the increase in toxicity [92]. GFPs can be fused with other proteins making those proteins fluorescent [93]. GFPs employed with microscopic applications such as Fluorescence Recovery After Photobleaching (FRAP) and Förster Resonance Energy Transfer (FRET) are used for imaging [94, 95]. GFPs can also be used in live-cell experiments as well [96]. Ranging from cyan to red natural sources such as hydrozoans, corals, and sea anemones yielded GFPs like proteins [97, 98, 99].

The most crucial part of the live-cell imaging technique is maintaining the cell in good condition after adding the FPs. The culture media is prepared based on the biochemical requirement of the cell. This usually contains amino acids, inorganic salts, vitamins, sugars, tricarboxylic acid cyclic intermediates, coenzymes, and lipids. However, surface tension, oxygen concentration, pH, buffering capacity, osmolarity, and viscosity should be controlled when a tissue culture media is prepared. In order to control dissolved gas concentrations, the cells have to be cultured in 5-7% of carbon dioxide, but this process will be difficult for live cell imaging [69]. The requirement of oxygen can differ for various cells. Maintaining the temperature of the culture is also an important part, particularly live-cell imaging. Customized chambers are designed for viewing live cell images using light or fluorescent microscopy [98]. Depending on the type of experiment, the conditions can be altered. Open chambers are exposed to the atmosphere, while closed chambers are sealed in order to avoid evaporation. Both are available commercially. A PS, a fluorescent molecule when excited, can quickly react with oxygen, producing free radicals which damage the cell. As these cells are fluorophores, they are susceptible to photodamage. Even if the fluorophore is absent, the cells are always sensitive to ultraviolet light [100]. Thus, with the introduction of FPs, detailed time-lapsed imaging of cells with fluorescent labels can be done.

Analysis workflow for cell imaging is depicted in Figure 5. [101] At first, cells are incubated at 37°C and 5% carbon dioxide is introduced at normal culture conditions. The conditions can be changed depending on the cell. These can be screened using wells with 96 or 384 well plates. Then the cell is treated with the compound under consideration. Depending on the mechanism of action and biological response, the compounds are added. In the third step, the live cells are stained with the fluorophore. This fluorophore is a fluorescent PS. However, staining is not necessary for live cell imaging, but visual detection will be more effective with staining. Then the stained cells are placed in the relevant imaging instrument with controlled temperature and humidity. These conditions can also be altered depending on the cell's nature. Imaging can be monitored continuously or discontinuously. Using continuous time-lapse acquisition, the image is recorded every hour for 24 hours, while in the discontinuous time-lapse acquisition, plates can be incubated and recorded at different time intervals. Finally, the cells can be analyzed [101].

Pretorius et al. had depicted a live cell image with a continuous time-lapse acquisition. Different phases of the cell cycle were demonstrated and the target cell was divided into daughter cells [102]. Hence, when the target tumor is destroyed, the fluorescence will be reduced. Thus, predicting the dependence of dosage and real-time monitoring effectively. As surgery is the frontline treatment for a large number of cancers [103, 104], NIR region fluorescence-guided tumor surgery can be effective for this [105]. In conclusion, progress in imaging technologies, targeting chemistry, and nanotechnology, combined with a thorough understanding of PDT-related molecular mechanisms, offer hope for the future of imaging in PDT. Imaging could be a key conduit for new PDT applications and the development of patient-specific treatments.

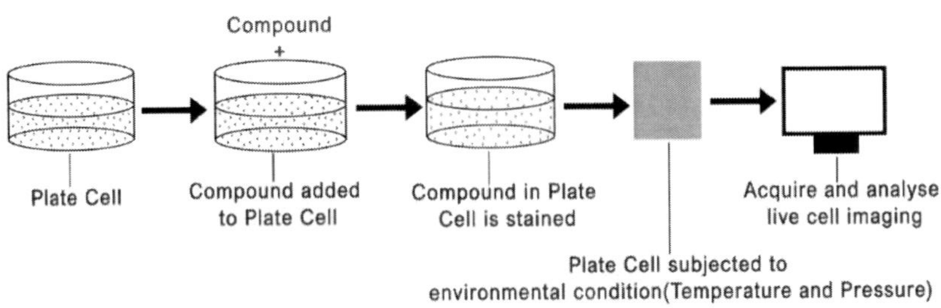

Figure 5. Analysis workflow for live cell imaging.

Conclusion

The relevance of photosensitization reactions is clearly understood with the significant increase in research across the globe. Cancer is one of the leading causes of mortality globally, and many research studies are focused in the last decade on developing novel therapeutics to lessen the adverse effects of existing treatments. As discussed, the combination of imaging and PDT can be the best tool in this field. Chemists are always looking for new compounds to design, synthesize, purify, and characterize that can be employed as fluorescent PS. The fluorescent PS design has advanced significantly in recent years, and new technologies are being developed to improve their efficiency. PDT and cell imaging are multimodal diagnostic and therapeutic techniques. This may be used in various disciplines, and researchers' imagination only limits its future application development.

References

[1] Ravelli, D., Protti, S., Albini, A., (2015). Energy and molecules from photochemical/photocatalytic reactions. An overview. *Molecules*. 20, 1527-1542.

[2] Dwivedi, A. H., Pande, U. C., (2012). Photochemical Degradation of Halogenated Compounds: a Review. *Sci Revs Chem Commun*. 2, 41-65.

[3] Abdel-Kader, M. H., (2016). *Photodynamic Medicine: From Bench to Clinic, The Journey of PDT Throughout History: PDT from Pharos to Present.* Royal Society of Chemistry.

[4] Hsieh, C. W., Chu, C. H., Lee, H. M., Yuan Yang, W., (2015). Triggering mitophagy with far-red fluorescent photosensitizers. *Sci. Rep.*, 5, 1-12.

[5] John, A. M., Thomas, R., Balakrishnan, S. P., Al-Zaqri, N., Alsalme, A., Warad, I., (2021), Diazo-pyrazole analogues as photosensitizers in dye sensitised solar cells: Tuning for a better photovoltaic efficiency using a new modelling strategy using experimental and computational data. *Zeitschrift fur Phys Chemie.*, 235, 1227-1245.

[6] McCullough, B. J., Neyhouse, B. J., Schrage, B. R., Reed, D. T., Osinski, A. J., Ziegler, C. J., (2018). Visible-Light-Driven Photosystems Using Heteroleptic Cu(I) Photosensitizers and Rh(III) Catalysts to Produce H_2. *Inorg Chem.*, 57, 2865-2875.

[7] Farrell, T. J., Hawkes, R. P., Patterson, M. S., Wilson, B. C., (1998). Modeling of photosensitizer fluorescence emission and photobleaching for photodynamic therapy dosimetry. *Appl Opt.* 37, 7168-7183.

[8] Spikes, J. D., (1990). New trends in photobiology:Chlorins as photosensitizers in biology and medicine. *J Photochem Photobiol B Biol.*, 6, 259-274.

[9] Josefsen, L. B., Boyle, R. W., (2008). Photodynamic therapy: Novel third-generation photosensitizers one step closer?. *Br J Pharmacol.*, 154, 1-3.

[10] Oleinick, N. L., Morris, R. L., Belichenko, I., (2002). The role of apoptosis in response to photodynamic therapy: What, where, why, and how. *Photochem Photobiol Sci.,* 1, 1-21.
[11] Malik, Z., Lugaci, H., (1987). Destruction of erythroleukaemic cells by photoactivation of endogenous porphyrins. *Br J Cancer.*, 56, 589-595.
[12] Babincová, M., Sourivong, P., Babinec, P., (2000). Gene transfer-mediated intracellular photodynamic therapy. *Med Hypotheses.*, 54, 180-181.
[13] Kou, J., Dou, D., Yang, L., (2017), Porphyrin photosensitizers in photodynamic therapy and its applications. *Oncotarget.*, 8, 81591-81603.
[14] Pouyet, B., Chapelon, R., (1987), Photochemical Mechanisms In Photosensitization. *J Phys Colloq.*, 48, 247-251.
[15] Dwivedi, A. H., Pillai, S. G., (2014). Photosensitized Reactions : an Important Route of Reaction in Photochemistry. *Int J Adv Technol Eng Sci.*, 2, 1-4.
[16] Ackroyd, R., Kelty, C., Brown, N., Reed, M., (2001). The History of Photodetection and Photodynamic Therapy. *Photochem Photobiol.,* 74, 656-669.
[17] Li, X., Lovell, J. F., Yoon, J., Chen, X., (2020). Clinical development and potential of photothermal and photodynamic therapies for cancer. *Nat Rev Clin Oncol.*, 17, 657-674.
[18] Levy, J. G., Obochi, M., (1996). New Applications in Photodynamic Therapy Introduction. *Photochem Photobiol.*, 64, 737-739.
[19] Yaroslavsky, A. N., Feng, X., Muzikansky, A., Hamblin, M. R., (2019). Fluorescence Polarization of Methylene Blue as a Quantitative Marker of Breast Cancer at the Cellular Level. *Sci Rep.*, 9, 1-10.
[20] Mamede, J. I., Hope, T. J. (2016). Detection and Tracking of Dual-Labeled HIV Particles Using Wide-Field Live Cell Imaging to Follow Viral Core Integrity. *HIV Protocols, 49-59.*
[21] Zotter, A., Bäuerle, F., Dey, D., Kiss, V., Schreiber, G., (2017). Quantifying enzyme activity in living cells. *J Biol Chem.* 292, 15838-15848.
[22] Kirchhausen, T., (2009). Imaging endocytic clathrin structures in living cells. *Trends Cell Biol.*, 19, 596-605.
[23] Fernandez, N. A., Liang, T., Gaisano, H. Y., (2011). Live pancreatic acinar imaging of exocytosis using syncollin-pHluorin. *Am J Physiol - Cell Physiol.* 300(6), C1513-1523.
[24] Wenus, J., Düssmann, H., Paul, P., Kalamatianos, D., Rehm, M., Wellstead, P. E., (2009). ALISSA: An automated live-cell imaging system for signal transduction analyses. *Biotechniques.*, 47, 1033-1040.
[25] Cimica, V., Reich, N. C., (2012). Nuclear trafficking of STAT proteins visualized by live cell imaging, *Methods Mol Biol.*, 967, 189-202.
[26] Pilz, G. A., Bottes, S., Betizeau, M., Jörg, D. J., Carta, S., Simons, B. D., (2018). Live imaging of neurogenesis in the adult mouse hippocampus. *Science.,* 359, 658-662.
[27] McCaw, D. L., Bryan, J. N., (2009), Photodynamic therapy. *Cancer Manag Small Anim Pract.*, 90, 163-166.
[28] Kapany, N. S., Peppers, N. A., Zweng, H. C., Flocks, M., (1963). Retinal photocoagulation by lasers. *Nature.* 199, 146-149.

[29] Goldman, L., (1966). A review: Applications of the laser beam in cancer biology. *Int J Cancer.* 1, 309-318.

[30] Hamblin, M. R., Hasan, T., (2004). Photodynamic therapy: A new antimicrobial approach to infectious disease. *Photochem Photobiol Sci.*, 3, 436-450.

[31] Lovell, J. F., Liu, T. W. B., Chen, J., Zheng, G., (2010). Activatable photosensitizers for imaging and therapy. *Chem Rev.*, 110, 2839-2857.

[32] Henderson, B. W., Dougherty, T. J., (1992). How Does Photodynamic Therapy Work? *Photochem Photobiol.*, 55, 145-157.

[33] Konan, Y. N., Gurny, R., Allémann, E., (2002). State of the art in the delivery of photosensitizers for photodynamic therapy. *J Photochem Photobiol B Biol.*, 66, 89-106.

[34] Ion R-M, (2021). Revisiting Tetra-*p*-Sulphonated Porphyrin as Antimicrobial Photodynamic Therapy Agent. *Coatings.*, 11, 393.

[35] Celli, J. P., Spring, B. Q., Rizvi, I., Evan, C. L., Samkoe, K. S., Verma, S., Pogue, B. W., (2010). Imaging and Photodynamic Therapy: Mechanisms, Monitoring and Optimization. *Chem Rev.*, 110, 2795-2838.

[36] Ni, J., Wang,Y., Zhang,H., Sun, J. Z., Tang, B. Z., (2021). Aggregation-Induced Generation of Reactive Oxygen Species: Mechanism and Photosensitizer Construction. *Molecules.*, 26, 268.

[37] Felsher, D. W., (2003). Cancer revoked: Oncogenes as therapeutic targets. *Nat Rev Cancer.*, 3, 375-380.

[38] Juarranz, Á., Jaén, P., Sanz-Rodríguez, F., Cuevas, J., González, S., (2008). Photodynamic therapy of cancer. Basic principles and applications. *Clin Transl Oncol.*, 10, 148-154.

[39] Sharman, W. M., Allen, C. M., VanLier J. E., (1999). Photodynamic therapeutics: Basic principles and clinical applications. *Drug Discov Today.*, 4, 507-517.

[40] Ethirajan, M., Chen, Y., Joshi, P., Pandey, R. K., (2011). The role of porphyrin chemistry in tumor imaging and photodynamic therapy. *Chem Soc Rev.*, 40, 340-362.

[41] Garland, M. J., Cassidy, C. M., Woolfson, D., Donnelly, R. F., (2009). Designing photosensitizers for photodynamic therapy: Strategies, challenges and promising developments. *Future Med Chem.*, 1, 667-691.

[42] Kamkaew, A., Lim, S. H., Lee, H. B., Kiew, L. V., Chung, L. Y., Burgess, K., (2013). BODIPY dyes in photodynamic therapy. *Chem Soc Rev.,* 42, 77-88.

[43] Li, X., Kim, J., Yoon, J., Chen, X., (2017). Cancer-Associated, Stimuli-Driven, Turn on Theranostics for Multimodality Imaging and Therapy. *Adv Mater.*, 29, 1-24.

[44] Baskaran, R., Lee, J., Yang, S. G., (2018). Clinical development of photodynamic agents and therapeutic applications. *Biomater Res.*, 22, 1-8.

[45] Bray, F., Ferlay, J., Soerjomataram, I., Siegel, R. L., Torre, L. A., Jemal, A., (2018). Global cancer statistics 2018: GLOBOCAN estimates of incidence and mortality worldwide for 36 cancers in 185 countries. *CA Cancer J Clin.*, 68, 394-424.

[46] Qumseya, B. J., David, W., Wolfsen, H. C., (2013). Photodynamic therapy for Barrett's esophagus and esophageal carcinoma. *Clin Endosc.*, 46, 30-37.

[47] Simone, C. B., Cengel, K. A., (2014), Photodynamic therapy for lung cancer and malignant pleural mesothelioma. *Semin Oncol.*, 41, 820-830.

[48] Wang, S., Bromley, E., Xu, L., Chen, J. C., Keltner, L., (2010). Talaporfin sodium. *Expert Opin Pharmacother.*, 11, 133-140.

[49] Biel, M. A., (2010). Photodynamic therapy of head and neck cancers. *Methods Mol Biol.*, 635, 281-293.

[50] Hopper, C., Niziol, C., Sidhu, M., (2004). The cost-effectiveness of Foscan mediated photodynamic therapy (Foscan-PDT) compared with extensive palliative surgery and palliative chemotherapy for patients with advanced head and neck cancer in the UK. *Oral Oncol.*, 40, 372-382.

[51] Banerjee, S. M., MacRobert, A. J., Mosse, C. A., Periera, B., Bown, S. G., Keshtgar, M. R. S., (2017). Photodynamic therapy: Inception to application in breast cancer. *The Breast.*, 31, 105-113.

[52] Fan, B. G., Andrén-Sandberg, Å., (2007). Photodynamic therapy for pancreatic cancer. *Pancreas*. 34, 385-389.

[53] Lee, T. Y., Cheon, Y. K., Shim, C. S., Cho, Y. D., (2012). Photodynamic therapy prolongs metal stent patency in patients with unresectable hilar cholangio-carcinoma. *World J Gastroenterol.*, 18, 5589-5594.

[54] Allison, R. R., Cuenca, R., Downie, G. H., Randall, M. E., Bagnato, V. S., Sibata, C. H., (2005). PD/PDT for gynecological disease: A clinical review. *Photodiagnosis Photodyn Ther.*, 2, 51-63.

[55] Zheng, H., (2005). A review of progress in clinical photodynamic therapy. *Technol Cancer Res Treat.,* 4, 283-293.

[56] Li, X., Fan, H., Guo, T., Bai, H., Kwon, N., Kim, K. H., (2019). Sequential Protein-Responsive Nanophotosensitizer Complex for Enhancing Tumor-Specific Therapy. *ACS Nano.*, 13, 6702-6710.

[57] Li, X., Kolemen, S., Yoon, J., Akkaya, E. U., (2017). Activatable Photosensitizers: Agents for Selective Photodynamic Therapy. *Adv Funct Mater.*, 27, 1-11.

[58] Fritsch, C., Backer-Wegerich, P. M., Menke, H., Ruzicka, T., Goerz, G., Olbrisch, R. R., (1997). Successful surgery of multiple recurrent basal cell carcinomas guided by photodynamic diagnosis. *Aesthetic Plast Surg.*, 21, 437-439.

[59] Stummer, W., Novotny, A., Stepp, H., Goetz, C., Bise, K., Reulen, H. J., (2000). Fluorescence-guided resection of glioblastoma multiforme by using 5-aminolevulinic acid-induced porphyrins: A prospective study in 52 consecutive patients. *J Neurosurg.*, 93, 1003-1013.

[60] Fritsch, C., Lang, K., Neuse, W., Ruziacka, T., Lehmann, P., (1999). Photodynamic Diagnosis and Therapy in Dermatology. *Skin Pharmacol Appl Skin Physiol.*, 11, 358-373.

[61] Awasthi, K., Yamamoto, K., Furuya, K., Nakabayashi, T., Li, L., Ohta, N., (2015). Fluorescence characteristics and lifetime images of photosensitizers of talaporfin sodium and sodium pheophorbide a in normal and cancer cells. *Sensors.*, 15, 11417-114130.

[62] Wilson, B. C., Patterson, M. S., Lilge, L., (1997). Review Article Implicit and Explicit Dosimetry in Photodynamic Therapy : a New Paradigm. *Lasers Med Sci.* 12, 182-199.

[63] Boere, I. A., Robinson, D. J., de Bruijn, H. S., van den Boogert, J., Tilanus, H. W., Sterenborg, H. J. C. M., (2003). Monitoring In Situ Dosimetry and Protoporphyrin IX Fluorescence Photobleaching in the Normal Rat Esophagus During 5-Aminolevulinic Acid Photodynamic Therapy. *Photochem Photobiol.*, 78, 271-277.

[64] De Bruijn, H. S., Kruijt, B., Van Der Ploeg - Van Den Heuvel A, Sterenborg, H. J. C. M., Robinson, D. J., (2007). Increase in protoporphyrin IX after 5-aminolevulinic acid based photodynamic therapy is due to local re-synthesis. *Photochem Photobiol Sci.*, 6, 857-864.

[65] Morgan, J., Oseroff, A. R., (2001). Mitochondria-based photodynamic anti-cancer therapy. *Adv Drug Deliv Rev.*, 49, 71-86.

[66] Ji, Z., Yang, G., Vasovic, V., Cunderlikova, B., Suo, Z., Nesland, J. M., (2006). Subcellular localization pattern of protoporphyrin IX is an important determinant for its photodynamic efficiency of human carcinoma and normal cell lines. J *Photochem Photobiol B Biol.*, 84, 213-220.

[67] Dysart, J. S., Patterson, M. S., (2006). Photobleaching kinetics, photoproduct formation, and dose estimation during ALA induced PpIX PDT of MLL cells under well oxygenated and hypoxic conditions. *Photochem Photobiol Sci.*, 5, 73-81.

[68] Rizvi, I., Anbil, S., Alagic, N., Celli, J. P., Zheng, L. Z., Palanisami, A., (2013). PDT dose parameters impact tumoricidal durability and cell death pathways in a 3D ovarian cancer model. *Photochem Photobiol.*, 89, 942-952.

[69] Mallidi, S., Anbil, S., Lee, S., Manstein, D., Elrington, S., Kositratna, G., (2014). Photosensitizer fluorescence and singlet oxygen luminescence as dosimetric predictors of topical 5-aminolevulinic acid photodynamic therapy induced clinical erythema. *J Biomed Opt.*, 19, 028001.

[70] Jensen, E. C., (2013). Overview of Live-Cell Imaging: Requirements and Methods Used. *Anat Rec.*, 296, 1-8.

[71] Sung, M. H., McNally, J. G., (2011). Live cell imaging and systems biology. *Wiley Interdiscip Rev Syst Biol Med.*, 3, 167-182.

[72] Lovell, J. F., Zheng, G., (2008). Activatable Smart Probes For Molecular Optical Imaging And Therapy. *J Innov Opt Heal Sci.*, 1, 45-61.

[73] Kobayashi, H., Choyke, P. L., (2019). Near-Infrared Photoimmunotherapy of Cancer. *Acc Chem Res.*, 52, 2332-2339.

[74] Mahmood, U., Tung, C. H., Bogdanov, A., Weissleder, R., (1999). Near-infrared optical imaging of protease activity for tumor detection. *Radiology.*, 213, 866-870.

[75] [74] live cell viability: Comparative study of fluorescence, oblique incidence reflection and phase contrast microscopy imaging techniques. *Opt Express.*, 12, 5754-5759.

[76] Furman, C. A., Chen, R., Guptaroy, B., Zhang, M., Holz, R. W., Gnegy, M., (2009). Dopamine and amphetamine rapidly increase dopamine transporter trafficking to the surface: live-cell imaging using total internal reflection fluorescence microscopy. *J Neurosci.*, 29, 3328-3336.

[77] Selinummi, J., Ruusuvuori, P., Podolsky, I., Ozinsky, A., Gold, E., Yli-Harja, O., (2009). Bright field microscopy as an alternative to whole cell fluorescence in automated analysis of macrophage images. *PLoS One.*, 4, e7497.

[78] Isherwood, B., Timpson, P., Mcghee, E. J., Anderson, K. I., Canel, M., Serrels. A., (2011). Live cell *in vitro* and *in vivo* imaging applications: Accelerating drug discovery. *Pharmaceutics.*, 3, 141-170.

[79] Burke, R. T., Orth, J. D., (2016). Through the looking glass: Time-lapse microscopy and longitudinal tracking of single cells to study anti-cancer therapeutics. *J Vis Exp.*, 2016, 1-14.

[80] Green, M, V., Pengo, T., Raybuck, J. D., Naqvi, T., McMullan, H. M., Hawkinson, J. E., (2019). Automated Live-Cell Imaging of Synapses in Rat and Human Neuronal Cultures. *Front Cell Neurosci.,* 13, 1-14.

[81] Kandel, M. E., He, Y. R., Lee, Y. J., Chen, T. H. Y., Sullivan, K. M., Aydin, O., Saif, M. T. A., Kong, H., Sobh, N., Popescu, G., (2020). Phase imaging with computational specificity (PICS) for measuring dry mass changes in sub-cellular compartments. *Nat Commun.*, 11, 6256

[82] Meng, H., Xue, M., Xia, T., Ji, Z., Tarn, D. Y., Zink, J. I., (2011). Use of size and a copolymer design feature to improve the biodistribution and the enhanced permeability and retention effect of doxorubicin-loaded mesoporous silica nanoparticles in a murine xenograft tumor model. *ACS Nano.*, 5, 4131-4144.

[83] Massignani, M., Canton, I., Sun, T., Hearnden, V., MacNeil, S., Blanazs, A., (2010). Enhanced fluorescence imaging of live cells by effective cytosolic delivery of probes. *PLoS One.*, 2010, 5, e10459.

[84] Dean, K. M., Palmer, A. E., (2014). Advances in fluorescence labeling strategies for dynamic cellular imaging. *Nat Chem Biol.*, 10, 512-523.

[85] Leem, J. W., Kim, S. R., Choi, K. H., Kim,Y. L., (2018). Plasmonic photocatalyst-like fluorescent proteins for generating reactive oxygen species. *Nano Converg.*, 5, 8.

[86] Zimmer, M., (2002). Green fluorescent protein (GFP): Applications, structure, and related photophysical behavior. *Chem Rev.*, 102, 759-781.

[87] Brown, E. B., Campbell, R. B., Tsuzuki, Y., Xu, L., Carmeliet, P., Fukumura, D., (2001). *In vivo* measurement of gene expression, angiogenesis and physiological function in tumors using multiphoton laser scanning microscopy. *Nat Med.* 7, 864-868.

[88] Hoffman, R. M., (2009). Dual-Color Imaging of Tumor Angiogenesis. In: Methods in Molecular Biology, Viral Applications of Green Fluorescent Protein. *Methods Mol Biol.*, 515, 45-61.

[89] Shimomura, O., Johnson, F. H., Saiga, Y., (1962). Extraction, purification and properties of aequorin, a bioluminescent. *J Cell Comp Physiol.*, 59, 223-239.

[90] Morise, H., Shimomura, O., Johnson, F. H., Winant, J., (1974). Intermolecular Energy Transfer in the Bioluminescent System of Aequorea. *Biochemistry.*,13, 2656-2662.

[91] Johnson, F. H., Shimomura, O., Saiga, Y., Gershman, L. C., Reynolds, G. T., Waters, J. R., (1962). Quantum efficiency of Cypridina luminescence, with a note on that of Aequorea. *J Cell Comp Physiol.*, 60, 85-103.

[92] Shimomura, O., (1979). Structure of the chromophore of Aequorea green fluorescent protein. *FEBS Lett.,* 104, 220-222.

[93] Song, Y. H., Kim, C. S., Seo, J. H., (2016). Noninvasive monitoring of environmental toxicity through green fluorescent protein expressing Escherichia coli. *Korean J Chem Eng.*, 33, 1331-1336.

[94] Su, W.W., (2005). Fluorescent protein as tools to aid protein production. *Microb Cell Fact.*, 4, 1-6.

[95] Zheng, C. Y., Petralia, R. S., Wang, Y. X., Kachar, B., (2011). Fluorescence recovery after photobleaching (FRAP) of fluorescence tagged proteins in dendritic spines of cultured hippocampal neurons. *J Vis Exp.*, 9, 9-11.

[96] Murakoshi, H., Shibata, A. C. E., Nakahata, Y., Nabekura, J., (2015). A dark green fluorescent protein as an acceptor for measurement of Förster resonance energy transfer. *Sci Rep.,* 5, 1-12.

[97] Glazachev, Y. I, Orlova, D. Y., Řezníčková, P., Bartova, E., (2018). Effective scheme of photolysis of GFP in live cell as revealed with confocal fluorescence microscopy. *Phys Biol.,* 15, 036008.

[98] Alieva, N. O., Konzen, K. A., Field, S. F., Meleshkevitch, E. A., Hunt, M. E., Beltran-Ramirez, V., (2008). Diversity and evolution of coral fluorescent proteins. *PLoS One.*, 3(7): e2680. https://doi.org/10.1371/journal.pone.0002680

[99] Khodjalov, A., Rieder, C.L., (2006). Imaging the division process in living tissue culture cells. *Methods.*, 38, 2-16.

[100] Deheyn, D. D., Kubokawa, K., Mccarthy, J. K., Murakami, A., Porrachia, M., Rouse, G. W., Holland, N. D., (2007). Endogenous Green Fluorescent Protein (GFP) in Amphioxus. *Biol Bull.,* 213, 95-100.

[101] Stephens, D. J., Allan, V. J., (2003). Light microscopy techniques for live cell imaging. *Science*, 300, 82-86.

[102] Wolf, S. G., Shimoni, E., Elbaum, M., (2020). *Molecular Devices, Cellular Imaging Insights.* https://www.moleculardevices.com/applications/live-cell-imaging#gref

[103] Pretorius, A. J., Khan, I. A., Errington, R. J., (2017). A Survey of Visualization for Live Cell Imaging. *Comput Graph Forum.*, 36, 46-63.

[104] Nagaya, T., Nakamura, Y. A., Choyke, P. L., Kobayashi, H., (2017). Fluorescence-guided surgery. *Front Oncol.*, 7, 314.

[105] Hill, T. K., Mohs, A. M., (2016). Image-guided tumor surgery: will there be a role for fluorescent nanoparticles? *Wiley Interdiscip Rev Nanomedicine Nanobiotechnology.*, 8, 498-511.

[106] Egloff-Juras, C., Bezdetnaya, L., Dolivet, G., Lassalle, H. P., (2019). NIR fluorescence-guided tumor surgery: New strategies for the use of indocyanine green. *Int J Nanomedicine.*, 14, 7823-7838.

Chapter 5

Emerging Trends in Nanomaterials for Photodynamic Therapy Applications: The Current Scenario

Jenny Jacob and Treesa Varghese*

School of Biosciences, Mar Athanasios College for Advanced Studies, Thiruvalla, Kerala, India

Abstract

Photodynamic therapy (PDT) is a promising and non-invasive therapeutic method, used as an alternative to chemotherapy or radiation therapy for disease treatment. The greater hydrophobicity, poor stability, low cell/tissue specificity, and minimal penetration capacity associated with the conventional photosensitizers could obstruct their therapeutic ability. Nanomaterials are one of the phenomenal materials with unique characteristics that make them striking candidates in the world of pharmacology. Physicochemical properties such as the ultra-small size in the dimensions of nanoscale, large surface area to mass ratio and high reactivity make them potent enough to replace the bulk materials having similar composition. The synthesis of nano-sized photosensitizers with their promising optical properties along with the enhanced biocompatibility and cytotoxicity make them an ideal candidate in the photodynamic therapy of tumors and other chronic diseases. Herein, the general features of nano-scale photosensitizers in terms of their classification, properties, applications and prospects in photodynamic therapy are reviewed. The focus of the chapter is mainly on describing the use of nanomaterials in photodynamic therapy whereby it would be

* Corresponding Author's Email: treesa@macfast.org.

In: Photosensitizers and Their Applications
Editors: Davor Margetic and Renjith Thomas
ISBN: 978-1-68507-880-5

useful in critical issues of human diseases with high efficiency and translational potential.

Keywords: photodynamic therapy, nanomaterials, photosensitizers

Introduction

Nanoparticles are the key and basic elements of nanotechnology that have different physicochemical properties than the bulk particles. They have surprising unique characteristics of being modified to different shapes (nanoparticles, nanosheets, nanorods) based on their dimensions. As far as the 1950's, pioneering work by Robert Dinegar and Victor La Mer in developing the process for synthesis of monodisperse colloidal materials has paved the foundations for nanocrystals as nanoparticles. Nanoscale research and colloidal research have continued much unappreciated until the great lecture by Richard Feynman in 1959 stating that there is plenty of room at the bottom. This was followed by a substantial interest at the nanoscale leading to the discovery of nanocrystalline semiconductor quantum dots in glass matrix by Alexei Ekimov in 1981. The breakthrough in nanocrystalline semiconductor quantum dots came with its subsequent discovery by Louis Brus in 1985 leading him to be the recipient of the 2008 Kavli Prize in Nanotechnology. The synthesis of these nanocrystals was then controlled by the method detailed by Moungi Bawendi in 1993 and further improved by Chris Murray and Louis Brus which has ultimately led to the present level of applications for these types of particles.

Nanotechnology has several advances in many fields including medicine and physiology both in scientific and technological aspects. Nanotechnology is an interdisciplinary area, and can be used to develop many enhanced diagnostic methods and treatments for various cancerous and non-cancerous diseases. The nanoparticles are used as fluorescent materials, contrast agents, drugs with targeting antibodies and molecular research tools [1]. Nanosomes, nanoshells, quantum dots and superparamagnetic nanoparticles are some examples of the nanoparticles used in medical science for diagnostic purposes [2]. Depending on the unique structural, optical and magnetic properties, the nanoparticles have been recommended in the development of many non-conventional treatments, especially photodynamic therapy (PDT) [3]. Photodynamic therapy is new emerging branch of science in medicine, first introduced by a medicine student, Oscar Raab, who was working in Munich,

Germany in the year 1900. He discovered that incubation of microorganisms such as paramecia with certain specific dyes could kill them when exposed to light, while they remain alive under dark condition [4]. Based on this observation, a phenomenon was proposed, in which atmospheric oxygen could exert a light dependent killing mechanism on microorganisms when incubated in the presence of some specific compound and this effect is called 'photodynamic action'. This method was first introduced in cancer therapy for tumor cell mapping. Modern era of PDT was initiated in 1970 by the effort of Thomas Dougherty. Together with coworkers he discovered a water-soluble photosensitizer (PS) which is a mixture of porphyrins collectively called 'hematoporphyrin derivative' (HpD). There are hundreds of PSs that have been used in medicine with improved application against malignant tumors, infectious diseases and many other diseases.

Photodynamic therapy is a form of photo-therapy used to prompt cellular damage (phototoxicity) with the aid of light and a photosensitizer in the presence of molecular oxygen and is also called photochemotherapy. PDT has proven ability to destroy microbial populations, like viruses, fungi and bacteria [5]. PDT is widely used to treat a variety of clinical conditions, like atherosclerosis, psoriasis and age-related macular degeneration. Additionally, it is also used for treating acne. This technology has also been used for treatment of malignant tumors [6].

Mechanism of Photodynamic Therapy

Photodynamic therapy represents a less invasive and most vibrant therapeutic methodology used to treat various non- carcinogenic and carcinogenic diseases. The working principles of PDT are typically based on the systemic or local application of a photosensitizing molecule or light absorbing molecules and light in an oxygen rich condition (Figure 1) [7].

In the presence of appropriate wavelengths, the light sensitized PSs accumulated in the affected tissues absorb the light leading to the development of selective photocytotoxic reactions in the pathological tissues. There are two basic mechanisms for PDT. The initial stages of both mechanisms are undistinguishable and both reactions are typically depending on the presence of oxygen in the biological system [8].

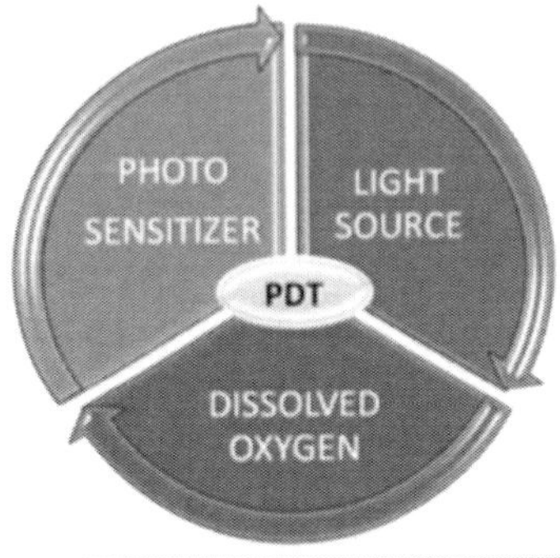

Figure 1. Working principles of PDT.

Type I Mechanism

In type I mechanism, after insertion into the cell, the accumulated PS irradiated with appropriate light of specific wavelength will absorb photons, and then convert from its singlet ground energy state (S°) into the singlet excited state (S^1) [9]. Because of the unstable excited energy state, a portion of the absorbed energy is then irradiated from the molecule as quantum of fluorescence. While the presence of energy left leads the PS molecule to enter an excited triplet state (T^1). At this stage, the PSs are able to release and transfer the energy into the biomolecules including nucleic acids, proteins and lipids present in the surrounding region. In other terms, releasing of electrons or hydrogen from the T^1 stage of PS then triggers the formation of reactive anion or free radical form of PS. Released electron then interact with oxygen in the surrounding region, hence induces the production of superoxide anion radical ($O_2^{\bullet -}$) and other reactive oxygen species (ROS) within the affected area. Accumulation of ROS then induces a strong oxidative stress in the biological system and finally cellular destruction [10, 11].

Type II Mechanism

In the second type of mechanism, after transition into the T^1 state, due to the same spins, energy from the PS will be directly transferred to the molecular oxygen in its basic triplet state form. At this time, the excited oxygen is said to be the 'singlet oxygen,' and is extreme powerful oxidizing agent [11, 12].

So, in this type of reaction, activated PSs never introduce a damage to the affected area, but activated oxygen species can be the only one culprit [13]. Functionality of these two mechanisms typically depends on many factors including structure of PSs, concentration of molecular oxygen, tissue specific pH and dielectric constant. Additionally, with low partial molecular oxygen in the tissue, type I mechanism will be the principal one while type II photodynamic mechanism has better therapeutic efficiency than type I [14]. The overall mechanism of photodynamic therapy is illustrated in Figure 2.

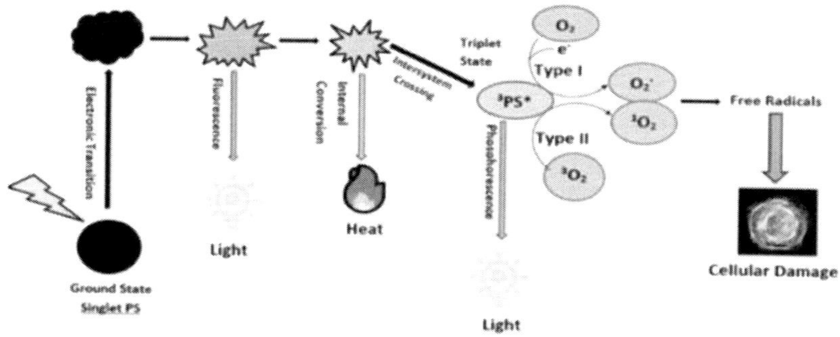

Figure 2. Mechanism of photodynamic reaction.

The major advantages of PDT are that it has improved selectivity indicating that if the tissue is exposed, the PDT will selectively damage the pathological cells but not the normal cell, thus the side effect is also lesser as compared to radiotherapy or chemotherapy techniques. Compared to any other traditional procedure, the peculiar characteristics of PDT gather this technique in the superior most level. Table 1 summarized important advantages of PDT in medicine.

Table 1. Advantages of photodynamic therapy in disease treatment

	Advantages
1	Effective site-specific drug delivery system
2	Prolongs the persistence of drug in the biological system
3	Reduces the possibility of drug tolerance
4	Less toxic to normal tissues with minimum side effects
5	Enhanced therapeutic efficacy
6	Economically viable
7	Simple technology to perform and needle free procedures

Development of nanotechnology spread out new opportunities for PDT, in which nanoparticles have been used as multi-functional and highly sophisticated medicines. More precisely, nanoparticles had been hired as: (i) photosensitizers; (ii) carriers of PSs; (iii) light harvesting molecules for PSs; (iv) carriers of magnetic nanoparticles.

Photosensitizers

Photosensitizers (PSs) are light absorbing molecules, which with the aid of absorbed light energy then generate reactive oxygen species (ROS) in an administered biological system. PSs used for therapeutic purposes must be non-toxic and their metabolic products should be less toxic. Majority of PSs exhibit a broad range of absorption spectra in the visible region and in particular, use of light radiation at ~600 nm (red light) is most effective and more potent for PDT. A good candidate of PSs must have a high photon absorption coefficient with high quantum yield for triplet formation. This triplet state forming molecules produce ROS and free radicals. Generally, these molecules have tricyclic or tetra-cyclic porphyrin-like backbone ring structures with conjugated double bonds. Table 2 summarizes the ideal characteristics of PSs for disease treatment in medicine.

Table 2. Characteristics of photosensitizers used in medicine

	Characteristics
1	Exhibit photobleaching (light-mediated destruction)
2	Ideal PSs have a strong absorption peak between 650 and 800 nm (in the red to near-infrared spectral region)
3	PSs should have a substantial triplet quantum yield to produce (ROS) under light irradiation.
4	It should not possess toxicity under dark condition
5	Should rapidly be cleared from the normal tissues
6	Have high photochemical reactivity
7	Exhibit cellular necrosis
8	Must show high level of chemical purity
9	Possess high stability at normal temperature
10	Should express high selectivity for neoplastic tissues
11	Synthesis must be inexpensive, simple and easy

Classic PSs face some disadvantages also. The most important is that many of the classic PSs display aggregation in the body, which then alters their functionality by shortening the half-life in triplet state and reducing the yield of ROS formation. Hence, introduction of a structural level modification can overpower the aggregation tendency of a PSs and thus their functionality could be improved. The synthesis of nano-based polymeric materials as PSs can be used to overcome this drawback of classic photosensitizers.

Application of Nanomaterials in PDT

Development of nanomaterial-based PDT offers a platform for modern medicine in terms of enhanced therapeutic efficacy. This is achieved through the peculiar characteristic nature of nanomaterials including enhanced solubility of hydrophobic PSs, large surface area, high drug loading capacity, site specific targeting ability, constant PS delivery rate, less cytotoxicity rate toward normal cells, greater permeability and retention (EPR) effects and multifunctional integration ability [15].

Nano-based drug delivery systems for PSs possess some potential benefits in PDT [15]: 1) due to the small size and large surface/volume ratio of nanoparticles, it is possible to deliver increased amount of PSs into the targeted tissues, 2) enhances the site-specific accumulation of reactive molecules by preventing the premature release before reaching the target tissue with minimum side effects, 3) presence of some hindrances in the bloodstream and an amphiphilic character of conjugated nanoparticles facilitate the tumor accumulation, 4) enhanced permeability and retention (EPR) effect of nanostructures allow the conjugated PS to accumulate in tumor affected area, 5) presence of pores in the nanoparticles, facilitate the intake or release of biologically active molecules and activation of absorbed therapeutics, 6) peculiar surface chemistry, cell uptake, biological distribution and pharmacokinetics of the PS due to the presence of functional groups facilitate their desired applications. Along with this, excellent biocompatibility, outstanding multifaceted practice including bio-sensing, diagnostics, therapeutics, and bio-imaging, has increased the potential use of these nanomaterials for PDT [16].

Both non-biodegradable and biodegradable types of nanomaterials can commonly be used for PDT. In the case of non-biodegradable nanostructures, the PSs are adhered onto the surface and the absorbed molecules will not be completely released for singlet oxygen generation. While, when the biodegradable nanoparticle (lipid based or polymeric materials) is used, the PSs are usually trapped into their interior core. Because of this reason, the PSs would release in a controlled manner and singlet oxygen generation typically depends on illumination with the light wave [17].

A wide range of nanomaterials including organic and inorganic particles have been widely used in PDT. Inorganic nanomaterials with PS capacity include carbon nanotubes (CNT), quantum dots (QD), mesoporous silica nanoparticles (MSNP), gold nanoparticles (AuNP), etc. [18, 19]. Furthermore, dendrimers, polymeric micelles, protein nanoparticles, viral nanoparticles, and liposomes, etc. are widely accepted and well-studied nano systems with enhanced photosensitization ability for PDT [20].

Traditionally, organic and inorganic nano scaled materials have functioned as drug carriers. Ideally, this drug carrier system must be biocompatible, and should be able to prolong the persistence of drug in the target tissues with minimum accumulation rate in normal cells. The organic nanomaterials including polymeric micelles, liposomes etc. were constructed through self-assembly method from its components. Most of these organic nanomaterials exhibit an amphipathic nature in aqueous medium, with a hydrophilic core in the interior region and is surrounded by a hydrophilic surface. The hydrophilic surface layer mainly consists of a surface monolayer with phospholipids, natural and/or synthetic polymers. Hydrophobic PSs along with drugs used in PDT permit their encapsulation into the interior core of the nano-based system [21].

Liposomes

In PDT, the liposomes function as an effective cargo system to deliver the PS molecule into the targeted tissues. Development of rose Bengal and gold nanoparticle liposomes has a potent application in the field of PDT [22]. Rapid and excellent drug release and accumulation characteristics of liposomes into the targeted cells give an extra qualification to function as a good candidate as a cargo system in cancer therapy.

Polymeric Nanomaterials

Polymeric nanomaterials are an emerging material to function as an efficient drug delivery system in cancer therapy. Along with an excellent stability and prolonged half-life in the circulatory system, the polymeric nanomaterials exhibit very good EPR capacity. Additionally, they are able to escape immune mediated recognition from macrophages. In PDT, the hydrophobic PSs can be packed in the interior core of the nano-scaled materials. Polylactide (PLA), polyglycolide (PGA), and their copolymer poly (D,L-lactide-co-glycolide) (PLGA) have been chiefly used. These particles are biocompatible in nature, exhibit an excellent physical robustness, versatility, high drug-loading capacity, and controlled drug release efficiency.

Viral Nanoparticles (VNPs)

Viral nanoparticle (VNPs) systems can also serve as an effective platform for PDT. VNPs have been considered as a drug delivery cargo in antimicrobial therapy and cancer treatment [23].

Carbon Nanotubes

Carbon nanotubes composed of pure carbon atoms, exhibits very good mechanical properties, high stability, excellent electrical conductivity and heat conductivity, minimum toxicity, environmental friendliness. Both single walled carbon nanotubes (SWNTs) and multiwalled carbon nanotubes (MWNTs) can be used as PS delivery agents in PDT. Within our body, the carbon nanotubes are internalized through endocytosis. Biological activities of carbon nanotubes on numerous cell types were studied. It also has a potential application in PDT as PSs towards tumor and photoinactivation of viruses, and bacteria, and carriers for controlled drug delivery and cytoprotective/antioxidative reagents [24- 26].

Fullerenes

Fullerenes are one type of carbon nanomaterials with closed-cage structure. They are equipped with some functional groups on its surface, making them soluble in nature, facilitating their easier interaction with the biological system. Additionally, based on the nature of functional groups incorporated, fullerenes can photo-inactivate both malignant cancer cells and pathogenic microbial cells effectively. Based on this peculiar property, a special consideration has been given to develop this nano system. Owing to their extended π-conjugation, fullerenes can absorb light from the visible range.

Therefore, upon light illumination leading to a high triplet yield, subsequent formation of reactive oxygen species is obtained [4].

Magnetic Nanoparticles

Magnetic nanoparticles are used widely in the field of bio magnetics. They have a multiple level of applications, including in magnetic hyperthermia therapy (MHT), PDT, photothermal therapy, immunotherapy and chemotherapy as well as in cell labelling and sorting, and drug delivery system. As a drug transporter, the superparamagnetic iron oxide nanoparticles (SPIONs) augment supplementary capabilities, like introduction of magnetic resonance imaging as image-based therapy and magnetic drug targeting, heating by an external magnetic field [27].

Quantum Dots

Quantum dots are considered as a central class of nano-based material, offering a multi-functional application ranging from biomedicine to energy conversion. Quantum dots have very good tunable surface chemistries and optical properties. Unique optical property of this nano system is due to the presence of cadmium-based semiconductors. Quantum dots possess a promising application in early detection and treatment of cancerous cells [28-33].

Silver Nanoparticles

Silver nanoparticles (NPs) are nano-scaled particles of silver with 1 nm to 100 nm in size, commonly exist in spherical, octagonal, diamond, or thin sheets in shapes. Silver nanoparticles function as a very good antimicrobial, anti-inflammatory agent and are a good candidate in the biomedical industry. Encapsulation of silver particles by a nano- based carrier enhances its efficiency by avoiding its agglomeration and helps in the delivery of the PSs together with it as a combination therapy. Under illumination by light, the supramolecular nano-carrier that is self-assembling, consists of the PSs, producing and releasing the reactive oxygen species (ROS) [34, 35].

Gold Nanoparticles

Gold nanoparticle Ps (AuNPs) are getting a supreme attention in the field of biomedical industry due to their multi-focused applications including drug delivery, cell labelling, and sensing. Additionally, localized surface plasmon resonances nature of gold nanoparticles may perhaps be used to increase the excitation efficacy of a photosensitizer. AuNPs with different shape and size have been widely used as PSs for cancer diagnostics, nanocarriers for drug delivery systems, PDT, plasmonic photothermal (PPT), and in laser-assisted pathogen killing. Gold nanoshells, a peculiar gold based nano system, can function as an excellent fluorescence imaging theranostic agent. The concentration of silver nanoparticles in tumors and their response to laser light have to be analyzed and improved [36-39].

Silica Nanoparticles

Silica nanoparticles have been considered as a promising molecule in PDT for the treatment of cancer. In this nano-based formulation, the PS and organically modified silica (ORMOSIL) nanoparticles are incorporated into each other covalently. Upon photoirradiation, this covalently incorporated PS molecule, strongly produces cytotoxic singlet oxygen molecules. Major advantages of this silica based nano system is that they do not release entrapped PSs, but can release reactive O_2 and $^1O_2$, thereby reducing the level of toxicity and degree of PSs agglomeration onto the blood vessels [40-42].

Application of PDT

Due to the non-invasive and rapidly developing nature, the PDT can be used for treating various malignant and non-malignant diseases. While the antimicrobial property offers an added benefits to its therapeutic efficiency [9, 43]. PDT is found to efficiently heal a vast range of ailments in many medicinal fields including, oncology, dentistry, ophthalmology, urology and dermatology, as well as it has a very good wound healing property.

Cancer Therapy

Surgical removal, radiation therapy, chemotherapy, and immunotherapy are the common therapeutic procedures offered to the cancer patients for their survival from the diseased condition. But major hurdles in cancer treatment include drug resistance and drug toxicity. PDT is being widely accepted as a successful practice in medicine for treating a number of tumors of the lung, breast, pancreas, prostate, and skin. PDT is used to treat many of the multi-drug resistant tumors, making it the most effective treatment procedure over the conventional cancer therapies. Furthermore, it expresses an excellent selectivity to the malignant tissues, whereas damage to the non-cancerous cells is limited [44, 45].

PDT induced tumor cell necrosis can be achieved through three major mechanisms: 1) ROS mediated direct cellular damage, 2) cessation of blood supply by shutting down the blood vessel to the tumor cell leading to an indirect damage, and 3) stimulation of immune system by tumor cell specific antigen induced activation of T lymphocytes [46, 47]. PDT is being used to treat both melanoma and non-melanoma type of skin cancer [48-50]. One of the top three cancers being diagnosed is lung cancer, which acts as the major culprit for the increased mortality rate in the USA [51]. Development of drug resistance to this cancer cell followed by nonreactivity to these drugs are the major hurdles for the conventional chemotherapy treatment and overcome by PDT [52]. PDT is being successfully used for treatment of microinvasive endobronchial non-small cell lung cancer (NSCLC) but is not apt to treat partially or completely obstructive endobronchial NSCLC [51]. Breast cancer is the most common and predominant form of mortality trait in females globally [52, 53]. Compared to the conventional therapies, PDT has a potent application and is simpler to repeat with minimum side effects [54]. Fourth most common cancer type developed among females is the cervical, affecting the mortality rate badly. Use of PDT helps to reduce the side effects of conventional therapeutic techniques such as bleeding and pain, can even keep unaffected the reproductive functionality among patients [55, 56]. Correspondingly, prostate tumor is one of the frequent cancer types in males across the world. Regrettably, in most cases, this was discovered in its advanced stages and in this stage, it has a poor prognosis due to both pathophysiological changes and inadequate response to treatment [57]. PDT has been used for treatment of both stage one as well as post radiotherapy prostate cancer in an effective and safe mode [58]. Brain cancer is considered as one of the most infiltrative and highly aggressive but poor prognostic types of malignancy. For the complete curative process, a repeated chemotherapy and radiation therapy treatment is necessary and most delicate. It is noticed that the PDT is useful and safe treating various types of malignant brain tumors [59]. PDT permits a synergic visualization of malignant cells and their selective destruction [60]. Still, a special attention must be given for careful monitoring of PDT-induced specific changes developed in the malignant tissues [59]. PDT was also found to be effective in the treatment of various types of malignancy associated with gastrointestinal tract [61]. Extensive indication and less invasive nature of PDT became the added rewards of PDT in treating GI cancer over surgery and endoscopic resection therapy, and is function as a complementary practice for radio-chemo-therapy [62]. PDT is being recommended as curative therapy for liver [63, 64], pancreatic [65-67], esophageal [62], gastric [68, 69], and biliary cancer [61].

Antimicrobial Photoinactivation Therapy

PDT is not confined to eliminating tumor cells, despite its reputation as an option in cancer treatment. The photodynamic effect was first proven against bacteria. The PDT has an ability to fight against and thereby cause the inactivation of bacteria, viruses, fungi, and protozoa [66, 70].

Antimicrobial photodynamic therapy (AMPDT), which is also known as photodynamic inactivation, is a cost-effective and safe way to treat a variety of infections. This is critical because multi-drug resistant microorganisms can easily infect lesions formed on the soft tissue causing delayed healing. Likewise, conventional therapy for these infectious diseases became ineffective and thus expensive. At this juncture, photodynamic therapy functions as a proposed therapy to treat infectious diseases in an effective and safe manner. Additionally, PDT has exhibited very good wound healing results, achieved through inactivation of bacteria and subsequent proliferation of fibroblasts [71].

PDT has a great ability to inactivate many of the virus in biological system including adenoviruses hepatitis A, hepatitis B, and hepatitis C viruses, human immunodeficiency virus (HIV), herpes viruses, human papillomavirus (HPV), enteroviruses, human cytomegalovirus, and human parvovirus B19 [72-74]. Most importantly, PDT shows a very good antiviral activity against SARS-CoV-2 [72, 75].

However, a secondary infection or nosocomial bacterial infection is also reported in most of the COVID-19 cases reported in the hospitals. Most predominant form of secondary infection associated with COVID-19 pandemic was Pneumococcal pneumonia caused by *Streptococcus pneumonia*. *Staphylococcus aureus* infection is also reported in some patients with flu- like infections, skin infections and bone infections. Therefore, PDT can be applied as a complementary strategy to treat SARS-CoV-2 related complications and can decrease the bacterial and viral load in the respiratory tract [76]. Mechanism of action of PDT is through structural level alteration of proteins, nucleic acids and membrane structures of infectious agents [77, 78].

Photodynamic antibacterial therapy (PDAT) is effective technology against many of drug resistant bacterial species [79]. In dental medicine, the pathogenic microorganism predominantly found in the periodontal plaque formation of the tooth including *Porphyromonas gingivalis*, *Fusobacterium nucleatum*, and *Staphylococcus* spp., have been effectively destroyed through PDT treatment [80, 81]. Hence, APDT has functioned as a prevailing alternative to antibiotic therapy, primarily to treat localized infections such as oral cavity and skin infections. PDT has been getting great attention on photodynamic anti-mycotic (fungal) therapy (PACT) particularly due to the lack of a wide spectrum of antifungal drugs for invasive infectious diseases. Additionally, efficiency of fungal treatment may vary from patient to patient and is typically dependent on the immune response of patient [82, 83].

Conclusion

The use of photodynamic therapy as a means to substitute chemotherapy is beyond question. When you combine this therapy with the use of nanotechnological materials, it leads to a vast segment of treatment options not limited to cancer. The disadvantages of PDT like hydrophobicity or stability can be effectively curbed and, in the process, create a nano-photosensitizer based therapeutic brilliance. The applications discussed above are indicative of the usefulness of nano-photosensitizers in diverse clinical scenarios. Photodynamic therapy using green nanomaterials can be explored further for a safer and more effective alternative to harmful chemotherapeutic drugs.

References

[1] Surendiran, A., Sandhiya, S., Pradhan, S. C., Adithan, C., (2009). Novel applications of nanotechnology in medicine. *Indian J. Med. Res.* 130, 689-701.

[2] Salata, O. V., (2004). Applications of nanoparticles in biology and medicine. *J. Nanobiotechnol.* 2, 1-6.

[3] Liu, R., Gao, Y., Liu, N., Suo, Y., (2021). Nanoparticles loading porphyrin sensitizers in improvement of photodynamic therapy for ovarian cancer. *Photodiagn. Photodyn. Ther.* 33, 102156.

[4] Hamblin, M. R., (2018). Fullerenes as photosensitizers in photodynamic therapy: pros and cons. *Photochem. Photobiol. Sci.* 17, 1515-1533.

[5] Saini, R., Lee, N. V, Liu, K. Y. P., Poh, C. F., (2016). Prospects in the application of photodynamic therapy in oral cancer and premalignant lesions. *Cancers.* 8, 83-97.

[6] Chen, J., Keltner, L., Christophersen, J., Zheng, F., Krouse, M., Singhal, A., Wang, S., (2002). New technology for deep light distribution in tissue for phototherapy. *Cancer J.* 8, 154-163.

[7] Allison, R. R., Moghissi, K., (2013). Photodynamic therapy (PDT): PDT mechanisms. *Clin. Endosc.* 46, 24-29.

[8] Celli, J. P., Spring, B Q., Rizvi, I., Evans, C. L., Samkoe, K. S., Verma, S., Pogue, B. W., Hasan, T., (2010). Imaging and photodynamic therapy: mechanisms, monitoring, and optimization. *Chem. Rev.* 110, 2795-2838

[9] Kwiatkowski, S., Knap, B., Przystupski, D., Saczko, J., Kędzierska, E., Knap-Czop, K., Kotlińska, J., Michel, O., Kotowski, K., Kulbacka, J., (2018). Photodynamic therapy - mechanisms, photosensitizers and combinations. *Biomed. Pharmacother.* 106, 1098-1107.

[10] Luksiene, Z., (2003). Photodynamic therapy: mechanism of action and ways to improve the efficiency of treatment. *Medicina (Kaunas, Lithuania),* 39, 1137-1150.

[11] Nowak-Stepniowska, A., Pergoł, P., Padzik-Graczyk, A., (2013). Photodynamic method of cancer diagnosis and therapy--mechanisms and applications. *Postepy Biochem.* 59, 53-63.

[12] Juzeniene, A., Moan, J., (2007). The history of PDT in Norway Part one: Identification of basic mechanisms of general PDT. *Photodiagn. Photodyn. Ther.* 4, 3-11.

[13] Fonseca, S. M., Pina, J., Arnaut, L. G., Seixas de Melo, J., Burrows, H. D., Chattopadhyay, N., Alcacer, L., Charas, A., Morgado, J., Monkman, A. P., Asawapirom, U., Scherf, U., Edge, R., Navaratnam, S., (2006). Triplet-state and singlet oxygen formation in fluorene-based alternating copolymers. *J. Phys. Chem. B.* 110, 8278-8283.

[14] Castano, A. P., Demidova, T. N., Hamblin, M. R., (2005). Mechanisms in photodynamic therapy: part two-cellular signalling, cell metabolism and modes of cell death. *Photodiagn. Photodyn. Ther.* 2, 1-23.

[15] Chizenga, E. P., Abrahamse, H., (2020). Nanotechnology in modern photodynamic therapy of cancer: a review of cellular resistance patterns affecting the therapeutic response. *Pharmaceutics.* 12, 632- 653.

[16] Yong, K. T., Roy, I., Swihart, M. T., Prasad, P. N., (2009). Multifunctional nanoparticles as biocompatible targeted probes for human cancer diagnosis and therapy. *J. Mater. Chem.* 19, 4655-4672.

[17] Bechet, D., Couleaud, P., Frochot, C., Viriot, M.-L., Guillemin, F., Barberi-Heyob, M., (2008). Nanoparticles as vehicles for delivery of photodynamic therapy agents. *Trends Biotechnol.* 26(11), 612-621.

[18] Maiti, D., Tong, X., Mou, X., Yang, K., (2018). Carbon-Based nanomaterials for biomedical applications: a recent study. *Front. Pharmacol.* 9, 1401-1441.

[19] Mokhtarzadeh, A., Eivazzadeh-Keihan, R., Pashazadeh, P., Hejazi, M., Gharaatifar, N., Hasanzadeh, M., Baradaran, B., de la Guardia, M., (2017). Nanomaterial-based biosensors for detection of pathogenic virus. *Trends Analyt. Chem.* 97, 445-457.

[20] Su, S., Kang, M. P., (2020). Recent advances in nanocarrier-assisted therapeutics delivery systems. *Pharmaceutics.* 12(9).837- 862

[21] Khalid, K., Tan, X., Mohd Zaid, H. F., Tao, Y., Lye Chew, C., Chu, D. T., Lam, M. K., Ho, Y. C., Lim, J. W., Chin Wei, L., (2020). Advanced in developmental organic and inorganic nanomaterial: a review. *Bioengineered.* 11, 328-355.

[22] Kautzka, Z., Clement, S., Goldys, E. M., Deng, W., (2017). Light-triggered liposomal cargo delivery platform incorporating photosensitizers and gold nanoparticles for enhanced singlet oxygen generation and increased cytotoxicity. *Int. J. Nanomed.*12, 969-977.

[23] Lin, S., Liu, C., Han, X., Zhong, H. Cheng, C., (2021). Viral nanoparticle system: an effective platform for photodynamic therapy. *Int. J. Mol. Sci.* 22, 1728-1743.

[24] Kostarelos, K., Lacerda, L., Pastorin, G., Wu, W., Wieckowski, S., Luangsivilay, J., Godefroy, S., Pantarotto, D., Briand, J. P., Muller, S., Prato, M., Bianco, A., (2007). Cellular uptake of functionalized carbon nanotubes is independent of functional group and cell type. *Nature Nanotechnol.* 2, 108-113.

[25] Liu, Z., Davis, C., Cai, W., He, L., Chen, X., Dai, H., (2008). Circulation and long-term fate of functionalized, biocompatible single-walled carbon nanotubes in mice probed by Raman spectroscopy. *Proc. Natl. Acad. Sci. USA.* 105, 1410-1415.

[26] Hirsch, A., (2002). Functionalization of single-walled carbon nanotubes. *Angew. Chem. Int. Ed.* 41, 1853-1859.

[27] Janko, C., Ratschker, T., Nguyen, K., Zschiesche, L., Tietze, R., Lyer, S., Alexiou, C., (2019). Functionalized superparamagnetic iron oxide nanoparticles (SPIONs) as platform for the targeted multimodal tumor therapy. *Frontiers Oncol.* 9, 1-14.

[28] Bawendi, M. G., Steigerwald, M. L., Brus, L. E., (1990). The quantum mechanics of larger semiconductor clusters ("Quantum Dots"). *Ann. Rev. Phys. Chem.* 41, 477-496.

[29] Pathak, S., Choi, S. K., Arnheim, N., Thompson, M. E., (2001). Hydroxylated quantum dots as luminescent probes for *in situ* hybridization. *J. Am. Chem. Soc.* 123, 4103-4104.

[30] Jaiswal, J. K., Mattoussi, H., Mauro, J. M., Simon, S. M., (2003). Long-term multiple color imaging of live cells using quantum dot bioconjugates. *Nat. Biotechnol.* 21, 47-51.

[31] Lim, Y. T., Kim, S., Nakayama, A., Stott, N. E., Bawendi, M. G., Frangioni, J. V., (2003). Selection of quantum dot wavelengths for biomedical assays and imaging. *Mol. Imag.* 2, 50-64.

[32] Bakalova, R., Ohba, H., Zhelev, Z., Nagase, T., Jose, R., Ishikawa, M., Baba, Y., (2004). Quantum Dot anti-CD Conjugates: Are they potential photosensitizers or potentiators of classical photosensitizing agents in photodynamic therapy of cancer? *Nano Lett.* 4, 1567-1573.

[33] Burda, C., Chen, X., Narayanan, R., El-Sayed, M. A., (2005). Chemistry and properties of nanocrystals of different shapes. *Chem. Rev.* 105, 1025-1102.

[34] Ramos, A. P., Cruz, M. A. E., Tovani, C. B., Ciancaglini, P., (2017). Biomedical applications of nanotechnology. *Biophys. Rev.* 9, 79-89.

[35] Liao, H., Hafner, J. H., (2005). Gold nanorod bioconjugates. *Chem. Mater.* 17, 4636-4641.

[36] Robertson, C. A., Evans, D. H., Abrahamse, H., (2009). Photodynamic therapy (PDT): a short review on cellular mechanisms and cancer research applications for PDT. *J. Photochem. Photobiol. B: Biol.* 96, 1-8.

[37] Kuo, W. S., Chang, C. N., Chang, Y. T., Yang, M. H., Chien, Y. H., Chen, S. J., Yeh, C.-S., (2010). Gold nanorods in photodynamic therapy, as hyperthermia agents, and in near-infrared optical imaging. *Angew. Chem. Int. Ed.* 49, 2711-2715.

[38] Camerin, M., Magaraggia, M., Soncin, M., Jori, G., Moreno, M., Chambrier, I., Cook, M. J., Russell, D. A., (2010). The *in vivo* efficacy of phthalocyanine-nanoparticle conjugates for the photodynamic therapy of amelanotic melanoma. *Eur. J. Cancer.* 46, 1910-1918.

[39] Reum, N., Fink-Straube, C., Klein, T., Hartmann, R. W., Lehr, C. M., Schneider, M., (2010). Multilayer coating of gold nanoparticles with drug-polymer coadsorbates. *Langmuir.* 26, 16901-16908.

[40] Murray, C. B., Norris, D J., Bawendi, M. G., (1993). Synthesis and characterization of nearly monodisperse CdE (E = sulfur, selenium, tellurium) semiconductor nanocrystallites. *J. Am. Chem. Soc.* 115, 8706-8715.

[41] Gao, X., Cui, Y., Levenson, R. M., Chung, L. W. K., Nie, S., (2004). *In vivo* cancer targeting and imaging with semiconductor quantum dots. *Nature Biotechnol.* 22, 969-976.

[42] Yan, F., Kopelman, R., (2003). The embedding of meta-tetra (hydroxyphenyl)-chlorin into silica nanoparticle platforms for photodynamic therapy and their singlet oxygen production and pH-dependent optical properties. *Photochem. Photobiol.* 78, 587-591.

[43] Sun, J., Kormakov, S., Liu, Y., Huang, Y., Wu, D., Yang, Z., (2018). Recent progress in metal-based nanoparticles mediated photodynamic therapy. *Molecules,* 23, 1704-1727.

[44] Montaseri, H., Kruger, C. A., Abrahamse, H., (2021). Inorganic nanoparticles applied for active targeted photodynamic therapy of breast cancer. *Pharmaceutics.* 13, 296-326.

[45] Mai, T. T., Yoo, S. W., Park, S., Kim, J. Y., Choi, K. H., Kim, C., Kwon, S. Y., Min, J. J., Lee, C., (2021). *In vivo* quantitative vasculature segmentation and assessment

for photodynamic therapy process monitoring using photoacoustic microscopy. *Sensors.* 21, 1776-1791.

[46] Kataoka, H., Nishie, H., Tanaka, M., Sasaki, M., Nomoto, A., Osaki, T., Okamoto, Y., Yano, S., (2021). Potential of photodynamic therapy based on sugar-conjugated photosensitizers. *J. Clin. Med.* 10, 841- 854.

[47] Mroz, P., Yaroslavsky, A., Kharkwal, G. B., Hamblin, M. R., (2011). Cell death pathways in photodynamic therapy of cancer. *Cancers.* 3, 2516-2539.

[48] Lee, E. H., Lim, S. J., Lee, M. K., (2019). Chitosan-coated liposomes to stabilize and enhance transdermal delivery of indocyanine green for photodynamic therapy of melanoma. *Carbohydr. Polym.* 224, 115143-115168.

[49] Serda, M., Szewczyk, G., Krzysztyńska-Kuleta, O., Korzuch, J., Dulski, M., Musioł, R., Sarna, T., (2020). Developing fullerene nanomaterials for better photodynamic treatment of non-melanoma skin cancers. *ACS Biomater. Sci. Eng.* 6, 5930-5940.

[50] de Albuquerque, I. O., Nunes, J., Figueiró Longo, J. P., Muehlmann, L. A., Azevedo, R. B., (2019). Photodynamic therapy in superficial basal cell carcinoma treatment. *Photodiagn. Photodyn. Ther.* 27, 428-432.

[51] Chhatre, S., Vachani, A., Allison, R. R., Jayadevappa, R., (2021). Survival outcomes with photodynamic therapy, chemotherapy and radiation in patients with stage III or stage IV non-small cell lung cancer. *Cancers.* 13, 803-814.

[52] Abe, C., Imai, T., Sezaki, A., Miyamoto, K., Kawase, F., Shirai, Y., Sanada, M., Inden, A., Kato, T., Shimokata, H., (2021). A longitudinal association between the traditional Japanese diet score and incidence and mortality of breast cancer-an ecological study. *Eur. J. Clin. Nutr.* 75, 929-936.

[53] Trindade, R., França, E. L., Laura, D., Fagundes, G., Fujimori, M., Gelli, P., Machi, F., Massmman, P.., Tozetti, I. A., Honorio-França, A. C., (2019). Biointerface Research in Applied Chemistry. *Biointerface Res. Appl.Chem.* 9, 4416-4423.

[54] Banerjee, S. M., El-Sheikh, S., Malhotra, A., Mosse, C A., Parker, S., Williams, N. R., MacRobert, A. J., Hamoudi, R., Bown, S. G., Keshtgar, M. R., (2020). Photodynamic therapy in primary breast cancer. *J. Clin. Med.* 9, 483-494.

[55] Chizenga, E. P., Chandran, R., Abrahamse, H., (2019). Photodynamic therapy of cervical cancer by eradication of cervical cancer cells and cervical cancer stem cells. *Oncotarget.* 10, 4380-4396.

[56] Afanasiev, M., Dushkin, A., Grishacheva, T., Afanasiev, S., Nesvizhsky, Y., Karaulov, A. V, Pylev, A., (2021). The multi-course approach of photodynamic therapy to treat invasive cervical cancer IB2: A Case Report. *Case Rep. Oncol.* 14, 506-519.

[57] Wang, Q., Zhang, X., Sun, Y., Wang, L., Ding, L., Zhu, W. H., Di, W., Duan, Y. R., (2019). Gold-caged copolymer nanoparticles as multimodal synergistic photodynamic/photothermal/chemotherapy platform against lethality androgen-resistant prostate cancer. *Biomater.* 212, 73-86.

[58] Gheewala, T., Skwor, T., Munirathinam, G., (2017). Photosensitizers in prostate cancer therapy. *Oncotarget.* 8, 30524-30538.

[59] Tanaka, K., Sasayama, T., Nakai, T., Kohmura, E., (2019). INNV-29. Photodynamic therapy for malignant brain tumors. *Neuro-Oncol.* 21, 136-136.

[60] Cramer, S. W., Chen, C. C., (2020). Photodynamic therapy for the treatment of glioblastoma. *Front. Surgery.* 6, 81-92.

[61] Yano, T., Wang, K. K., (2020). Photodynamic Therapy for gastrointestinal cancer. *Photochem. Photobiol.* 96, 517-523.

[62] Inoue, T., Ishihara, R., (2021). Photodynamic therapy for esophageal cancer. *Clin. Endosc.* 54(4), 494-498.

[63] Zou, H., Wang, F., Zhou, J. J., Liu, X., He, Q., Wang, C., Zheng, Y. W., Wen, Y., Xiong, L., (2020). Application of photodynamic therapy for liver malignancies. *J. Gastr. Oncol.* 11, 431-442.

[64] Satiya, J., Schwartz, I., Tabibian, J. H., Kumar, V., Girotra, M., (2020). Ablative therapies for hepatic and biliary tumors: endohepatology coming of age. *Transl. Gastroent. Hepatol.* 5, 14-38.

[65] Wang, Y., Wang, H., Zhou, L., Lu, J., Jiang, B., Liu, C., Guo, J., (2020). Photodynamic therapy of pancreatic cancer: Where have we come from and where are we going? *Photodiagn. Photodyn. Ther.* 31, 101876-101912.

[66] Yang, H., Liu, R., Xu, Y., Qian, L., Dai, Z., (2021). Photosensitizer nanoparticles boost photodynamic therapy for pancreatic cancer treatment. *Nano-Micro Lett.* 13(1), 35-51.

[67] Quilbe, A., Moralès, O., Baydoun, M., Kumar, A., Mustapha, R., Murakami, T., Leroux, B., de Schutter, C., Thecua, E., Ziane, L., Colombeau, L., Frochot, C., Mordon, S., Delhem, N., (2020). An Efficient photodynamic therapy treatment for human pancreatic adenocarcinoma. *J. Clin. Med.* 9, 192- 205.

[68] Mae, Y., Kanda, T., Sugihara, T., Takata, T., Kinoshita, H., Sakaguchi, T., Hasegawa, T., Tarumoto, R., Edano, M., Kurumi, H., Ikebuchi, Y., Kawaguchi, K., Isomoto, H., (2020). Verteporfin-photodynamic therapy is effective on gastric cancer cells. *Mol. Clin. Oncol.* 13, 10-16.

[69] Afzal, A., Qayyum, M. A., Shah, M. H., (2020). Comparative assessment of trace elements in the blood of gastric cancer patients and healthy subjects. *Biointerf. Res. Appl. Chem.* 11, 10824-10843.

[70] Dai, T., Huang, Y. Y., Hamblin, M. R., (2009). Photodynamic therapy for localized infections--state of the art. *Photodiagn. Photodyn. Ther.* 6, 170-188.

[71] Kharkwal, G. B., Sharma, S. K., Huang, Y. Y., Dai, T., Hamblin, M. R., (2011). Photodynamic therapy for infections: clinical applications. *Lasers Surg. Med.* 43, 755-767.

[72] Svyatchenko, V. A., Nikonov, S. D., Mayorov, A. P., Gelfond, M. L., Loktev, V. B., (2021). Antiviral photodynamic therapy: Inactivation and inhibition of SARS-CoV-2 *in vitro* using methylene blue and Radachlorin. *Photodiagn. Photodyn. Ther.* 33, 102112- 102130.

[73] Namvar, M. A., Vahedi, M., Abdolsamadi, H. R., Mirzaei, A., Mohammadi, Y., Azizi Jalilian, F., (2019). Effect of photodynamic therapy by 810 and 940 nm diode laser on Herpes Simplex Virus 1: An *in vitro* study. *Photodiagn. Photodyn. Ther.* 25, 87-91.

[74] Owczarek, W., Slowinska, M., Walecka, I., Ciazynska, M., Nowicka, D., Walczak, L., Paluchowska, E., (2021). Correlation of the ALA-PDT treatment efficacy and

the HPV genotype profile of genital warts after cryotherapy failure and podophyllotoxin therapy in male patients. *Life.* 11, 146-154.

[75] Kipshidze, N., Yeo, N., Kipshidze, N., (2020). Photodynamic therapy for COVID-19. *Nature Photonics.* 14, 651-652.

[76] Dias, L. D., Blanco, K. C., Bagnato, V. S., (2020). COVID-19: Beyond the virus. The use of photodynamic therapy for the treatment of infections in the respiratory tract. *Photodiagn. Photodyn. Ther.* 31, 101804- 101807.

[77] Conrado, P. C. V., Sakita, K. M., Arita, G. S., Galinari, C. B., Gonçalves, R. S., Lopes, L. D. G., Lonardoni, M. V. C., Teixeira, J. J. V., Bonfim-Mendonça, P. S., Kioshima, E. S., (2021). A systematic review of photodynamic therapy as an antiviral treatment: Potential guidance for dealing with SARS-CoV-2. *Photodiagn. Photodyn. Ther.* 34, 102221-102243.

[78] Bhapkar, S., Jadhav, U., Kumbhar, N., Gacche, R., Jagtap, S., (2021). Photodynamic therapy (PDT): An alternative approach for combating COVID-19. *Biointerf. Res. Appl. Chem.* 11, 12808-12830.

[79] Pérez-Laguna, V., García-Luque, I., Ballesta, S., Rezusta, A., Gilaberte, Y., (2021). Photodynamic therapy combined with antibiotics or antifungals against microorganisms that cause skin and soft tissue infections: a planktonic and biofilm approach to overcome resistances. *Pharmaceuticals.* 14, 603-624.

[80] Amos-Tautua, B. M., Songca, S. P., Oluwafemi, O. S., (2019). Application of porphyrins in antibacterial photodynamic therapy. *Molecules.* 24, 2456- 2494.

[81] Silvestre, A. L. P., Di Filippo, L. D., Besegato, J. F., de Annunzio, S. R., Almeida Furquim de Camargo, B., de Melo, P. B. G., Rastelli, A. N. de S., Fontana, C. R., Chorilli, M., (2021). Current applications of drug delivery nanosystems associated with antimicrobial photodynamic therapy for oral infections. *Int. J. Pharmaceut.* 592, 120078- 120139.

[82] Bapat, P., Singh, G., Nobile, C. J., (2021). Visible lights combined with photosensitizing compounds are effective against *Candida albicans* biofilms. *Microorganisms.* 9, 500-519.

[83] Liang, Y. I., Lu, L. M., Chen, Y., Lin, Y. K., (2016). Photodynamic therapy as an antifungal treatment. *Exp. Ther. Med.*12, 23-27.

Chapter 6

Synthesis of Pheophorbide a and Chlorin e_6 Photosensitizers

Akmaral Kussayeva[1], Zhi-Long Chen[2] and Davor Margetić[1,*]

[1]Division of Organic Chemistry and Biochemistry, Ruđer Bošković Institute, Zagreb, Croatia
[2]Department of Pharmaceutical Science & Technology, College of Chemistry and Biology, Donghua University, Shanghai, China

Abstract

Porphyrins and their derivatives are an important class ofphotosensitizers which are applied in photodynamic therapy (PDT). In this review, synthetic routes to pheophorbide a and chlorin e_6 photosensitizers from natural sources and their functional derivatizations are described.

Keywords: porphyrins, synthesis, pheophorbide a, chlorins, photodynamic therapy

Introduction

Photodynamic therapy (PDT) is nowadays a well-accepted treatment for the tumor elimination. This method exploits the photochemical activation with tissue-penetrating light of photosensitizer molecules (PS) into an efficient and selective photodynamic reaction [1]. Advantageous characteristics of PDT methodology include significantly diminished toxic adverse effects as well as

* Corresponding Author's Email: margetid@irb.hr.

In: Photosensitizers and Their Applications
Editors: Davor Margetic and Renjith Thomas
ISBN: 978-1-68507-880-5

body deformation caused by the current medical treatment practices against tumors.

Several porphyrinic molecules such as Veterporfin, Talaporfin and Photofrin are clinically available for PDT. In order to get around their disadvantages, and with the aim of increasing selective tumor accumulation and targeting, many efforts were undertaken. This resulted in development of photosensitizers of second and third generations. Derivatives of chlorins or bacteriochlorins belong to second generation PSs and conjugated PSs on carrier are the third generation [2].

There are specific requirements for an ideal PS in tumor PDT: low dark toxicity and at the same time strong photocytotoxicity, good tissue infiltration, high extinction coefficient, fast removal from the organism and multiple routes of application. In comparison to Photofrin, derivatives of chlorophyll display in most cases small dark toxicity, very quick elimination from healthy tissue, and low skin phototoxicity. The main impediment of the chlorophyll derivatives which restricts their PDT clinical application is poor solubility in water. Among novel and promising PSs is pheophorbide a - chlorophyll metabolic breakdown product, with tetrapyrrole core which structurally resembles natural porphyrins [3]. It has a strong absorption in the tissue penetrating wavelength range (650-700 nm), significant ROS generation, low dark toxicity and when applied as a photosensitizer can generate considerable antiproliferative effects on cancer [4].

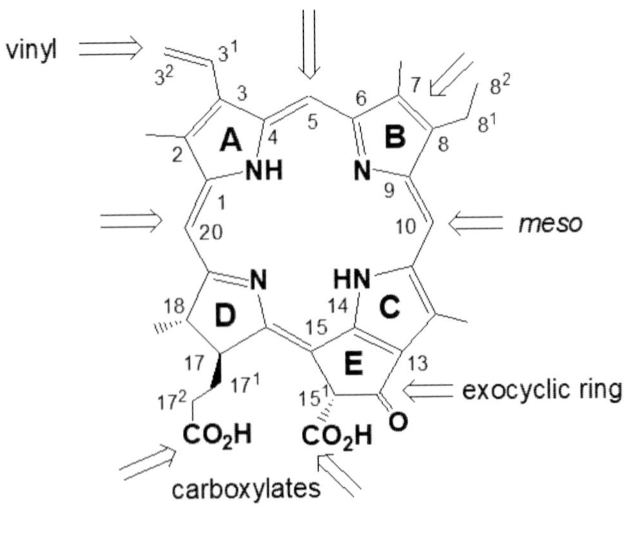

Figure 1. Functionalization sites of pheophorbide a.

Developmental efforts toward novel chlorophyllic PSs with better characteristics are based on the preparation of molecules, which are then tested for PDT. Basically, these molecules could be obtained by applying total, multistep synthesis of chlorophyll such as the one achieved by Woodward [5, 6]. However, semisynthesis using natural products as a starting point for further chemical modifications is synthetically more viable approach to structurally complex molecules often including many rings and chiral centers [7-10]. Besides the fact that isolation from natural sources significantly shortens chemical synthesis, it is more ecologically friendly to use renewable natural materials, which are ambudant in nature.

In this chapter, the most common synthetic approaches towards chlorophyllic PSs (pheophorbides and chlorins) will be presented, and chemical derivatizations illustrated only by molecules which were evaluated in PDT.

Pheophorbide a **1** has available several functionalization sites (Figure 1) [11], due to its substituents which differ in reactivity. Carboxylic groups at 15^1 and 17^2 positions are highly reactive, as well as 13^1-exocyclic ring carbonyl. Reduction of the vinyl group usually by catalytic hydrogenation is another way of structural modification. The *meso* positions could be modified in similar way as porphyrins, whereas reduction of pyrrole B ring at C^7 and C^8 leads to bacteriochlorophylls. Some specific chlorophyll chemistry could also take place, such as epimerization of 13^1-exocyclic ring chiral center [12] or allomerization, the oxygen-mediated cleavage of the isocyclic ring.

Isolation from Natural Sources and Semisynthesis

The foundation of semisynthesis of pheophorbide a and chlorin derivatives is the isolation of chlorophyll from abundant natural sources, such as green algae, *Spirulina maxima* or *Chlorella*, nettle leaves or spinach leaves (Scheme 1) [13]. The acidolysis treatment of natural materials with sulphuric acid in methanol led to the removal of magnesium from the chlorophyll core and subsequent transesterification of phytyl chain to methyl ester, thus forming methyl pheophorbide a. When chlorophyll is treated with hydrochloric acid, pheophorbide a could be obtained [14].

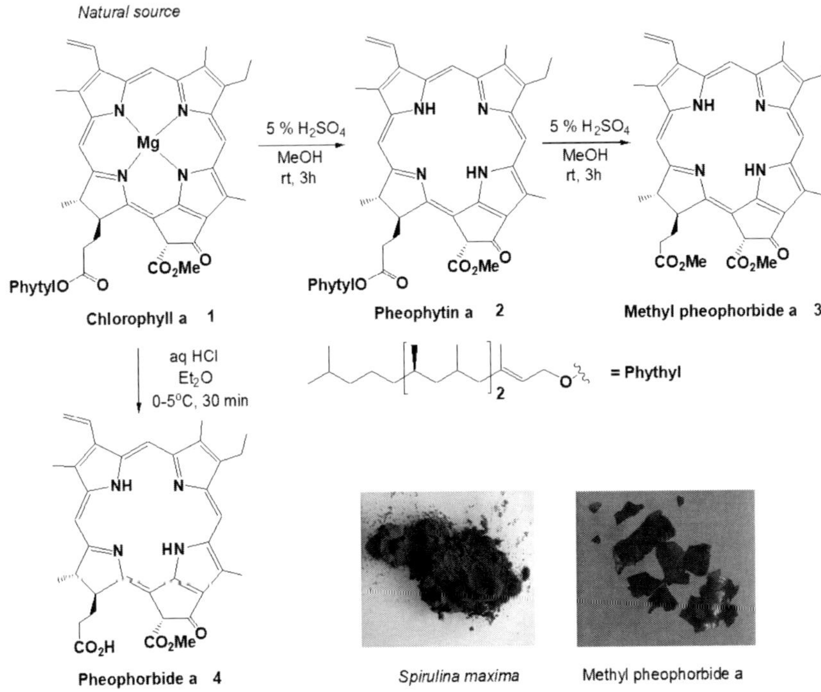

Scheme 1. Isolation of methyl pheophorbide a from natural sources.

Methyl pheophorbide a has some specific chemical properties such as unstable exocyclic E ring susceptible to nucleophilic attack. Therefore, for further derivatization **3** is decarboxylated to pyropheoporbide or hydrolysed to chlorin e_6 (Scheme 2). This transformation could be achieved by thermal decarboxylation of **3** by reflux at 175°C in *sym*-collidine providing methyl pyropheophorbide **5** in high yield [15], whereas treatment with base affords chlorin e_6 [16]. The change of hydrolysis conditions leads to different products. Purpurin-18 **7** is easily prepared from pheophorbide a and its esters as a result of oxidative cleavage of isocyclic ring E with LiOH [17] or KOH (1-propanol, pyridine, air, 2 h) [18]. Its anhydride functionality could be readily transformed to different imides. For instance, synthetic route from highly reactive **7** was employed in preparation of 13^1-lysine amide [19]. Simple treatment of purpurin e_6 with aqueous solution of lysine and pyridine afforded chemoselectively 13^1-amide product **9** in 80% yield. The evaluation of efficacy of the 13^1-lysyl chlorin p_6 **9** in PDT against 9L rat glioma cells *in vivo* revealed significant inhibition of tumour growth and was shown to be 10 to 100 times more toxic than Photofrin-II.

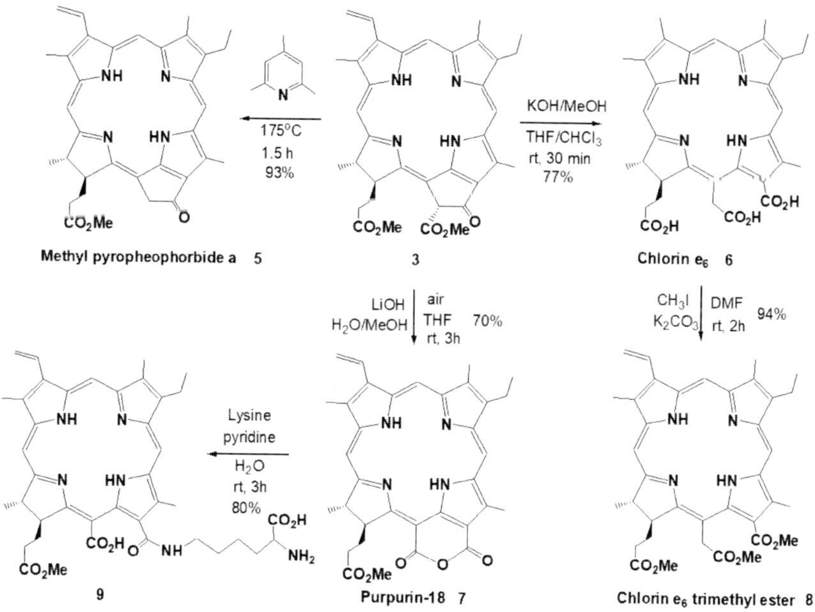

Scheme 2. Chemical transformations of methyl pheophorbide a.

Reactions on E Ring of Pheophorbide a

Nucleophilic attack on the E ring of pheophorbide a leads to formation of the 13^1-amide derivatives [20]. The reaction outcome is twofold, depending on the reaction conditions, and this is one of the peculiar behaviors of pheophorbide a. At elevated temperature and with the equimolar amount of amines pheophorbide a **3** forms carboxamides at 15^2-carbomethoxy group (amidation) instead of the cleavage of the exocycle. In contrast, at room temperature with an excess of amine nucleophilic attack is preferred on the exocyclic carbonyl with concomitant formation of 13^1-carboxamides. This type of reaction was used for instance in synthesis of guanidine derivatives **11** and **12** (Scheme 3) [21]. Room temperature reaction of dimethyl pheophorbide a with ethylenediamine provides chlorin-e_6 13^1-diaminoethylcarboxamide **10**. Guanidine/chlorin conjugate **11** was obtained when amide **10** was subjected to guanidinylation with 1H-pyrazole-1-carboxamidine reagent. This guanidine conjugate can also be prepared in two synthetic steps, by the guanidinylation/deprotection sequence. Guanidinylation reaction of amide **10** with *N,N'*-di-Boc-1H-pyrazole-1-carboxamidine reagent afforded *N,N'*-di-

tert-butyloxycarbonylguanidine product **12** in 70% yield. Boc deprotection was achieved by the treatment with trifluoroacetic acid. Molecules **11** and **12** exhibited lower cytotoxicity and higher *in vitro* photocytotoxicity in comparison with the anticancer drug *m*-THPC in MTT assay in A549 lung cancer cell lines. Both conjugated derivatives have induced severe intracellular, oxidative photodynamic damage in lung cancer cells by *in vivo* PDT.

Reactions on Vinyl Position of Pheophorbide a

Vinyl functionalization of methyl pheophorbide a was performed in several synthetic steps (Scheme 4) [22]. The vinyl group was firstly oxidized ($NaIO_4/OsO_4$) to a 13^2-*R/S* epimeric mixture of aldehydes **13**. Then, the 3^1-formyl group was reduced to alcohol **14** with *tert*-butylamine/borane complex [23]. The condensation of acetylacetone and alcohol **14** catalyzed by $Zn(OAc)_2$ provides product **15** after treatment with hydrochloric acid (for demetallation of porphyrinic core). Acetylacetonate functionality on **15** readily cyclized with hydrazine derivatives (nucleophilic reagents) to provide corresponding pyrazoles **16** as epimeric mixtures. The heterocycle formation proceeded easily at room temperature without the employment of catalysis.

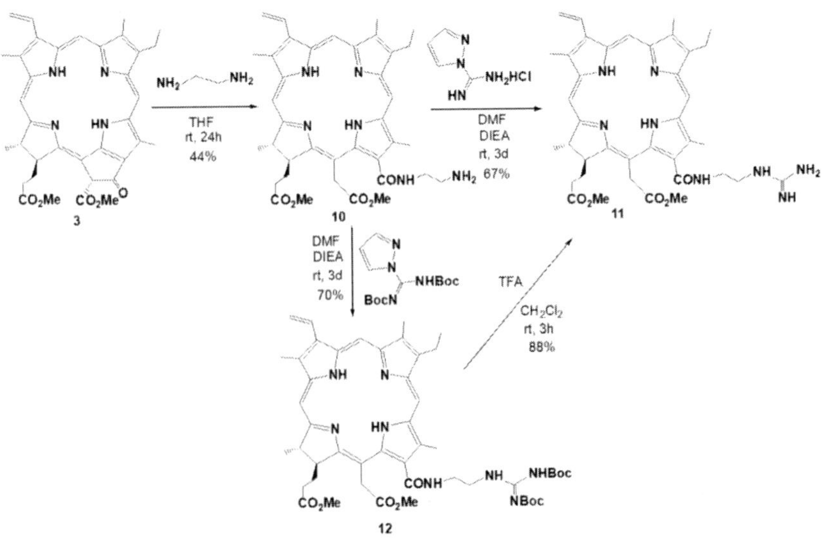

Scheme 3. Synthesis of guanidine derivatives **11** and **12**.

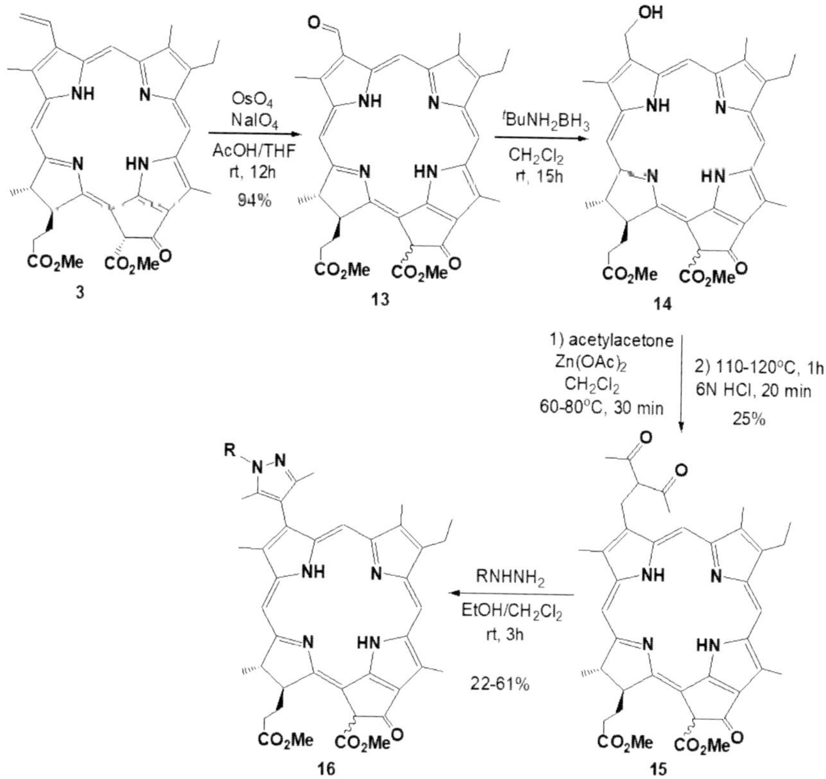

Scheme 4. Functionalization of vinyl group of methyl pheophorbide a.

The examination of the PDT effects *in vitro* of these new derivatives on A549 cancer cells showed that compound **15** and pyrazole **16** (with R=CH_2CH_2OH substituent) display promising effects. More polar substituents showed more pronounced effect than the nonpolar or lipophilic aromatic ring substituents.

The extension of the π-system of chlorin was achieved by functional modification of vinyl substituent (Scheme 5) [24]. Esterification of chlorin e6 to trimethyl ester **8** with methyl iodide was carried out in order to facilitate the separation and purification. Regioselective bromination of **8** at low temperature (-97°C) led to addition reaction, which was followed by thermal HBr elimination (at 80°C). Substituted product **18** was obtained as the major product, which was accompanied by formation of a small amount of 3^1,20-dibrominated product **17**. The later product was obtained as the major one when bromination was carried out at 80°C. Standard palladium-catalyzed Suzuki-Miyaura cross coupling reaction of bromide **18** with arylboronic esters efficiently provided chlorins **19** with extended π-system possessing *E*-configuration. Final synthetic step was ester hydrolysis with NaOH to chlorins **20**. This functionalization caused change of spectral characteristics and *in vitro* PDT activity was found to be higher than chlorin e_6 and for compound **20** (with Ar = *p*-OMePh) strong phototoxicity against human hepatoma HepG2 cells was observed in *in vivo* studies [25].

The 3^1-(heptyloxyethyl)pyropheophorbide a **22** was prepared from dimethyl pheophorbide a in two synthetic steps, for which details were not given in the paper (Scheme 6) [26]. Compound **22** was found to be quite effective in PDT and in comparison with Photofrin, the *n*-heptyl ether of analogue **5** showed better *in vivo* sensitizing activity on SMT/F tumors.

Cycloaddition reactions were also employed for functionalization of vinyl substituents of chlorins and pheophorbides. While 1,3-dipolar cycloaddition led to formation of five-membered heterocycle on the chlorin C^3 atom [27, 28], when vinylic group was involved in Diels-Alder reactions, it acted as the part of 1,3-diene moiety, together with the π-system of pyrrole ring A, as in reactions of DMAD with chloroporphyrins **8**, **25** and **26** (Scheme 7) [29]. Esterification of chlorin e_6 **6**, chlorin e_4 **23** and chlorin p_6 **24** was achieved by treatment with diazomethane in THF. These esters were subsequently oxidized by 2,3-dichloro-5,6-dicyano-1,4-benzoquinone (DDQ) to pheophorphyrin compounds **8**, **25** and **26** in high yield. Oxidation was necessary since chlorins **6**, **23** and **24** or their methyl esters were not reactive in Diels-Alder reactions. Thermal reactions of chloroporphyrins **8**, **25** and **26** with DMAD after prolonged heating in toluene provided the corresponding Diels-Alder adducts **27-29** in moderate yields. By treatment with triethylamine cycloadducts **27-29** were rearranged quantitatively to conjugated benzochloroporphyrin *trans*-isomers **30-32**. In the final reaction step, these benzochloroporphyrins were subjected to base rearrangement with DBU leading to benzochloroporphyrin *trans*-isomers **33-35**. Rearrangement of **30** to **33** was accompanied with isomer **36** possessing exocyclic F ring as a minor product and the same product mixture was obtained by treatment of **9** with DBU.

The evaluation of *in vitro* dark cytotoxicity and photodynamic efficacy of benzochloroporphyrins **33-36** on human hepatoma BEL-7402 cells has shown that **36** and **35** are more effective PDT photosensitizers than **33** and **34**.

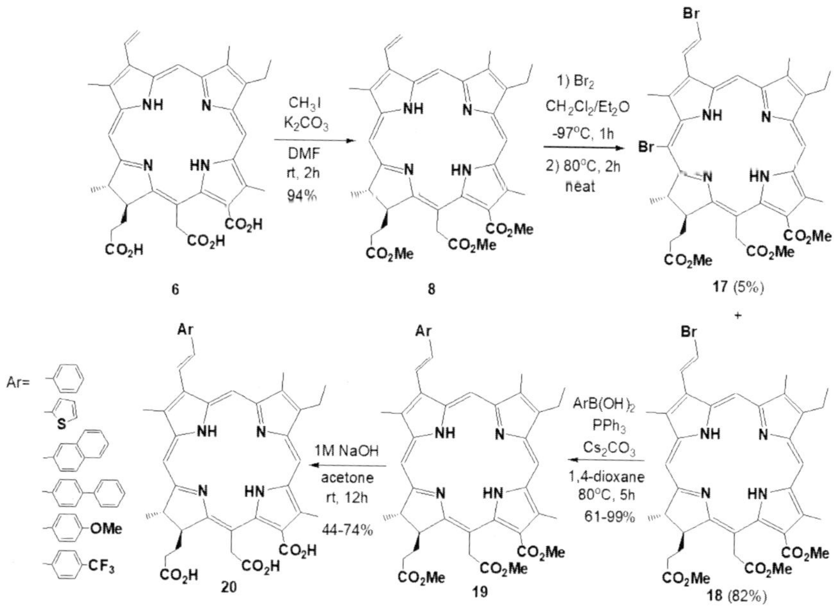

Scheme 5. The extension of the π-system of chlorin.

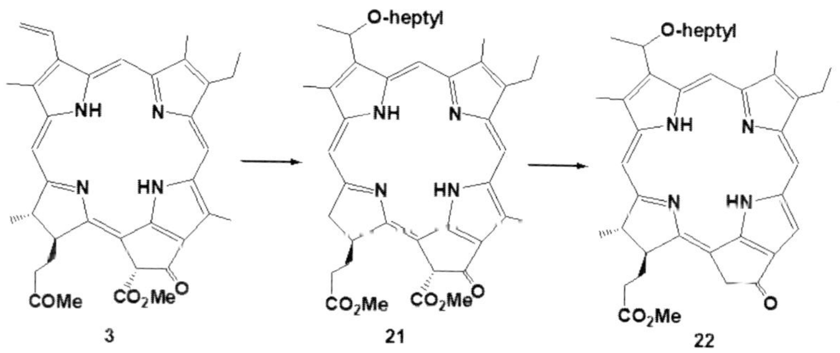

Scheme 6. Ether functionalization of dimethyl pheophorbide a.

Reactions on *Meso* Positions of Pheophorbide a

Four *meso* positions of pheophorbide a are generally available for structural modifications, but it was found experimentally that the porphyrin ring position 20 is the reactive one. As an illustration, the reaction of pyropheophorbide a with pyridinium perbromide in dichloromethane at room temperature gave 20-bromo-chlorin **37** regioselectively (Scheme 8) [30]. This 20-bromopyropheophorbide a **37** was used in Migita-Kosugi-Stille Pd-catalyzed cross-coupling: tributylvinyltin in refluxing THF by catalytic action of tetrakis(triphenylphosphine) palladium(0) to afford 20-vinyl-chlorin **38** [31]. Acetylenic substitution could be achieved by application of tributylstannyl (trimethylsilyl)acetylene and TMS removal with tetrabutylammonium fluoride afford **42**. Under similar conditions, the acetylation reaction of **37** with tributyl(1-ethoxyvinyl)tin affords 20-acetyl-pyropheoporbide **39**. Fluorination of **5** with *N*-fluorobenzenesulfonimide (NFSI) provides 20-flouro-pyropheoporbide **40** [32].

Functional Derivatizations of Chlorins

Different chemoreactivity of 13^1, 15^2 and 17^3 carboxylic groups of chlorin e_6 was used for selective functionalizations and preparation of formic, acetic and propionic amino acid derivatives. This property was employed to introduce amino acids and enhance water solubility, membrane affinity and penetrability. For instance, unsubstituted 17^3-carboxylic acid group of pheophorbide a **4**, was joined to an amino acid, such as lysine(O^{t}Bu) methyl ester by standard peptide coupling protocol. Acid activation with DCC/DMAP in the presence od amino acid formed lysine pheophorbide a derivative **43**. Treatment with strong base caused isocyclic ring-opening to 17^3-lysine chlorin e_6 product **44**, which could be easily deprotected to **45** (Scheme 9) [33]. Highly efficient nucleophilic ring-opening of the isocyclic ring of pheophorbide a with ethylenediamine was employed as the first reaction step to synthesize 13^1 amino acid functionalized chlorin e_6 **48**. Amide **48** was then subjected to peptide coupling protocol with protected lysine, and the obtained lysine derivative was deprotected in the final step to **49**.

When lysine was replaced by (L)-aspartic acid dimethyl ester hydrochloride, corresponding 17^3-aspartate pheophorbide a **46** was obtained. The hydrolysis of F-ring with NaOMe/MeOH led to formation of **47**, an 17^3-aspartate conjugate of dimethyl chlorin e_6 [34].

Significant differences in the reactivity of three carboxylic acids of chlorin e_6 enable selective methylation of the 15^2-and 17^3-carboxylic groups (Scheme 10). When chlorin e_6 was esterified with methanol in acidic conditions (5% H_2SO_4) dimethyl ester **50** was obtained quantitatively. This chemical behaviour gives an entry to 13^1 derivatization, which is an alternative synthesis to methyl pheophorbide a route described in Scheme 9 (isocyclic ring opening). Chemical activation of 13^1 carboxylic acid position of **50**

(HOBt/TBTU) and coupling with L-Asp(OtBu)2 afforded a 13[1]-aspartate diester conjugate **51**, which was easily deprotected to final aspartyl conjugate **52**. Analogous reagents were used for coupling of **50** β-alanine *tert*-butyl ester. Almost quantitative deprotection of **53** with TFA provided 13[1]-alanyl diester conjugate **54**, which was subsequently coupled with aspartate di(*tert*)butyl ester in previously described conditions, followed by removal of the *tert*-butyl protecting group to obtain conjugate **55**. The most reactive 15[2]-carboxylic acid was activated with DCC/DMAP and coupled with protected amino acids (Asp and Lys). These conjugates of chlorin e6 were esterified with diazomethane, while TFA was used for deprotection of Boc groups. This synthetic sequence led to the 15[2]-aspartyl and 15[2]-lysyl conjugates **56** and **57** in medium to high yields. Analogously, the employment of (L)-Asp(OMe)2 in conjunction with HBTU/triethylamine in dichloromethane (rt, 2h), followed by esterification with diazomethane provided in 61% yield 15[2]-(L)-aspartylchlorin-e6 tetramethyl ester **58**, the dimethyl ester of **56.** Further elaboration of these conjugates was the insertion of palladium into the chlorin core by treatment with palladium acetate. The Pd(II)-metallochlorins **59** and **60** were obtained in quantitative yield.

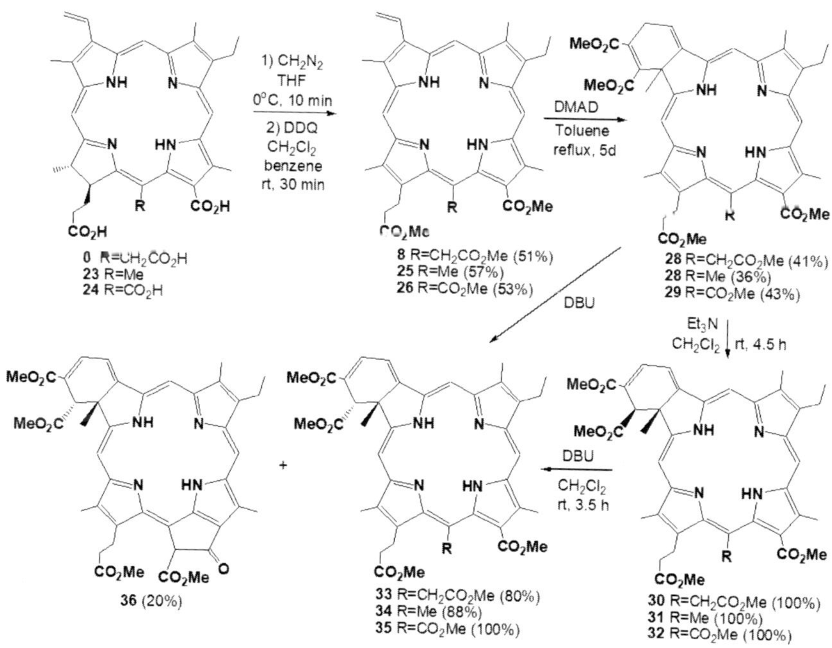

Scheme 7. Cycloaddition reactions of chlorins.

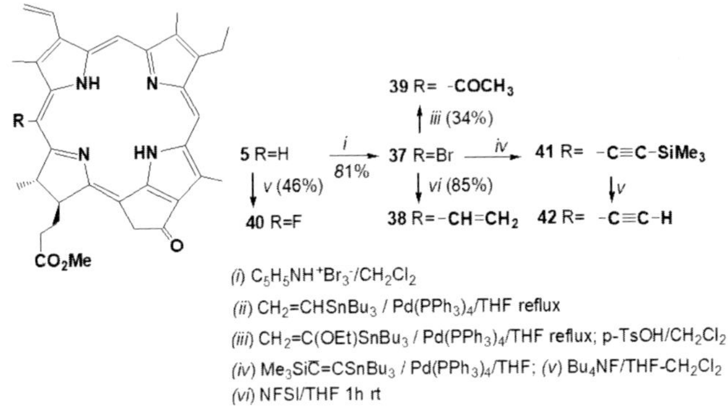

Scheme 8. Synthesis of *meso*- deriviatives of pyropheophorbide a.

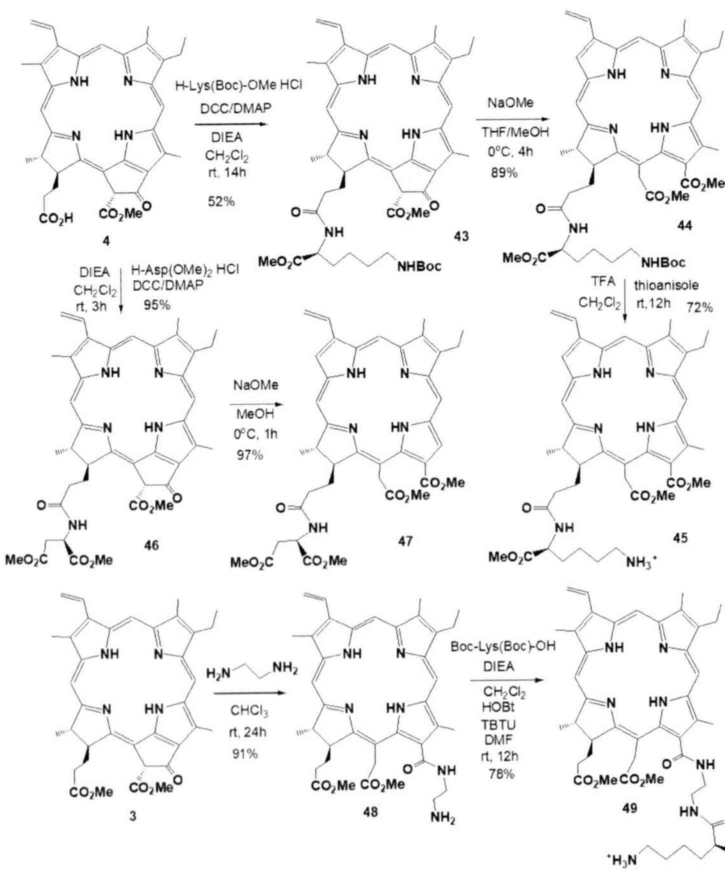

Scheme 9. Synthesis of amino acid derivatives of chlorin e_6 and pheophorbide a.

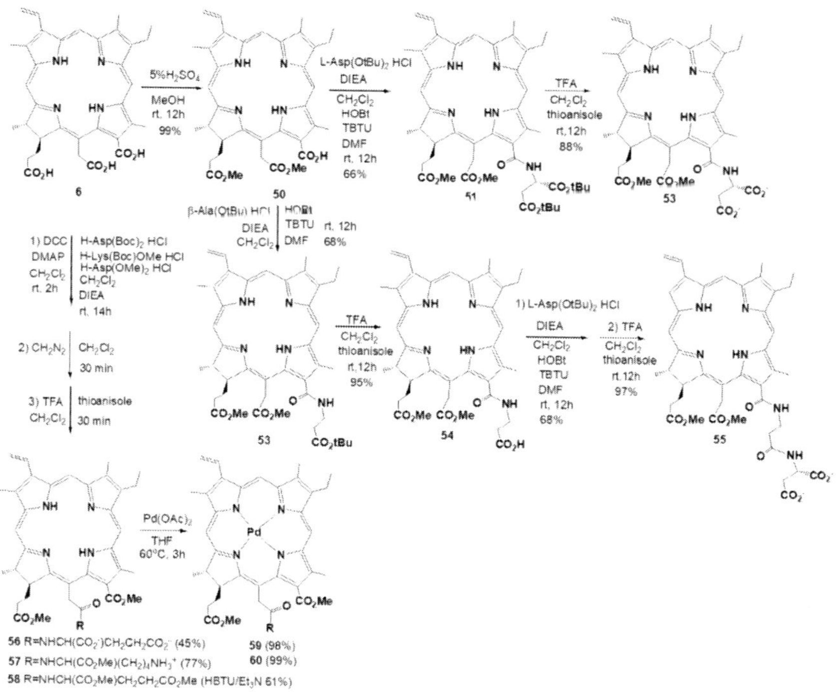

Scheme 10. Synthetic utilization of differences in the reactivity of three carboxylic acids of chlorin e_6.

The evaluation of phototoxicity of chlorin e_6 and these derivatives in HEp2 cells has shown that all derivatives are highly toxic. The highest phototoxicity exhibited conjugates **52**, **55** and **56**, and the presence of palladium(II) decreased phototoxicity.

The differences in reactivity of three carboxylic groups in chlorin e_6 were also employed in selective synthesis of derivatives with artesunate side chain (ART) at 15^2, 17^3 and 13^1 positions with the aim of increasing ROS generation in PDT (Scheme 11) [35]. The use of EDCI formed a stable intermediate of $13^1/15^2$ seven membered ring anhydride via dehydration of carboxyl groups. Nucleophilic attack of ethylene diamine due to the different reactivity of carbonyls at 13^1 and 15^2 positions, the anhydride intermediate selectively provides a 15^2 condensation product **61**. Boc deprotection (**62**) was followed by condensation with artesunate *N*-hydroxysuccinimide (NHS) ester afforded 15^2 artesunate derivative **63** in 44% total yield of three synthetic steps. The treatment of chlorin e_6 with an excess of *N*-Boc-ethylenediamine in the presence of EDCI provided the triple condensation product **64**. Employment of the same reaction sequence (deprotection and condensation) afforded the triple artesunate conjugate **65**. When pheophorbide was used as a starting material, the 17^3 condensation product was obtained from *N*-Boc-ethylenediamine and EDCI, thus the exocyclic E ring effectively acted as 13^1 and 15^2 carboxylic group protection. Then base hydrolysis was used to open the ring to final product **67**. Again, Boc deprotection and condensation were applied to obtain chlorin e6-17^3-ART conjugate **69**. Direct nucleophilic attack of ethylenediamine onto methyl pheophorbide a led to 13^1 amide product **70** through ring opening and condensation. Esters were then hydrolyzed and **71** was amidated with ART to chlorin e_6-13^1-ART conjugate **72**.

These chlorin e_6-artesunate conjugates showed better phototoxicity against HepG2 cells than chlorin e_6 in the *in vitro* evaluation of photodynamic effect. The intracellular ROS production was significantly increased by the introduction of ART substituents and tris-conjugate **65** showed much more potent phototoxicity than the mono conjugates.

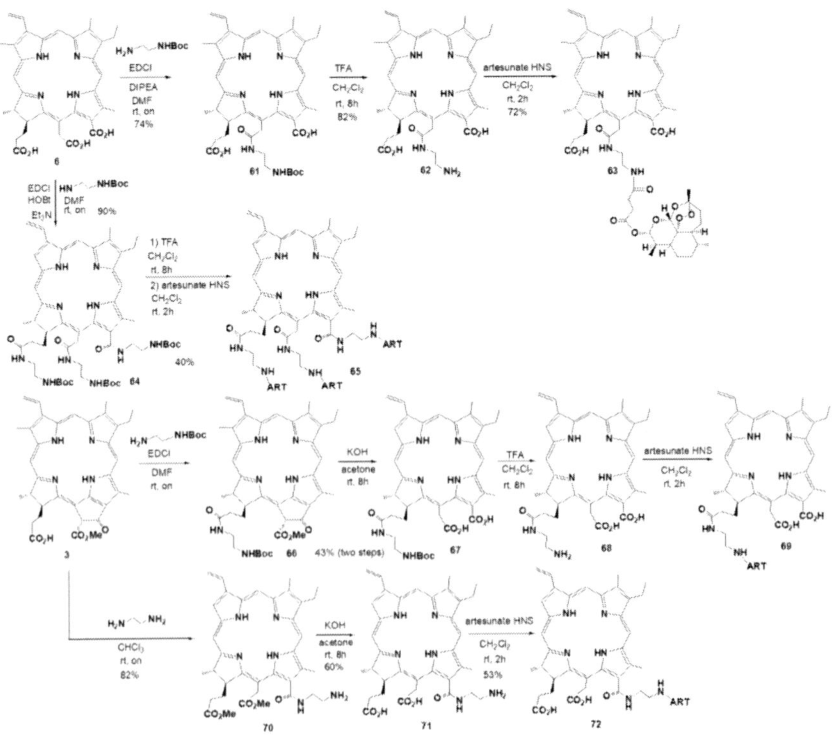

Scheme 11. Synthesis of synthesis of artesunate derivatives of chlorin e_6.

Direct Amide Preparations from Chlorins

The maleimide group was introduced in the photosensitizers with the assumption that it would inhibit the ROS scavenging by deactivating intracellular glutathione and cysteine through Michael addition of thiol group, thus elevating the cytotoxicity of photosensitizers [36]. The syntheses of 15^2-maleimide compounds **74** and **75** starts with the conversion of chlorin e_6 to a 13^1,15^2-anhydride by intramolecular dehydration with EDCI as a condensation reagent (Scheme 12). This *in situ* formed anhydride was then treated with *N*-(2-aminoethyl)maleimide to chemoselectively provide 15^2-amide. Subsequent esterification using methyl iodide led to formation of amide **73**. Michael reaction of thiol groups of cysteine or glutathione onto maleimide led to the introduction of amino acids onto the maleimide side chain without need for the functional group protection of amino acids. Applying similar synthetic sequence, 13^1-maleimide compounds **78** and **79** were prepared from 15^2,17^3-dimethyl ester of chlorin e_6, which was produced by esterification of chlorin e_6. Finally, pheophorbide 17^3-maleimide compounds **81** and **82** were prepared from pheophorbide a, however, due to the susceptibility of exocyclic E ring, the Michael reaction step provided lower yields in comparison to chlorins **73** and **77**.

PDT study revealed that all maleimide compounds have increased photocytotoxic activity in HepG2 cell compared with chlorin e_6 and its trimethyl ester. The follow-up evaluation of the PDT efficacy revealed that the maleimide derivative **77** exhibited more potent photocytotoxicity against HepG2 cells than compounds **78** and **79** [37].

Synthesis of Metal Complexes

Modulation of phototoxicity of photosensitizers by insertion of metals is not fully explored yet. Insertion of metal into porphyrinic core is a straightforward and high-yielding synthetic step which offers additional way to modulate physico-chemical properties of PSs. As an example, metallation of porphyrine core of chlorins to obtain metal complexes **M83-86** is given in Sheme 13 [38]. Complexes of zinc and nickel were prepared by refluxing of the corresponding metal acetylacetonate in toluene, while acetylacetonates were prepared *in situ* by reaction of acetylacetone with metal acetates. Preparation of copper complexes was achieved by reaction of chlorins with copper acetate in chloroform/methanol mixture.

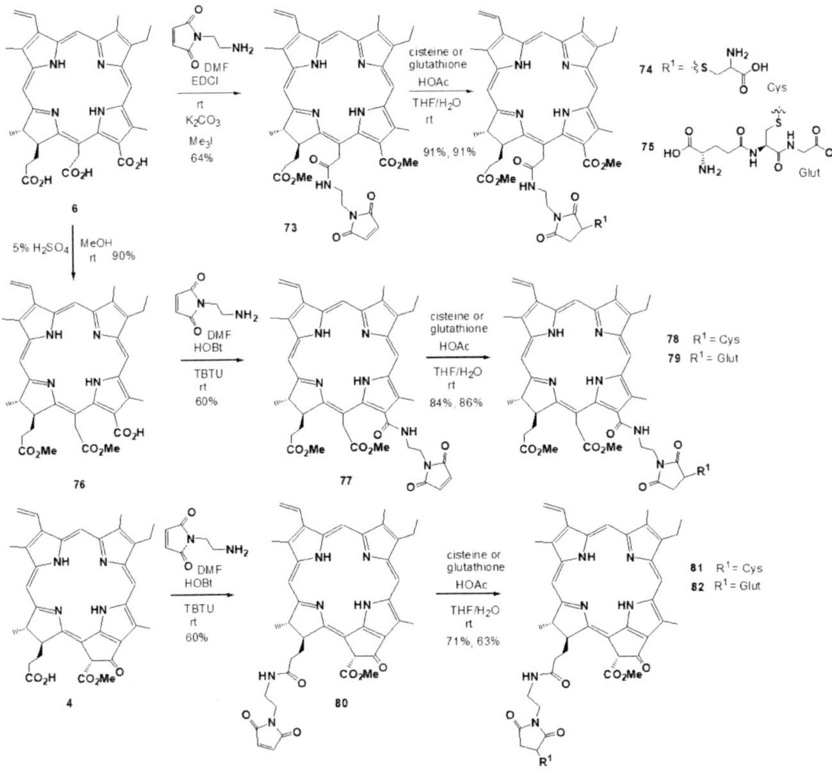

Scheme 12. Employment of maleimides in synthesis of amino acid derivatives of chlorin e_6.

The introduction of the metal into the chlorin core has an important effect on the cytotoxicity of the compounds on Hela cells. Among them, zinc complexes have the highest dark cytotoxicity, and they are significantly more toxic than ligands **83-86**.

Zinc methylpheophorbide a was easily obtained in an analogous manner from methylpheophorbide a via direct metalation with zinc acetate in methanol [39]. The PDT studies on human adenocarcinoma cell line A549 tumor revealed that Zn-pheophorbide a showed a strong photodynamic effect, which is larger in comparison to its Mg analogue (chlorophyllide a) and Photofrin [40].

Synthesis of nickel complex **89** [41] was carried out in an analogous way to copper complex **88** (Scheme 14) [42]. Treatment of methyl pheophorbide a with hydrobromic acid in acetic acid, and subsequent addition of benzyl alcohol provided benzyl ether **87**. Metallation was carried out with the corresponding metal acetates.

Complex **89** has demonstrated good PDT efficacy, which was better in human ovarian endometrioid adenocarcinoma MDAH-2774 than in human breast adenocarcinoma cells MCF-7.

Pyropheophorbide Derivatizations

Poor water solubility of pyropheophorbide a was resolved by simultaneous introduction of amino acid at 17^3-position and ethers at 3^1-position. A series of water-soluble amino acid conjugates of pyropheophorbide a ethers **92** was synthesized in five synthetic steps starting from pheoporbide a (Scheme 15) [43]. Thermal decarboxylation of pheophorbide a was achieved in milder conditions (118°C) in comparison with collidine decarboxylation of methylpheophorbide a described earlier. Ether chains at vinylic 3^1 position were introduced by hydrobromination (in analogy to procedure in Scheme 14), followed by substitution with alcohols. Standard peptide coupling protocols (EDCI) were applied to merge Boc-protected amino acids with 17^3-carboxylic group. This synthetic sequence was completed by acidic removal of Boc protection groups (TFA).

The results of *in vitro* photodynamic antitumor activity of those compounds showed that molecules exhibit higher phototoxicity and lower dark toxicity against three types of tumor cell lines compared to veterporfin.

Additional F ring on pyropheophorbide a was constructed by chemical reactions on both vinyl and *meso* positions. It was found out that the presence of bromine in *meso* position is essental for ring closure of F ring (Scheme 16) [44]. Firstly, the 3-vinyl group of **4** was hydrated to the 3-(1-hydroxyethyl) group by the addition of hydrogen bromide and successive hydrolysis to afford methyl bacteriopheophorbide-d **93** [45]. Regioselective bromination at the C20-position using pyridinium perbromide in dichloromethane at room temperature led to methyl 20-bromobacteriopheophorbide-d **94**. Treatment of **94** with pTsOH led to several reactions: acidic dehydration, cyclization and debromination and four products were obtained. The first product is methylpyropheophorbide a **5**, which was formed by dehydration of the C3-substituent and debromination at the C20-position of **94**. The other product is methyl 3-devinyl-3,5-ethylene-pyropheophorbide-a **96**, where **94** has the ethylene linkage at the C3- and C5-positions and it is at the same time debrominated. Minor products **95** and **97** were also obtained.

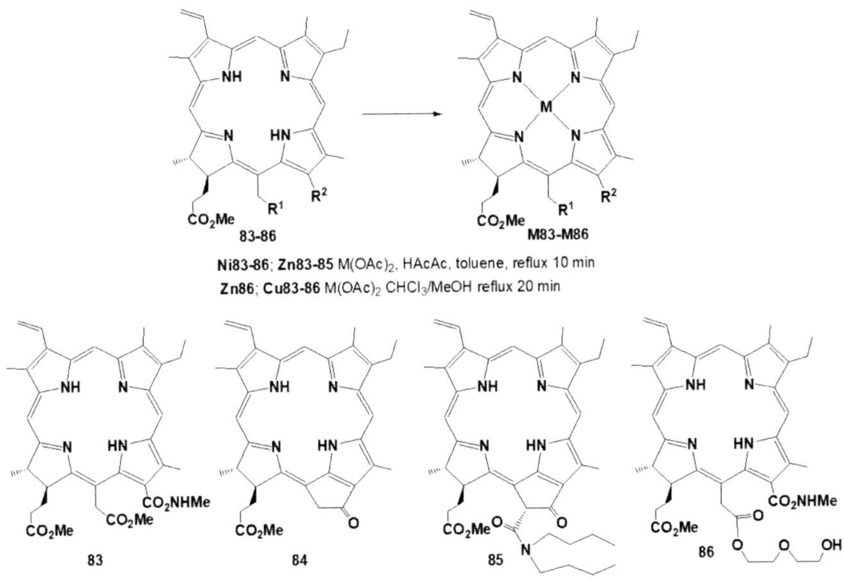

Scheme 13. The insertion of the metal into the chlorin core.

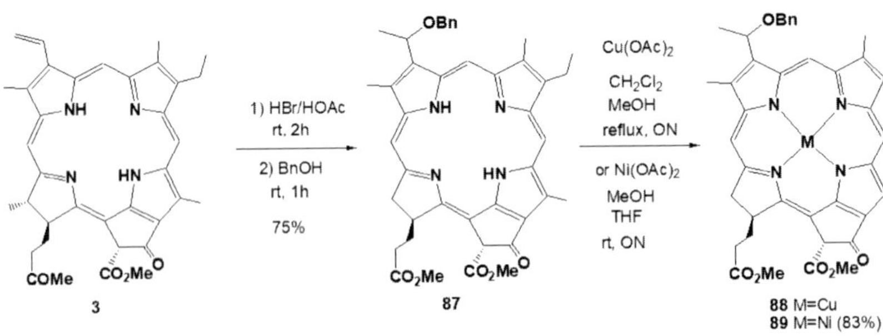

Scheme 14. Synthesis of metal complexes of pheophorbide a.

Simple synthetic route was applied to prepare 17^3-2-deoxyglucosamide conjugate of pyropheophorbide a (Scheme 17) [46]. Firstly, the carboxy group was activated with DCC and *N*-hydroxysuccinimide in DMF. In subsequent reaction, pyropheophorbide succinimidyl ester **98** was treated with D-glucosamine hydrochloride in a solution of sodium methoxide in DMSO providing conjugate **99**. PDT efficacy of conjugate **99** was evaluated in a 9L glioma rat model, which efficiently caused selective mitochondrial damage to the region of a tumor.

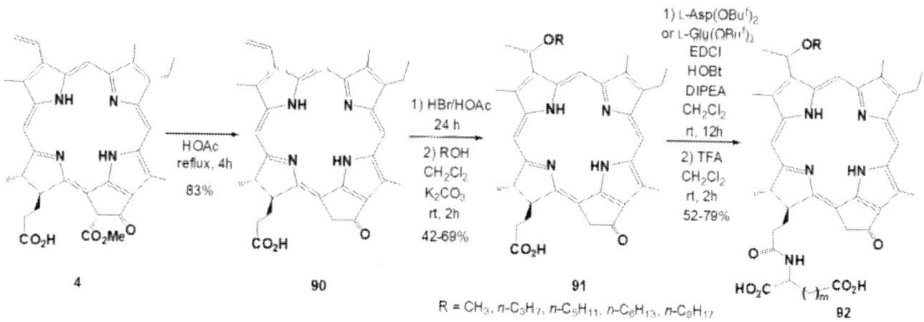

Scheme 15. Water solubilization of pyropheophorbide a.

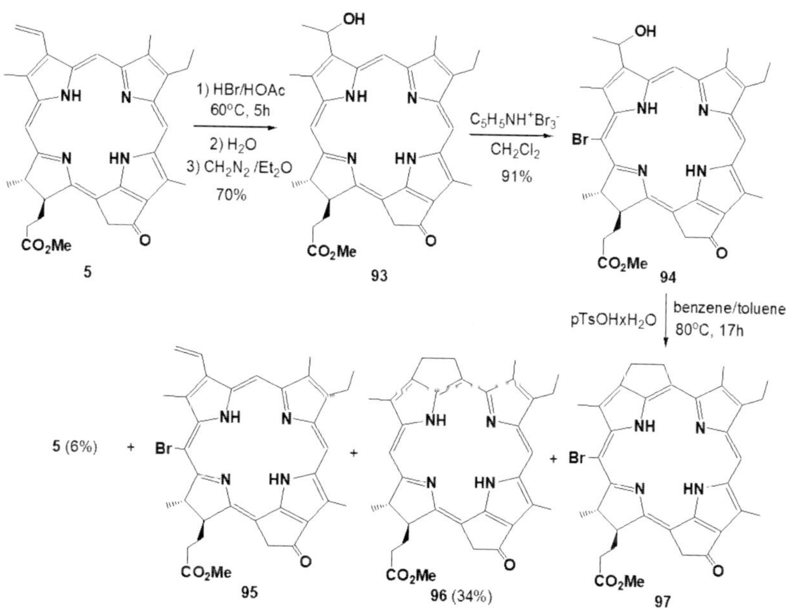

Scheme 16. The construction of F ring on pyropheophorbide a.

Exemplary functionalization reactions of pyropheophorbide a are given in Scheme 18. The 17^3-aminolevulinic acid conjugate **100** of pyropheophorbide a was obtained in nearly quantitative yield by EDCI coupling of **4** with methyl aminolevulinate (Scheme 18) [47]. Subsequently, the 3^1-vinyl position of **100** was functionalized in three reaction steps: bromination with 30% HBr/acetic acid to afford bromide **101**, which was without isolation treated with alcohols to obtain corresponding ethers. In the final synthetic step, methyl ester was hydrolyzed with LiOH to afford 3^1-pyrophephorbide - aminolevulinic acid conjugates **102** in moderate yields.

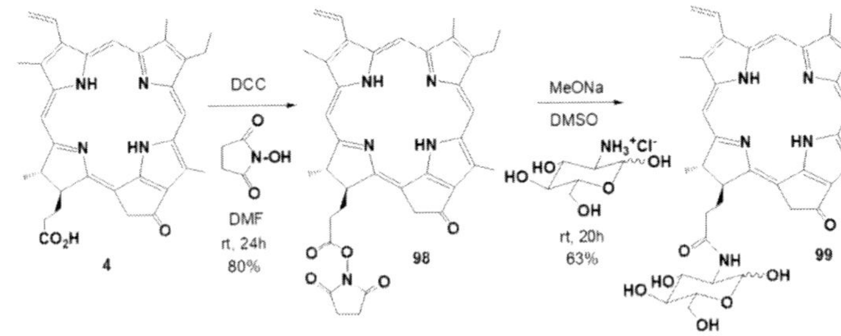

Scheme 17. Synthesis of 17^3-2-deoxyglucosamide conjugate of pyropheophorbide a.

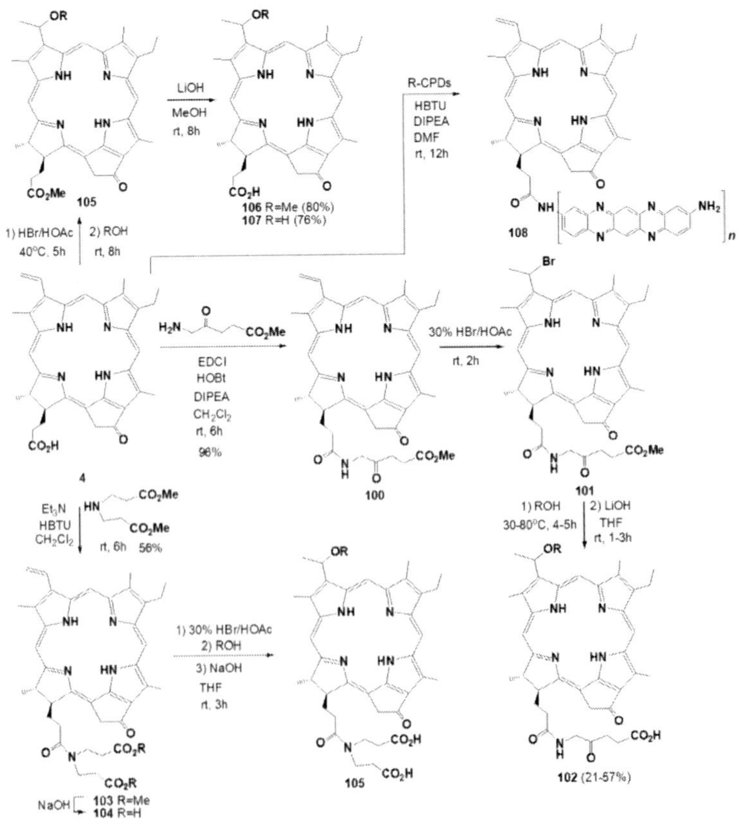

Scheme 18. Functionalization reactions of pyropheophorbide a.

Within this series, compounds **102** (with R = *n*-Bu, ethoxyl) have shown outstanding photodynamic anti-tumor effects on human non-small cell lung cancer cell line A549 without obvious skin phototoxicity. In addition, remarkable inhibition of tumor growth was observed for the amide **104** in the investigation of the *in vivo* PDT effects against human cholangiocarcinoma xenograft nude mouse model [48].

Thc 17^3-dicarboxyethyl pyropheophorbide a amide derivatives **105** were prepared in an analogous manner starting with coupling of 3,3'-azanediyldipropionate with pyropheophorbide a (Scheme 18) [49]. Derivative **105** (R = *n*-pentyl) exhibited significant photocytotoxicity and after PDT treatment, the growth of Eca-109 tumor was significantly inhibited.

Already established synthetic sequence encompassing bromination, etherification of vinylic bond and hydrolysis of methyl ester provided 3-(1-methoxyethyl) pyropheophorbide a **106** in 80% overall yield (Scheme 18) [50]. When water was used instead of methanol, 3-(1-hydroxyethyl) pyropheophorbide a **107** was obtained in 76% yield [51]. *In vitro* and *in vivo* photocytotoxicity studies indicated that **106** is prospective photosensitizer in PDT for age-related macular degeneration, whereas **107** displayed excellent antitumor PDT activity on human bile duct carcinoma cells (QBC-939).

In the final example of pyropheophorbide functionalizations, covalent linking of pyropheophorbide a with red emission polymer dots (R-CPDs) was constructed by standard HBTU/DIPEA peptide coupling protocol in DMF, and the 17^3- pyropheophorbide a R-CPDs conjugate **108** was obtained (Scheme 18) [52]. Photo toxicity and dark toxicity of **108** was evaluated in human esophageal cancer cell line (Eca-109) showing that the incorporation of p-dots into porphyrin moiety improves biocompatibility and enhances the photodynamic effect.

Conslusion

Various functional derivatives of pheophorbide a and chlorin e_6 were prepared by modern synthetic methods. The main route to obtain these compounds is semisynthesis, starting from naturally occurring molecules and their chemical modification. Novel compounds have shown enhanced activity in cancer photodynamic therapy.

References

[1] Pandey, R. K. Recent advances in photodynamic therapy. *J. Porph. Phthaloc.*, **2000**, *4*, 368-373.

[2] Yoon, I.; Li, J. Z.; Shim, Y. K. Advance in Photosensitizers and Light Delivery for Photodynamic Therapy. *Clin. Endosc.*, **2013**, *46*, 7-23.

[3] Smith, K. M., Porphyrins, Clorrins and Phthalocyanines. in *Comprehensive Heterocyclic Chemistry*, Katritzky, A. R.; Rees, C. W. (Eds.), Elsevier, Oxford **1984**, vol. 4, Chapter 3.07, pp. 377-442.

[4] Saide, A.; Lauritano, C.; Ianora, A. Pheophorbide a: State of the Art. *Mar. Drugs*, **2020**, *18*, 257-268.

[5] Woodward, R. B.; Ayer, W. A.; Beaton, J. M.; Bickelhaupt, F.; Bonnett, R.; Buchschacher, P.; Closs, G. L.; Dutler, H.; Hannah, J.; Hauck, F. P.; Itô, S.; Langemann, A.; Le Goff, E.; Leimgruber, W.; Lwowski, W.; Sauer, J.; Valenta, Z.; Volz, H. The total synthesis of chlorophyll. *J. Am. Chem. Soc.*, **1960**, *82*, 3800-3802.

[6] Woodward, R. B. The total synthesis of chlorophyll. *Pure Appl. Chem.*, **1961**, *2*, 383-404.

[7] Majhi, S.; Da, D. Chemical derivatization of natural products: Semisynthesis and pharmacological aspects - A decade update. *Tetrahedron*, **2021**, *78*, 131801.

[8] Hamann, M. T. Enhancing marine natural product structural diversity and bioactivity through semisynthesis and biocatalysis. *Curr. Pharm. Des.*, **2003**, *9*, 879-889.

[9] Dupuis, S. N.; Robertson, A. W.; Veinot, T.; Monro, S. M. A.; Douglas, S. E.; Syvitski, R. T.; Goralski, K. B.; McFarland, S. A.; Jakeman, D. L. Synthetic diversification of natural products: semi-synthesis and evaluation of triazole jadomycins. *Chem. Sci.*, **2012**, *3*, 1640-1644.

[10] Sasaki, S.; Hashimoto, Y.; Kinoshita, Y.; Tamiaki, H.; Duan, S.; Wang, X.-F.; Saga, Y.; Yamamoto, H.; Ikeuchi, T.; Shishioh, N. Synthesis of C3/C13-Substituted Semi-Synthetic Bacteriochlorophyll-a Derivatives and Their Properties as Functional Dyes. *ChemPhotoChem*, **2020**, *4*, 5399-5407.

[11] Briš, A.; Marinić, Ž.; Chen, Z.-L.; Margetić, D. Synthesis of Chlorins by Diels-Alder Cycloadditions of Pheophorbide a and its Derivatives. *Synlett*, **2015**, *26*, 991-994.

[12] Katz, J. J.; Norman, G. D.; Svec, W. A.; Strain, H. H. Chlorophyll diastereoisomers. Nature of chlorophylls a' and b' and evidence for bacteriochlorophyll epimers from proton magnetic resonance studies. *J. Am. Chem. Soc.*, **1968**, *90*, 6841-6845.

[13] Nifantiev, N. E.; Yashunsky, D. V. *Water-Soluble Poprhyrin Derivatives for Photodynamic Therapy, their Use and Manufacture*. Patent WO 200209882, PCT/US2002/016992, Dec 12, 2002.

[14] Schanderl, S. H.; Chichester, C. O.; Marsh. G. L. Degradation of Chlorophyll and Several Derivatives in Acid Solution. *J. Org. Chem.*, **1962**, *27*, 3865-3868.

[15] Smith, K. M.; Goff, D. A.; Simpson, D. J. The *meso* substitution of chlorophyll derivatives: direct route for transformation of bacteriopheophorbides d into bacteriopheophorbides c. *J. Am. Chem. Soc.*, **1985**, *107*, 4946-4954.

[16] Bui, H. T. H.; Pham, T. T.; Nguyen, H. T. T.; Do, T. M.; Nga, V. T.; Bac, N. D.; Huyen, V. T. B.; Le, H. M.; Tran, Q. C. Transformation Chlorophyll a of *Spirulina*

platensis to Chlorin e6 Derivatives and Several Applications. *Open Acc. Maced. J. Med. Sci.*, **2019**, *7*, 4372-4377.

[17] Wang, J. J.; Yin, Y. F.; Yang, Z. Synthesis of purpurin-18 imide derivatives from chlorophyll-a and -b by modifications and functionalizations along their peripheries. *J. Iran. Chem. Soc.*, **2013**, *10*, 583-591.

[18] Hoober, J. K.; Sery, T. W.; Yamamoto, N. Photodynamic sensitizers from chlorophyll: purpurin-18 and chlorin p6. *Photochem. Photobiol.*, **1988**, *48*, 579-582.

[19] Lee, S.-J. H.; Jagerovic, N.; Smith, K. M. Use of the chlorophyll derivative, purpurin-18, for syntheses of sensitizers for use in photodynamic therapy. *J. Chem. Soc., Perkin Trans., 1* **1993**, *19*, 2369-2377.

[20] Belykh, D. V.; Pushkareva, E. I. Amidation of the Ester Group of Methylpheophorbide a with Sterically Nonhindered Primary and Secondary Amines. *Russ. J. Gen. Chem.*, **2011**, *81*, 1216-1211.

[21] Gao, Y.-H.; Lovreković, V.; Kussayeva, A.; Chen, D.-Y.; Margetić, D.; Chen, Z.-L. The photodynamic activity of dimethyl 13^1-[2-(guanidinyl)ethylamino] chlorin e6 photosensitizers in A549 tumor. *Eur. J. Med. Chem.*, **2019**, *177*, 144-152.

[22] Galindev, O.; Badraa, N.; Dalantai, M.; Sengee, G.-I.; Dorjnamjin, D.; Shim, Y. K. Synthesis of pyrazole substituted methyl pheophorbide-a derivatives and their preliminary in vitro cell viabilities. *Photochem. Photobiol. Sci.*, **2008**, *7*, 1273-1281.

[23] Oba, T.; Tamiaki, H. Why do chlorosomal chlorophylls lack the 13^2-methoxycarbonyl moiety? An *in vitro* model study. *Photosynth. Res.*, **1999**, *61*, 23-31.

[24] Cao, L.; Guo, X.; Wang, L.; Wang, S.; Li, Y.; Zhao, W. Synthesis and *in vitro* phototoxicity of novel π-extension derivatives of chlorin e6. *New J. Chem.*, **2017**, *41*, 14279-14287.

[25] Dong. Y.; Guangzhe Li, G.; Wang, L.; Cao, L.; Li, Y.; Zhao, W. Anti-tumor evaluation of a novel methoxyphenyl substituted chlorin photosensitizer for photodynamic therapy. *J. Photochem. Photobiol. B*, **2020**, *211*, 112015.

[26] Pandey, R. K.; Constantine, S.; Goff, D. A.; Kozyrev, A. N.; Dougherty, T. J.; Smith, K. M. Chlorophyll-a derivatives in photodynamic therapy: effect of position of heptyl ether side-chains on *in vivo* photosensitizing activity. *Bioorg. Med. Chem. Lett.*, **1996**, *6*, 105-110.

[27] Shasha, Y.; Xisen, X.; Yang, L.; Jiahzu, L.; Yingxue, J.; Caixia, Q.; Jinjun, W. Synthesis of Chlorophyllous Chlorins Derivatives Substituted by Aromatic Groups on Their Periphery. *Chin. J. Org. Chem.*, **2014**, *34*, 362-370.

[28] Liu, Y.; Xu, X.; Li, J.; Yin, J.; Qi, C.; Wang, J. Cyclopropanation of Chlorophyllous Degradation Products and Synthesis of Chlorin Derivatives. *Chin. J. Org. Chem.*, **2014**, *34*, 552-560.

[29] Yu, S.; Xu, X.; Liu, Y.; Li, Y.; Jin, Y.; Qi, C.; Wang, J. Design, synthesis, and *in vitro* photodynamic activities of benzochloroporphyrin derivatives as tumor photosensitizers. *Bioorg. Med. Chem. Lett.*, **2008**, *18*, 293-297.

[30] Kamatani, Y.; Kitagawa, Y.; Tamiaki, H. 20-(*N*-Methylpyridiniumyl)ethynylated chlorophyll-a derivative with an intense Qx absorption band at a green to orange region. *Tetrahedron Lett.*, **2018**, *59*, 978-981.

[31] Tamiaki, H.; Matsunaga, S.; Taira, Y.; Wada, A.; Kinoshita, Y.; Kunieda, M. Synthesis of zinc 20-substituted bacteriochlorophyll-d analogs and their self-aggregation. *Tetrahedron Lett.*, **2014**, *55*, 3351-3354.

[32] Ogaswara, S.; Nakano, K.; Tamiaki, H. Synthesis of Fluorinated Chlorophylls-a and Their Bio/Physico-Chemical Properties. *Eur. J. Org. Chem.*, **2020**, 5537-5543.

[33] Jinadasa, R. G. W.; Hu, X.; Vicente, M. G. H.; Smith, K. M., Syntheses and Cellular Investigations of 17^3-, 15^2-, and 13^1-Amino Acid Derivatives of Chlorin e_6. *J. Med. Chem.*, **2011**, *54*, 7464-7476.

[34] Hargus, J. A.; Fronczek, F. R.; Vicente, M. G. H.; Smith, K. M. Mono-(L)-aspartylchlorin-e_6. *Photochem. Photobiol.*, **2007**, *83*, 1006-1015.

[35] Guo, X.; Wang, L.; Wang, S.; Li, Y.; Cao, L.; Cai, R.; Zhao, W. Synergistic antiproliferative effect of chemo-phototherapy: Synthesis and photodynamic activity evaluation of novel Chlorin e_6-artesunate conjugates as antiproliferative agents. *Bioorg. Med. Chem. Lett.*, **2017**, *27*, 4548-4551.

[36] Guo, X.; Wang, L.; Wang, S.; Li, Y.; Zhang, F.; Song, B.; Zhao, W. Syntheses of new chlorin derivatives containing maleimide functional group and their photodynamic activity evaluation. *Bioorg. Med. Chem. Lett.*, **2015**, *25*, 4078-4081.

[37] Guo, X.; Wang, S.; Zhang, F.; Li, G.; Li, Y.; Zhao, W. Derivatization of chlorin e_6 with maleimide enhances its photodynamic efficacy in HepG2 cells. *J. Porph. Phthaloc.*, **2020**, *24*, 1093-1098.

[38] Tarabukina, I. S.; Pylina, Y. I.; Velegzhaninov, I. O.; Startceva, O. M.; Shadrin, D. M.; Belykh, D. V. Synthesis and cytotoxic activity of novel transition metal porphyrinates based on chlorophyll a derivatives. *Butlerov Commun.*, **2015**, *43*, 18-23.

[39] Szczygieł, M.; Urbańska, K.; Jurecka P.; Stawoska, I.; Stochel, G.; Fiedor, L. Central Metal Determines Pharmacokinetics of Chlorophyll-Derived Xenobiotics. *J. Med. Chem.*, **2008**, *51*, 4412-4418.

[40] Jakubowska, M.; Szczygieł, M.; Michalczyk-Wetula, D.; Susz, A.; Stochel, G.; Elas, M.; Fiedor, L.; Urbanska, K. Zinc-pheophorbide a - Highly efficient low-cost photosensitizer against human adenocarcinoma in cellular and animal models. *Photodiagnosis Photodyn. Ther.*, **2013**, *10*, 266-277.

[41] Er, O.; Lambrecht. F. Y.; Ocakoglu, K.; Kayabasi, C.; Gunduz, C: Primary evaluation of a nickel-chlorophyll derivative as a multimodality agent for tumor imaging and photodynamic therapy. *J. Radioanal. Nucl. Chem.*, **2015**, *306*, 155-163.

[42] Ocakoglu, K.; Er, Ö.; Yurt Lambrecht, F.; Yılmaz Süslüer, S.; Kayabasi, C.; Gündüz, C.; Yılmaz, O. Evaluation of cancer imaging potential and photodynamic therapy efficacy of copper (II) benzyloxypheophorbide-a. *J. Drug. Target.*, **2015**, *23*, 89-95.

[43] Meng, Z.; Shan, B.; Zhang, L.; Han, G.-Y.; Liu, M.-H.; Jia, N.-Y.; Miao, Z.-Y.; Zhang, W.-N.; Sheng, C.-Q.; Yao, J.-Y. Amino acid derivatives of pyropheophorbide-a ethers as photosensitizer: Synthesis and photodynamic activity. *Chin. Chem. Lett.*, **2016**, *27*, 623-626.

[44] Doi, M.; Tamiaki, H. Synthesis of a chlorophyll-a derivative fused with an additional exo-five-membered ring and its optical properties. *Tetrahedron Lett.*, **2019**, *60*, 150934.

[45] Tamiaki, H.; Ariki, N.; Sugiyama, H.; Taira, Y.; Kinoshita, Y.; Miyatake, T. Synthesis of 3,20-disubstituted chlorophyll-a derivatives and reactivity of the substituents. *Tetrahedron*, **2013**, *69*, 8412-8421.

[46] Zhang, M.; Zhang, Z.; Blessington, D.; Li, H.; Busch, T. M.; Madrak, V.; Miles, J.; Chance, B. C.; Glickson, J. D.; Zheng, G. Pyropheophorbide 2-Deoxyglucosamide: A New Photosensitizer Targeting Glucose Transporters. *Bioconj. Chem.*, **2003**, *14*, 709-714.

[47] Gao, Y. H.; Zhu, X.-X.; Zhu, W.; Wu, D.; Chen, D.-Y.; Yan, Y.-J.; Wu, X.-F.; O'Shea, D. F.; Chen, Z.-L. Synthesis and evaluation of novel chlorophyll a derivatives as potent photosensitizers for photodynamic therapy. *Eur. J. Med. Chem.*, **2020**, *187*, 111959.

[48] Wu, Z.-M.; Wang, L.; Zhu, W.; Wang, L.-X.; Gao, Y.-H.; Wu, H.-M.; Wang, M.; Hu, T.-S.; Yan, Y.-J.; Chen, Z.-L. Preparation of a chlorophyll derivative and investigation of its photodynamic activities against cholangiocarcinoma. *Biomed. Pharmacother.*, **2017**, *92*, 285-292.

[49] Zhu, W.; Wang, L.-X.; Chen, D.-Y.; Gao, Y.-H.; Yan, Y.-J.; Wu, X.-F-; Wang, M.; Han, Y.-P.; Chen, Z.-L. Synthesis and biological evaluation of 17^3-dicarboxylethylpyropheophorbide-a amide derivatives for photodynamic therapy. *Bioorg. Med. Chem. Lett.*, **2018**, *28*, 2784-2788.

[50] Bao, L.-L.; Bian, J.; Yan, Y.-J.; Zhang, L.-J.; O'Shea, D. F.; Chen, Z.-L. *In vitro* and *in vivo* evaluation of a pyropheophorbide-a derivative as a potential photosensitizer for age-related macular degeneration. *Biomed. Pharmacother.*, **2017**, *88*, 1220-1226.

[51] Zhang, X.-H.; Zhang, L.-J.; Sun, J.-J.; Yan, Y.-J.; Zhang, L.-X.; Chen, N.; Chen, Z.-L. Photodynamic efficiency of a chlorophyll-a derivative *in vitro* and *in vivo*. *Biomed. Pharmacother.*, **2016**, *81*, 265-272.

[52] Sajjad, F.; Shao, Z.-Y.; Chen, T.; Wu, X.-F.; Ruan, J.-W.; Chen, L.-J.; Yan, Y.-J.; Margetić, D.; Chen, Z.-L. Synthesis and *in vitro* PDT evaluation of red emission polymer dots (R-CPDs) and pyropheophorbide-α conjugates. *Sci. Rep.*, **2021**, *11*, 10013.

Chapter 7

Photosensitizers and Their Applications in China

Yi-Jia Yan[1], Danye Chen[1], Davor Margetić[2,*] and Zhi-Long Chen[1,3,†]

[1]Department of Pharmaceutical Science and Technology, Donghua University, Shanghai, China
[2]Division of Organic Chemistry and Biochemistry, Ruđer Bošković Institute, Zagreb, Croatia
[3]Department of Pharmacy, Huadong Hospital affilated to Fudan University, Shanghai, China

Abstract

Since domestically produced hematoporphyrin derivative (HpD) became available, photodynamic therapy (PDT) was rapidly evolved and has been widely applied in the treatment of a large variety of diseases. This chapter provides an overview on the discovery and development of new photosensitizers, and the clinical applications of PDT in China.

Keywords: photosensitizer, photodynamic therapy, China

Introduction

Novelty and high potential of use of photosensitizer molecules in photodynamic therapy have been recognized in China four decades ago.

* Corresponding Author's Email: margetid@irb.hr.
† Corresponding Author's Email: zlchen1967@qq.com.

In: Photosensitizers and Their Applications
Editors: Davor Margetic and Renjith Thomas
ISBN: 978-1-68507-880-5

Scientific efforts were focused on synthesis of novel functional derivatives of photosensitizer molecules, study of their physico-chemical properties and *in vitro* and *in vivo* evaluation of their efficacy against different diseases, and introduction to clinical practice.

The Discovery of New Photosensitizers in China

The operation of photodynamic therapy (PDT) in China was first performed in 1981 to treat basal cell carcinoma. Since then, the studies and research on PDT grew rapidly. In the past four decades, large number of photosensitizers were developed and evaluated. Some of them have passed through the clinical trials and been approved by National Medical Product Administration of China (NMPA) for the therapy of both oncological and non-oncological conditions. Meanwhile, numerical reports on their clinical manifestations and *in vivo* or *in vitro* experimental results emerged, which allows Chinese physicians to gain profound understanding on the application of PDT in clinics to nurse patients. In this chapter, different photosensitizers developed and manufactured in China will be described, including those adopted by NMPA, in clinical stage and in research.

Photosensitizers Approved by NMPA

HiPorfin

HiPorfin [1], also known as Hematoporphyrin Injection or BHpD, which was modified from Photofrin II, is the sodium salt of various porphyrin mixtures prepared by semi chemical synthesis. The main ingredients are hematoporphyrin, porphyrin polymer, hydroxyethyl-vinyl hypo porphyrin and minor amount of protoporphyrin. 1,3,5,8-tetramethyl-2,4-bis(α-hydroxyethyl) porphine-6,7- dipropionic acid (hematoporphyrin) functions as the main active chemical, while the structures of some other components remain unpublished. Prepared in 1983, HiPorfin was the first photosensitizer developed by Chinese researchers of Beijing Institute of Pharmaceutical Industry.

HiPorfin received the official approval of National Medical Product Administration of China in 2001 to locate and treat various superficial cancers, like bladder cancer and oral cancer. Meanwhile, it was discovered that it can cure some other precancerous lesions as well as port wine stains (PWS). The results of phase III clinical trials demonstrated its feasibility in treating malignant tumor. Among the 100 cases of upper gastrointestinal cancer, in which 28 cases were esophageal cancer, 56 cases were cardia cancer, 16 cases were gastric cancer, the overall effective rate of therapy was 84.8% and complete reaction and partial reaction took 69.8% under the condition of the drug dosage of 5 mg/kg, the excitation wavelength of 514.5 nm and emission wavelength of 630 - 690 nm. Moreover, in the clinical research of HiPorfin on the treatment of port wine stains, the overall effective rate of drug reached 96.15% in the 26 cases, which was significantly higher than that of the group of patients treated by traditional therapy of CO_2 laser when the dosage was 0.01 mg/mL and the laser light was 532 nm.

In the pharmacokinetics study of HiPorfin, the content of drug distributed in various tissues of experimental mice reached its maximum in 1 hour after the intravenous injection. It was found that 73.29% of drug would be excreted from the body of mice in 5 hours after injection and it reached 91% on the 11th day.

The known disadvantage is that due to the complexity of the chemical constitution for HiPorfin, it could hardly be metabolized from skin tissue, which made the patients have to be protected from light source for a month to prevent some photosensitivity side effects like skin redness and pigmentation on the exposed area. HiPorfin is now marketed by Chongqing Maile Modern Biopharmaceutics Co. Ltd. The Chinese national approval number was H20064306.

Hemoporfin (Hematoporphyrin Monomethyl Ether, HMME)

Hematoporphyrin monomethyl ether (HMME), known as Hemoporfin for injection and PSD-044, is a chemical mixture of two constitutional isomers of compound 8-(1-hydroxyethyl)-13-(1-methoxyethyl)-3,7,12,17-tetramethyl-1H, 23H-porphine-2,18-dipropanoic acid and compound 8-(1-methoxyethyl)-13-(1-hydroxyethyl)-3,7,12,17-tetramethyl-21H,23H-porphine-2,18-dipropanoic acid (Figure 1). It was first synthesized by Hao Zhang under the supervision of De-Yu Xu in 1988 [2]. The determination of its structure was finished by Ping Lu and Zhi-Long Chen in 1996 [3]. Ying Gu was responsible for the evaluation towards its anti-port wine stains activity [4]. HMME has been licensed to Fudan Zhangjiang BioPharmaceutical Co. Ltd.

In UV-vis spectrum, HMME exhibited multiple absorption peaks at 395, 497, 531, 567 and 620 nm (Figure 1) and when excited by 385 nm light, it showed emission peak at 615 nm in fluorescence spectrum.

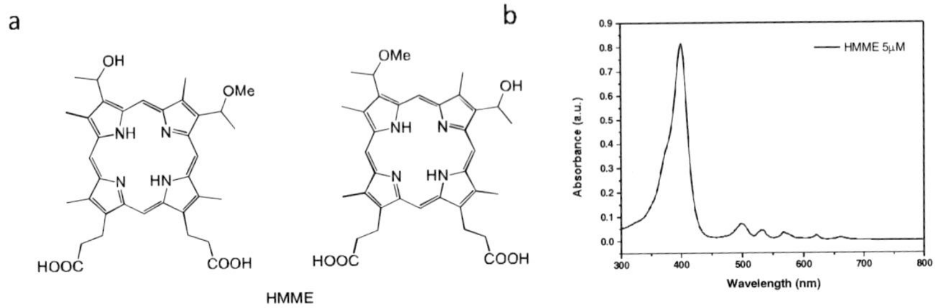

Figure 1. Chemical structure (a) and UV-vis spectrum (b) of HMME.

Table 1. Pharmacokinetics parameters of HMME

Dosage (mg/kg)	P (μg/ml)	A (μg/ml)	B (μg/ml)	π(min^{-1})	α (min^{-1})	β (min^{-1})	$T_{1/2\pi}$ (min)	$T_{1/2\alpha}$ (min)	$T_{1/2\beta}$ (min)
10	11.184	1.715	0.301	0.227	0.044	0.0046	3.051	15.77	149.84
20	265.970	3.666	0.788	0.376	0.034	0.0043	1.845	20.46	159.71
40	945.560	13.949	1.203	0.265	0.032	0.0020	2.617	21.57	353.01

Dosage (mg/kg)	K_{12} (min^{-1})	K_{21} (min^{-1})	K_{13} (min^{-1})	K_{31} (min^{-1})	K_{10} (min^{-1})	V_c (L/kg)	Cl (L·kg^{-1}·min^{-1})	AUC (μg·min^{-1}·ml^{-1})
10	0.0611	0.0698	0.0511	0.0077	0.086	0.757	0.0652	153
20	0.0429	0.0387	0.0561	0.0053	0.271	0.074	0.0200	998
40	0.0214	0.0354	0.0300	0.0022	0.210	0.040	0/0084	4764

Note: P, A, B, experience constant; π, α, distribution rate constant; β, elimination rate constant; $T_{1/2\pi}$,$T_{1/2\alpha}$, distribution half-life; $T_{1/2\beta}$, distribution half-life; K_{12}, K_{21}: apparent first-order intercopartmental transfer rate constants between the shallow peripheral and central compartments; K_{13}, K_{31}, apparent first-order intercopartmental transfer rate constants between the deep peripheral and central compartments; K10, apparent first-order elimination rate constant from the central compartment; Vc, apparent volume of distribution of the central compartment; Cl, clearance; AUC, area under the plasma drug concentration-time curve.

Approved by NMDA in 2017, HMME-PDT has become the most effective measurement to manage the symptom of port wine stains in China and gradually being introduced to the world. In the phase III clinical trials, the overall effective rate of HMME-PDT reached 97.4%, excellence rate was 60.0%, and the cure rate was 28.1% in 440 port wine stains patients under the suggested dosage of 5 mg/kg and 532 nm exciting laser. After the treatment,

168 cases of side effects were reported in total, among which the skin pigmentation was the most general one. All other types of adverse reactions were moderate, and the patients recovered soon after treatment. The results also indicated that among the three types of port wine stains, the curative effect of HMME was most obvious on treating red port wine stains and least on thickening port wine stains.

Some other reports also revealed the anti-tumor effects of HMME [5]. In 2004, the researchers in Harbin Medical University applied HMME-PDT to the treatment of cerebral gliomas. One-year survival rate of all 34 cases was 76.5%, which was significantly higher than that of patients with the treatment of microsurgery and radiotherapy.

After intravenous injection of HMME, the pharmacokinetic process in experimental rabbits fitted to a three-compartment model and the elimination of the drug was found to be in accord with linear kinetics [6]. Some pharmacokinetics parameters are listed in Table 1. The first and the second distribution phases showed that the drug could be distributed from the central chamber to the peripheral chambers rapidly after intravenous injection. Thereafter, the drug would be metabolized quickly, which means it can hardly accumulate. These features help to alleviate the long-lasting phototoxicity of normal tissues, especially in skin and eye tissues. Patients treated with HMME only need to be kept from light for 2 weeks after treatment, while those treated with HiPorfin needed 4 weeks. In addition, the probability of getting infection for patients after HMME-PDT was also lower than that of HiPorfin as thick scans were less likely to form after HMME-PDT.

Aminolevulinic Acid Hydrochloride for Topical Powder (5-ALA)

5-ALA (5-Aminolevulinic Acid Hydrochloride for Topical Powder) is approved by NMPA of China in 2007. It was widely adapted to manage the symptoms of urethral condylomata acuminata in China. Compared with traditional therapy of CO_2 laser, ALA-PDT possesses lower recurrence rate and higher efficiency. It was developed and currently licensed under Shanghai Fudan-zhangjiang Bio-Pharmaceutical Co., Ltd. The Chinese national approval number is H20070027.

Photosensitizers in Clinical Trials

Photocyanine

Photocyanine (Suftalan zinc) is the market name for the photosensitizer 2,9-dipotassium-sulfonate-16,23-dimethylenephthalimido- phthalocyanine zinc ($ZnPcS_2P_2$) and its isomers (Figure 2a). It was synthesized by Jin-Ling Huang et al. in 2000 [7, 8], passed through the stage I clinical trials targeting tumor in 2013 and entered phase II clinical trials in 2015 for primary esophageal cancer with complete or incomplete obstruction [9].

Photocyanine is embodied with special optical properties which made it a good photosensitizer. Its absorption peak locates on 675 nm in DMF solution (Figure 2b), which possesses notable penetration into body tissues and the fluorescence emission peak around 685 nm could be observed. Furthermore, this compound shows minimal absorption in the range of visible light, meaning that the skin phototoxicity caused by Photocyanine under natural light could be maintained at a low level.

In vitro photodynamic anti-tumor experiments showed that Photocyanine managed to eliminate human leukaemia HL60 cell under dosage of 10 μg/mL. For *in vivo* anti-tumor activity, the inhibition rate of murine transplanted solid tumor U_{14} and S_{180} with Photocyanine reached 89.8% and 90.8% respectively, which is significantly higher than that of HiPorfin. The mechanistic study demonstrated that Photocyanine, due to its amphiphilicity, was capable of penetrating cell membrane and was accumulated in lysosome and mitochondria. Its concentration in cytoplasm and nuclear membrane is lower. The light irradiation caused decrease of mitochondrial membrane potential and increase of caspase-3 activity to induce cell apoptosis. In acute toxicology study, Photocyanine had no obvious effect on experimental mice under 100 mg/kg, which was 20 times higher than the therapeutic dose. Some side effects were noted, in long-term toxicity test, most experimental animals suffered from dysphoria and a few experienced slight edemas on the faces. Neither mortality nor abnormity were observed, and no obvious phototoxicity was discovered [9].

Deuteporfin

Deuteporfin, known as PSD-007 in literature, was first synthesized by De-Yu Xu et al. in 1984 [10]. It is composed of seven porphyrins, in which 3-(or 8-)-(1-methoxyethyl)-8-(or 3-)-hydroxyethyl-deuteroporphyrin (MHD), 3-(or 8-)-(1-methoxyethyl)-8-(or 3-)-vinyldeuteroporphyrin (MVD) and 3,8-di-(1-methoxyethyl)-deuteroporphyrin (DMD) are the main anti-tumor active components. Deuteporfin has earned the permission to enter phase I and phase II clinical trials in 2009 to target cancer. However, its phase II clinical trials were terminated in 2019.

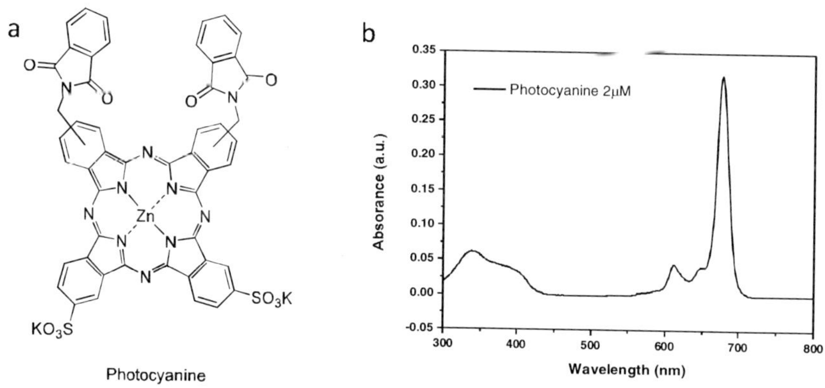

Figure 2. Chemical structure (a) and UV-vis spectrum (b) of Photocyanine.

Table 2. The IC_{50} values of the main active components for Deuteporfin and YHpD in Hela cells

Compound	IC_{50}(μg/ml)	95% Confidence limit
MVD	8.92	8.06-9.87
DMD	15.51	4.15-16.97
MHD	6.30	5.56-7.11
YHpD	11.70	10.16-12.73

Table 3. The proportion of mice recovered from tumor after photodynamic treatment

Tumor type	Drug		Control
	PSD-007	Photofrin II	
Lewis lung cancer	8/10	6/12	0/10
U_{14} cervical carcinoma	8/10	9/14	0/10
Sarcoma S_{180}	8/10	6/10	0/10

The ratio refers to the proportion of recovered mice among all mice model

Deuteporfin exhibited anti-tumor photodynamic activity on multiple types of cancer [10]. The *in vitro* studies of Deuteporfin against human liver cancer cells SMMC-7721 and human gastric cancer cells NGCC-8310 showed that the IC_{50} of drug was around 10 μg/mL under laser energy of 6 J while the cell death rate achieved by Photofrin II under the same drug and light dosage was only 20%. The IC_{50} values of the main active components for Deuteporfin in Hela cells was detected and YHpD (a type of hematoporphyrin derivative) was taken as reference (Table 2). Its anti-tumor activity was further confirmed on animal models. The photodamaging efficacy of Deuteporfin-PDT on different types of animal tumors was listed in table 3 and Photofrin II was used as a control drug.

The phototoxicity of Deuteporfin was evaluated on mice and the resulted syndromes include erythema, edema and necrosis of the ears. Death cases were discovered under high dosage group, indicating that Deuteporfin acquired a certain degree of toxicity. However, as a photosensitizer with definite chemical structure, its light toxicity is milder compared with HpD. The results of acute toxicity test showed the minimal lethal dosage of Deuteporfin was 171 mg/kg and the approximate lethal dosage was 114 mg/kg, which were far beyond the suggested dosage (2 mg/kg), indicating that Deuteporfin was safe to use.

Sinoporphyrin Sodium

Sinoporphyrin Sodium (DVDMS-2, also known as DVDMS) is a porphyrin derivative, 1-(6,7-sodium dipropionate-1,3,5,8-tetramethyl-2-vinyl-4-porphin) ether (Figure 3), isolated and purified from Photofrin. It was first obtained by Qi-Cheng Fang [11] and it has been approved for phase II clinical trials to treat esophageal cancer in 2015.

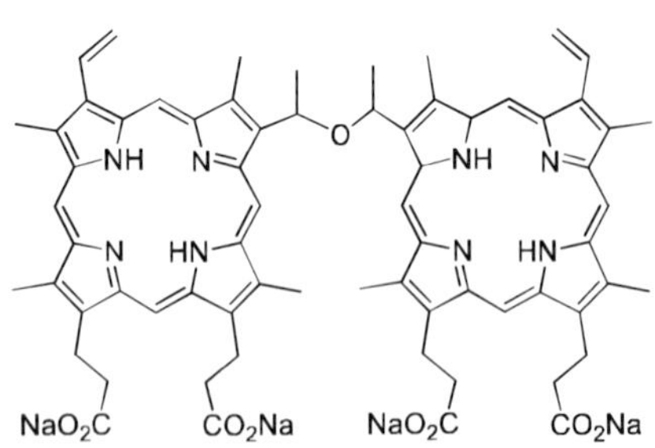

Figure 3. Chemical structure of DVDMS-2.

Similar to Photofrin, DVDMS-2 has five distinctive absorption peaks in UV spectrum at 359, 514, 548, 580 and 631 nm and the strongest one is located around 358 nm. When excited by 359 nm laser, the compound showed fluorescence emission peak at 642 nm. DVDMS-2 embodied strong anti-tumor activity on multiple types of cancer. *In vitro* study showed that the inhibition rates of DVDMS-2 PDT (0.5 μg/ml, 30 mW/cm^2, 180 s) towards breast cancer cells MCF-7, liver cancer cells HepG2, kidney cancer cells Ketr3 and lung cancer cells H406 all exceeded 85.5%. The IC_{50} value of DVDMS-2 PDT on 12 types of human cancer cells examined by MTT assay was around 0.207-0.584 μg/mL, while that of the reference drug Photofrin II was 0.3-0.5 μg/ml. *In vivo* studies demonstrated that the 90% of the transplanted tumor of human esophageal cancer cells CaEs-17 and human liver carcinoma cells HepG2 could be inhibited by DVDMS-2 PDT (2 mg/kg). This therapeutic effect could only be obtained by Photofrin PDT under much higher dosage of 20 mg/kg [12]. These results proved that DVDMS-2 possesses higher anti-tumor activity than Photofrin. Other *in vivo* and *in vitro* studies on different cancers provided similar results (Table 4).

The mechanism studies showed that the reactive oxygen species (ROS) generated by DVDMS-2 could damage DNA and change mitochondrial membrane potential (MMP) to induce apoptosis, lower the expression of proliferation cell nuclear antigen (PCNA) and inhibit the growth of new blood vessels.

Table 4. *In vivo* and *in vitro* studies on various cancers of DVDMS-2

Targets	*in vivo/in vitro*	Reference
Eca-109 Cells	*in vitro*	[13]
4T1 Cells	*in vivo & in vitro*	[14, 15]
MDA-MB-231 Cells	*in vitro*	[16]
U87MG Cells	*in vivo & in vitro*	[17]
PC9 Cells	*in vivo & in vitro*	[18]
MCF-7 Cells	*in vivo & in vitro*	[19]
Human colorectal cancer (CRC)	*in vivo & in vitro*	[20]
SGC7901 Cells, Caco2 Cells, SW480 Cells, SW620 Cells	*in vitro*	[21]
HSC-4 Cells	*in vitro*	[22]
SCC-15 cells	*in vitro*	[23]
Cholangiocarcinoma Cell	*in vitro*	[24]
CX-1	*in vitro*	[25]

The pre-clinical evaluations of the pharmacokinetics and safety of DVDMS-2 were also finished [26, 27]. The viability of normal tissues incubated with DVDMS-2 were not significantly affected, which suggested the low cytotoxicity of DVDMS-2. No observed adverse effect level (NOAEL) for DVDMS-2 with or without were 2 and 6 mg/kg in SD rats, respectively. Among the 270 test rats, liver injury was the most prominent side effect and only one rat died when given 60 J of 630 nm laser light and 18 mg/kg DVDMS-2. In the experiment on Beatle dogs, skin swelling, ulceration and changes in blood coagulation were observed in middle dose group, and yellow pigmentation was observed to deposit in the liver, spleen, local lymph nodes and marrow of dogs in high dose group. Most of these side effects would go away on their own soon. The pharmacokinetics of DVDMS-2 in tumor-bearing mice conformed to a two-compartment model, and some pharmacokinetics parameters are listed in Table 5.

Photosensitizers in Research

A large number of photosensitive compounds have been synthesized and evaluated *in vitro* and *in vivo* for photodynamic anti-tumor effect in China (Figure 4).

Table 5. Pharmacokinetics parameters of DVDMS-2

Dosage (mg/kg)	C_0 (mg L^{-1})	$T_{1/2\alpha}$ (min)	$T_{1/2\beta}$ (min)	K_{10} (h^{-1})	K_{12} (h^{-1})	K_{21} (h^{-1})
1	14.16	0.11 ± 0.01	7.68 ± 0.54	0.61 ± 0.08	4.64 ± 0.43	0.89 ± 0.15
2	21.07	0.13 ± 0.02	4.31 ± 0.56	0.59 ± 0.09	3.78 ± 0.42	1.59 ± 0.13
4	31.47	0.42 ± 0.08	6.82 ± 1.53	0.27 ± 0.05	0.86 ± 0.12	0.60 ± 0.07

Dosage (mg/kg)	Cmax (mg L^{-1})	$AUC_{0\to\infty}$ (mg h^{-1} L^{-1})	$T_{1/2}$ (h)	Cl (mg$L^{-1}$$kg^{-1}$)	MRT (h)	V_{ss} (mg kg^{-1})
1	14.43 ± 1.91	23.51± 2.72	1.13 ± 0.12	0.04	10.12 ± 1.91	0.43
2	21.05 ± 1.38	45.93 ± 5.08	1.18 ± 0.19	0.06	5.77 ± 1.02	0.32 ± 0.03
4	31.47 ± 3.32	113.07 ± 10.12	2.49 ± 0.56	0.04	8.77 ± 1.73	0.31 ± 0.05

In recent years, with the development of photodynamic antimicrobial therapy, photosensitizers with anti-microbial effects were also designed and synthesized.

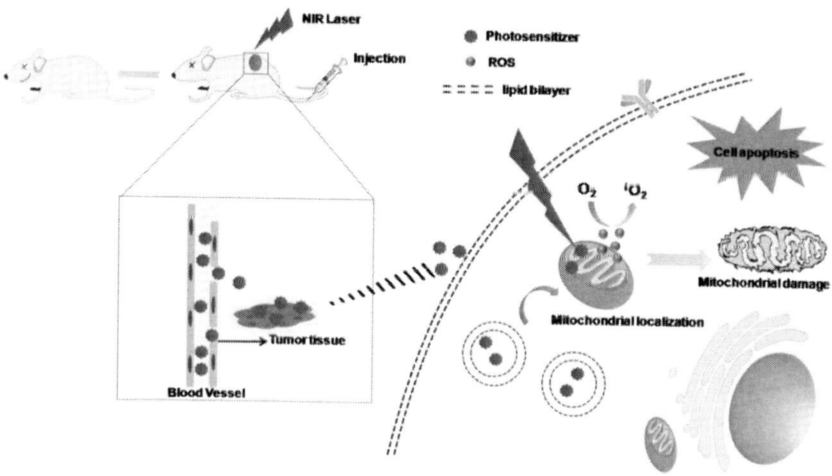

Figure 4. Schematic diagram of photodynamic antitumor activity *in vivo*.

Porphyrin-Type Photosensitizers

Porphyrins are a group of conjugated macrocyclic compounds formed by four pyrrole molecules together connected by methylene. Porphyrin is one of the most intensively studied chemical structure in the research of photodynamic therapy, as both the β positions and *meso* positions on the skeleton ring could be substituted by other functional groups, which allows it to be modified in the reaction. It has strong absorption around 400-450 nm (Soret band) and several weaker absorption peaks at 500-650 nm. It possesses strong oxygen generation ability and therefore received great attention. Photosensitizers HMME, ALA, etc. mentioned above all belong to this type of molecules.

Jing-Jing Chen et al. developed a novel photosensitizer named *meso*-5-[*p*-diethylene triamine pentaacetic acid (DTPA)-aminophenyl]-10,15,20-triphenyl-porphyrin (compound **1**, Figure 5) and investigated its anti-tumor activities against human gastric cancer cell line HGC27 and SNU-1 [28]. Despite the low absorbance, 650 nm was the optimal wavelength for its *in vitro* evaluation. Under the light doses of 6 and 12 J/cm^2, the IC_{50} for **1**-PDT was 4.75 and 2.98 μM separately for HGC27 cells, and 4.75 and 6.38 μM for SNU-1 cells. Further mechanism investigation presented that **1**-PDT could induce cell apoptosis through mitochondria apoptotic pathway via **1** localized in lysosome. Their further research specialized on the function of PDT in combating multidrug resistance (MDR), and the results showed that **1**-PDT could eliminate all human gastric cancer cells with or without MDR [29].

Shuai Meng et al. synthesized porphyrin **2** bearing four lysine moieties (Figure 6). It possesses low oil water partition coefficient, high singlet oxygen quantum yield and thus the high photoinactivation property against methicillin-resistant *Staphylococcus aureus* (MRSA), *Escherichia coli* and *Pseudomonas aeruginosa* with a low light dose (6 J/cm^2). Its MIC value was 3.91 mM, 1.95 mM and 7.81 mM against MRSA, *E. coli* and *P. aeruginosa*, respectively. Meanwhile, compound **2** was stable in serum and lower cytotoxicity to normal cell (A929 cells), indicating that it could be safe in the antibacterial practice [30].

Many porphyrin-type photosensitizers contain metal centers like ruthenium(II), iridium(III), and platinum(II). Wei-Jun Xuan et al. reported a *meso*-substituted tetramethylpyridinium porphyrin (**3**) and its Pd-derivative (**4**) for antimicrobial therapy (Figure 6). Compound **4** possesses higher singlet oxygen quantum yield and better photodynamic antimicrobial activity than **3**. The eradication of methicillin-resistant *Staphylococcus aureus*, *Escherichia coli* and *Candida albicans* could be obtained using 405 nm light dose of 10 J/cm^2 and concentrations of 10 nM, 50 nM and 500nM, respectively. The *in vivo* study on mouse model of *E. coli* infection also presented the therapeutic effect of **4** was somewhat superior compared to **3** [31].

Fluorinated photosensitizers received great attention in past decades as fluorinated compounds showed favorable physiological activities, which enhance the interaction of photosensitizer with biological targets to show anti-tumor effects. Zhi-Long Chen's group reported a series of fluorinated hematoporphyrin ether derivatives (Figure 7). Among them compound **5** shows the longest absorption wavelength at 622 nm and highest singlet oxygen quantum yield of 0.0957 min^{-1}. Its anti-tumor activity on human lung carcinoma cells A549 *in vivo* and *in vitro* was proved to be better than its homologous compound HMME [32].

Chen's group also designed a series of porphyrin derivatives, among them 5,10,15,20-tetrakis(5-morpholinopentyl)-21H,23H-porphin (compound **6**, Figure 7) which possesses strong near infrared absorption at 648 nm and characteristic large fluorescence emission at 653 nm. It showed low dark toxicity and optimal photodynamic activity *in vitro* and *in vivo*. Notably, the *in vivo* fluorescence image on nude mice displayed that this compound possessed high selectivity on tumor cells at 6 h post injection, indicating its potential use in fluorescence diagnosis [33].

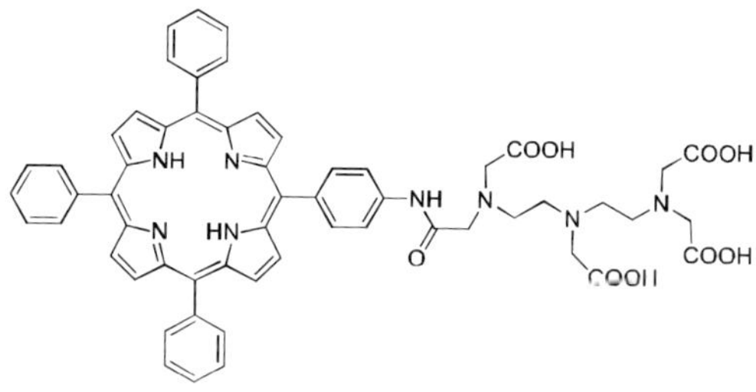

1 by Chen J.

Figure 5. Chemical structure of porphyrin **1**.

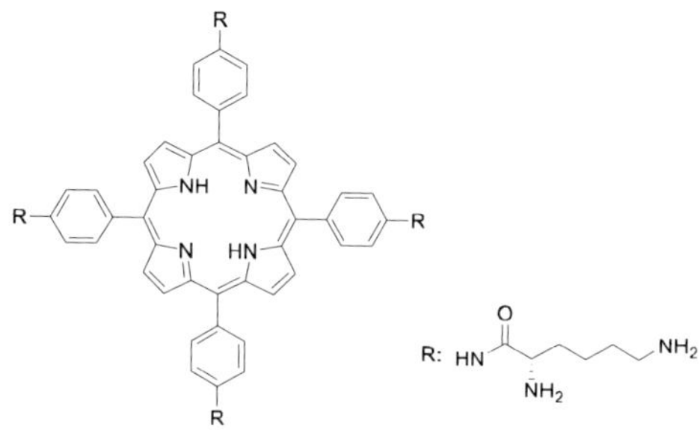

2 by Meng S.

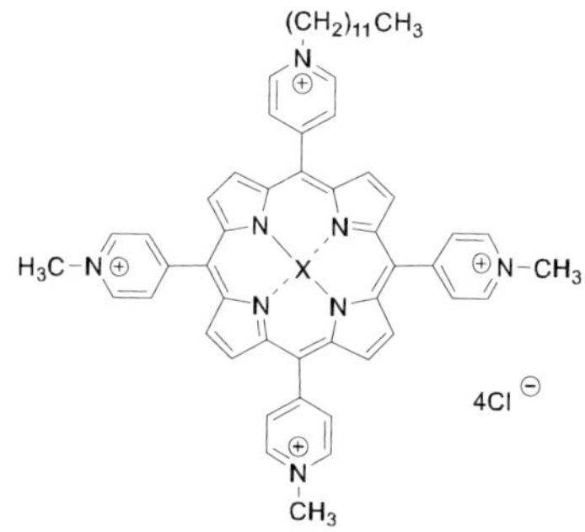

3 X=2H
4 X=Pd by Xuan W.

Figure 6. Chemical structures of porphyrins **2-4**.

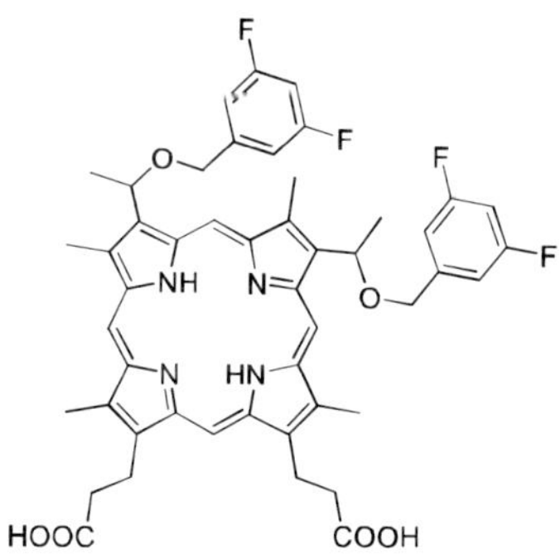

5 by Chen ZL

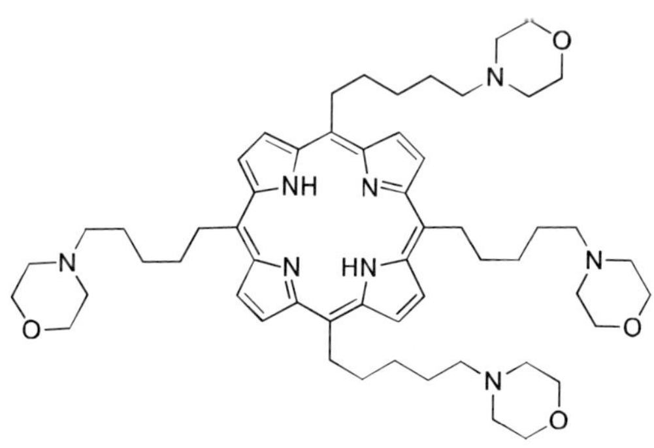

6 by Chen ZL

Figure 7. Chemical structures of porphyrins **5** and **6**.

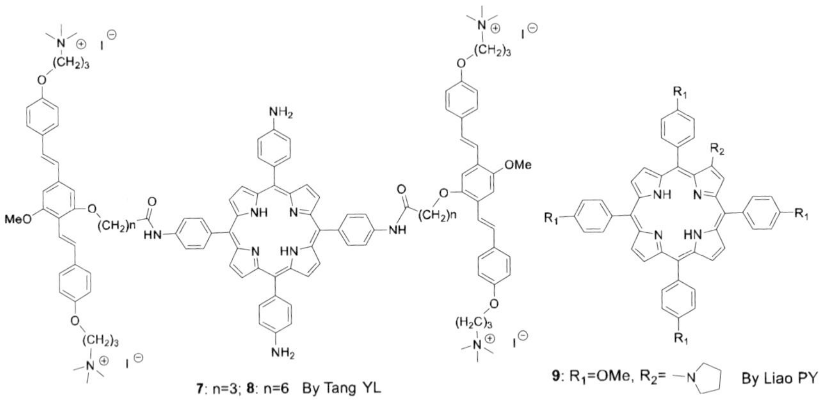

Figure 8. Chemical structures of porphyrins **7-9**.

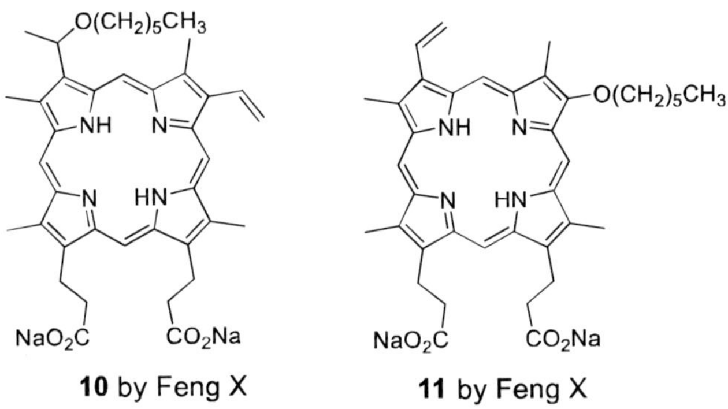

Figure 9. Chemical structures of porphyrins **10** and **11**.

The group of Yan-Li Tang reported two anti-bacterial photosensitizers **7** and **8** with donor-acceptor system in which cationic conjugated oligomer acts as donor and porphyrin acts as acceptor (Figure 8). This system brought the compounds strong fluorescence resonance energy transfer (FRET) and hence greatly enhances the singlet oxygen generation. The singlet oxygen quantum yield reached 121% when taking rose bengal as the standard reference. The half-inhibitory concentration was as low as 33 and 88 nM for methicillin-resistant *Staphylococcus aureus* and *Escherichia coli*, respectively. The IC_{50} value for two compounds on MRSA were 34 and 33 nM separately under 60 min irradiation, but both of them showed lower effect with IC_{50} of 92 and 88 nM on *E. coli* probably due to its unique outer membrane [34].

Ping-Yong Liao et al. reported a series of tetraphenylporphyrins with different substituents at 4-phenyl position and beta position. Compound **9** (Figure 8) with OMe substituents on phenyl position and pyrrolidine on beta position possessed the longest wavelength absorption at 668 nm and highest absorption. Its singlet oxygen yield was 3 times higher than HMME and the IC_{50} value against Hela cells was 4.38 μM and 386.9 μM with and without light irradiation respectively. The results of *in vitro* and *in vivo* studies suggested that this compound embodied higher phototoxicity and lower dark toxicity than HMME, making it a promising photosensitizer [35, 36, 37].

A novel class of water-soluble porphyrin derivative and its isomers, named photohexer-1 (**10**) and photohexer-2 (**11**) (Figure 9) were synthesized by Xiao-Lan Feng and her group. They extended the conjugated system of porphyrin and added carboxylic group to the ring structure. These modifications turned out to embody the two compounds with good absorption in phototherapeutic window and enhanced hydrophilicity. Porphyrin **11** exhibited stronger phototoxicity against breast cancer cells (4T1 murine breast cancer cells, human breast cancer cells MCF-7, and MDA-MB-231) and weaker dark toxicity compared to porphyrin **10** while minor cell damaging was observed on the normal cell lines HUVEC. Moreover, both of them showed better therapeutic effect than their homologous compound HMME [38].

Chlorin-Type Photosensitizers

Compared with porphyrin-type photosensitizers, chlorin-type photosensitizers possess strong absorption in NIR region, which means chlorin photosensitizer-based PDT could be applied to the treatment of deep-seated tumors. This special property attracts the great attention of researchers and several chlorin-type photosensitizers have been used in clinics like *m*-THPC and Talaporfin.

One of the main obstacles restraining the development of chlorin photosensitizers was their low solubility. Some researchers synthetized conjugated chlorins with biological molecules like amino acid and sugar to resolve this problem. Zhi Meng et al. reported a series of chlorin based amino acid conjugates, which introduced L-glutamic acid and L-aspartic acid to pheophorbide a and chlorin p_6 to enhance their water solubility and bioactivity. Aspartylchlorin p_6 dimethyl ester (**12**) (Figure 10) exhibited strong absorption at long wavelength above 650 nm and the highest molar absorption coefficient (3.73×10^5 M^{-1} cm^{-1}). Its dark toxicity *in vitro* could almost be neglected but both of its IC_{50} values against B16-F10 murine melanoma cells and human melanoma A375-S2 cells were the lowest one in all tested compounds (0.68

and 0.98), including reference drug verteporfin. The mechanism study showed that **12**-PDT mainly causes cell necrosis and late period apoptosis. *In vivo* experiment on B16-F10 tumor bearing mice demonstrated that the photodynamic antitumor effect of 1 mg/kg **12** was similar to that of verteporfin under 2 mg/kg, suggesting **12** as a potent photosensitizer. Amino acid and alkoxy groups were also conjugated with pyropheophorbide a ethers and all the resulting compounds showed low dark toxicity and higher photocytotoxicity than verteporfin *in vitro* [39].

Xing-Jie Zhang et al. designed eight new amino acid conjugates of chlorin p_6 ethers. Among them some compounds possessed the photocytotoxicity against A549 cells and B16-F10 cells *in vitro* with IC_{50} value 0.20 mmol/L and 0.41 mmol/L separately for these two cells lines, which represented 31- and 24-fold increase of antitumor activity in comparison to Talaporfin. The structure-activity relationship analysis revealed the increase of carbon chain length of alkoxy ether bonds at 3(1)-position and the substitution of L-glutamic acid with L-aspartic acid could enhance the anti-tumor activity of compounds. 3(1)-hexyloxy chlorin e_6-based 15(2)- or 13(1)-amino acid derivatives were synthesized to improve the biological activity and pharmacokinetic property of chlorin e_6. All compounds displayed stronger anti-tumor activity than Talaporfin against A549 cells, HeLa cells and B16-F10 cells *in vitro*. Compounds **13** and **14** (Figure 11) showed better therapeutic effects *in vivo* than other prepared compounds and Talaporfin, suggesting them as potent photosensitizers [40, 41].

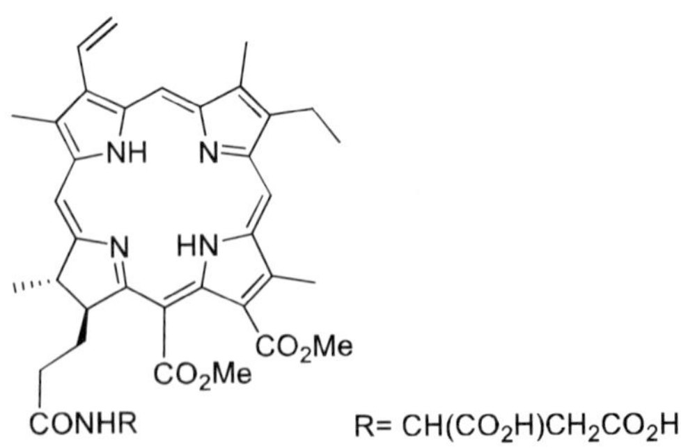

12: By Meng Z

Figure 10. Chemical structures of chlorins **12**.

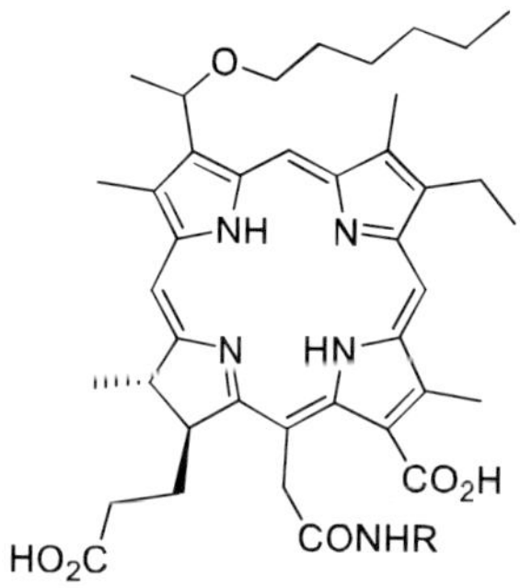

13: R=$CH(CO_2CH_3)CH_2CH_2CH_2CH_2NH_2$
By Zhang X

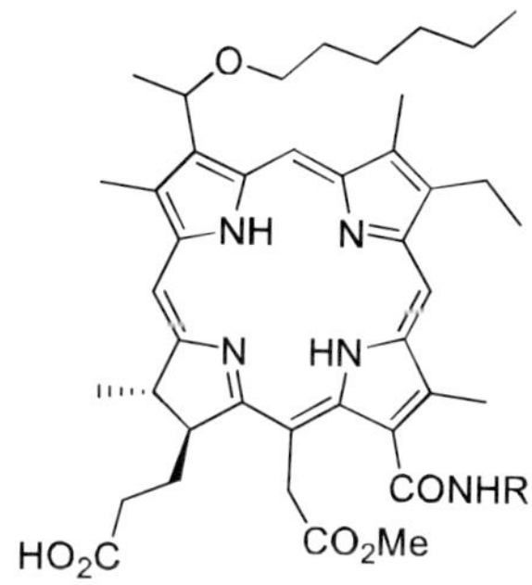

14: R= $CH(CO_2H)CH_2CO_2H$
By Zhang X

Figure 11. Chemical structures of chlorins **13** and **14**.

A series of pyropheophorbide a derivatives were synthesized by Ying-Hua Gao et al. through incorporating 5-ALA in the structure to enhance the photodynamic activity. All compounds showed strong absorption in 660-670 nm. Among them, compounds **15** and **16** (Figure 12) were considered to be potent, as they had more photodynamic antitumor activity than HPPH under the same conditions, and exhibited lower skin phototoxicity. Further mechanism studies revealed that the intracellular distribution of **15** and **16** was mainly in lysosome, mitochondria and endoplasmic reticulum, and they could induce cell apoptosis and necrosis efficiently after PDT [42].

Purpurin is a special group of chlorin derivatives which could be obtained from the degrading product of chlorophyll a. It incorporates an additional imide ring fused at the *meso*-position of chlorin macrocycle in the structure, which greatly extends the absorption wavelength of compound. Several photosensitizers were developed based on purpurin due to this special photophysical property. Yi-Jia Yan, et al. reported the synthesis of chlorin p_6-13,15-*N*-(cyclohexyl)imide (**17**) and its anti-tumor activity (Figure 12). Purpurin **17** shows characteristic long wavelength absorption at 698 nm and significant therapeutic effect on multiple human tumor cell lines (AGS, SGC7901, A549, SW480). Only human leukemia HL60 cells were resistant to **17**-PDT. The mechanism studies indicated that **17**-PDT induced apoptosis in A549 cells and caused the significant up-regulation of TNFSF13, XIAP, Bcl-2, Smac gene expression and down-regulation of Survivin, Livin, BID, caspase 9 gene expression [43].

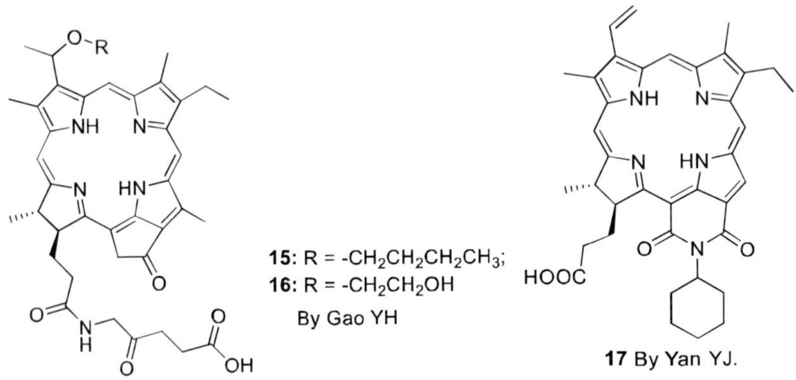

Figure 12. Chemical structures of chlorins **15**, **16** and purpurin **17**.

Chun-Ye Zhang et al. reported 5,10,15,20-tetrakis[(5-*N*-morpholino) pentyl] chlorin (**18**) (Figure 13) as a novel photosensitizer. The NIR absorption at 644 nm, rapid cellular uptake, low dark toxicity and high phototoxicity against both Hela cells and Eca-109 cells *in vivo* and *in vitro* make this compound a promising photosensitizer. It was found that chlorin **18** was mostly distributed in cytoplasm and nuclear membranes of Eca-109 cells, and the biodistribution studies showed that it mainly accumulated in tumor after 12 h post injection [44].

Meso-tetra(3-morpholinomethyl-4-methoxyphenyl) chlorin (**19**) was designed and evaluated by Zhi-Long Chen's group. Its NIR absorption peak located on 655 nm and its IC_{50} value for Eca-109 cells and KB cells irradiated with 4 J/cm^2 light was around 3.67 μM and 5.25 μM, respectively. The significant therapeutic effect on nude mice bearing Eca-109 tumors and no side effects were observed [45].

Yi Dong et al. introduced aromatic groups to the vinyl of Ce_6 skeleton to extend the conjugation system of chlorin e_6. The resulting series of compounds showed favorable optical properties and among them, (*E*)-3(2)-(4-methoxyphenyl)-chlorin e_6 (**20**) (Figure 14) displayed the best photobiological porperties. It could generate large amount of singlet oxygen in HepG2 cells under light irradiation. The absence of dark toxicity, strong phototoxicity against HepG2 cells, rapid cellular uptake and high PDT efficacy *in vivo* suggested the compound as a potent photosensitizer [46].

Xian-Sheng Ke et al. introduced multiple types of beta-hydrophilic conjugates with different charge to the macrocycle core of zinc(III) porpholactol to synthesize a series of chlorin like photosensitizers and investigated whether this could influence the NIR absorption and water solubility of compounds. The maximum absoraption wavelength of the six compounds $ZnL_{1\text{-}6}$ is located at 618 nm, which indicated a red-shift for about 12 nm compared with zinc(III) porpholactol. These compounds would self-assemble into nanosized structures under aqueous environment. The cellular uptake rate as well as photodynamic anti-tumor ability *in vitro* followed the order of $ZnL_{4\text{-}6}$ (positive charge conjugats) - $ZnL_{1\text{-}3}$ (negative charge or neutral conjugates), but all displayed IC_{50} values in the sub-micromolar range against HepG2, HeLa and MDA-MB-231 cells. The further mechanism studies on ZnL_4 (**21**) (Figure 14) revealed that the compound was mainly distributed in lysosome, mitochandira and nuclei to induce cell apoptosis [47].

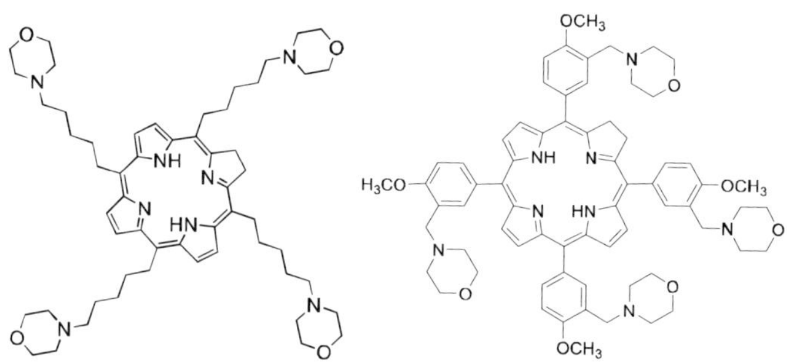

18 By Zhang CY

19 By Zhang LJ

Figure 13. Chemical structures of chlorins **18** and **19**.

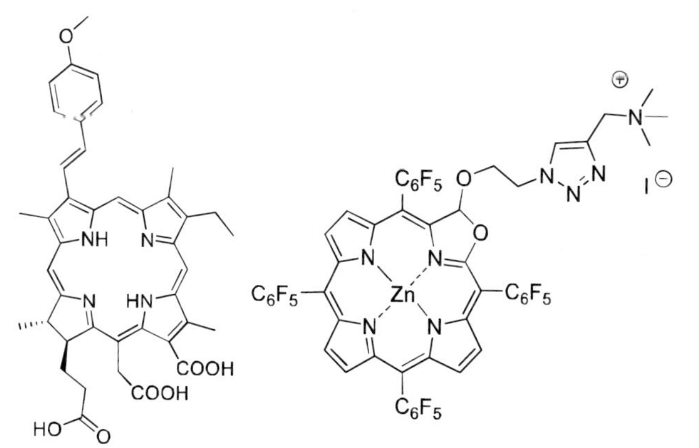

20 By Dong Y.

21 By Ke XS

Figure 14. Chemical structures of chlorin **20** and porpholactol **21**.

Bacteriochlorin

Li-Jun Zhang et al. synthesized a bacteriochlorin-based compound **22** with characteristic long wavelength absorption peak at 732 nm (Figure 15). Its intensive singlet oxygen generation, low dark toxicity, rapid cellular uptake and strong photodynamic anti-tumor efficacy in Eca-109 cells (IC_{50} = 2.64 μM when light dosage was 12 J/cm^2) and mice bearing S180 murine sarcoma suggested **22** as a potent photosensitizer. It mainly distributed in cytoplasm and nuclear membrane to induce cell apoptosis and necrosis [48].

Wei Zhu et al. conducted a comparative study between different novel porphyrin, chlorin and bacteriochlorin derivatives. Among them bacteriochlorin derivatives showed the absorption wavelength at 736 nm and compound **23** (Figure 15) possesses the strongest molar extinction coefficient of 11.02×10^4 $M^{-1}cm^{-1}$. Bacteriochlorin **23** was also the compound with lowest dark toxicity (IC_{50} > 500 μM) and strongest phototoxicity (IC_{50}= 2.43 μM) against Eca-109 cells *in vitro*. The *in vivo* experiment on Eca-109 tumor bearing mice further demonstrated the therapeutic effect of **23** and suggested it as a promising photosensitizer [49].

Phthalocyanine

Phthalocyanine is a macrocyclic molecule composed of four isoindoles. Compared with porphyrin, the maximum absorption wavelength of phthalocyanine is significantly red-shifted and its molar extinction coefficient in NIR region is far higher than that of porphyrin, which could imply strong therapeutic effect at small dosage.

Xiong-Jie Jiang, et al. introduced a series of silicon (IV) phthalocyanine polyamine conjugates with pH dependent manner, whose fluorescence quantum yield and singlet oxygen quantum yield would be significantly elevated under acidic environment. All compounds showed high phototoxicity against HT29 cells and HepG2 cells with IC_{50} of 0.03 μM. Among them **24** (Figure 16) exhibited strongest difference in singlet oxygen generation and absorption in pH range of 6.5 to 7.4, allowing this compound to differentiate normal tissues from tumor cells and embodying them selectivity. They also axially substituted silicon (IV) phthalocyanines with different polyamine groups to give nine hydrophilic compounds whose florescence quantum yields generally increased when pH decreased. The most potent compounds **25** and **26** (Figure 16), possessed stronger phototoxicity on both HT29 cells and CHO cells with IC_{50} value of 1.1 nM against prior one. Furthermore, they exhibited inhibition toward tumor growth in nude mice bearing-xenografted HT-29 cells [50, 51].

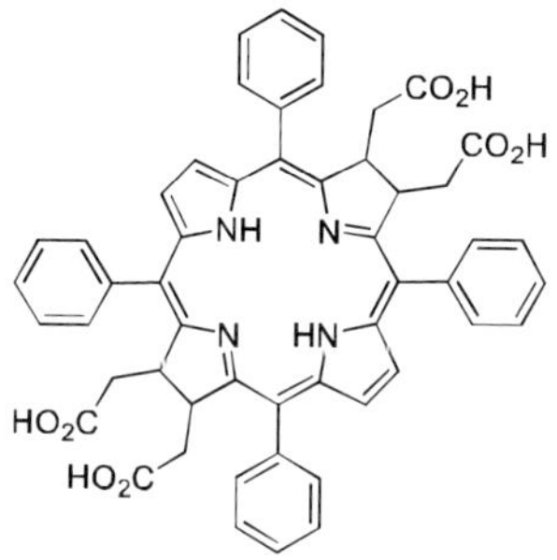

22 By Zhang LJ

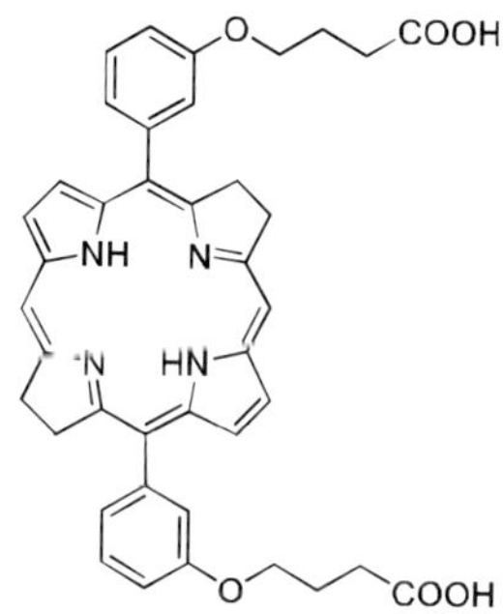

23 By Zhu W

Figure 15. Chemical structures of bacteriochlorins **22** and **23**.

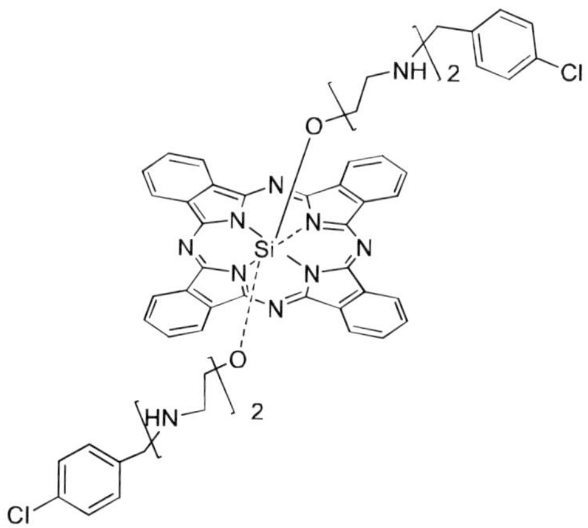

24 By Jiang XJ

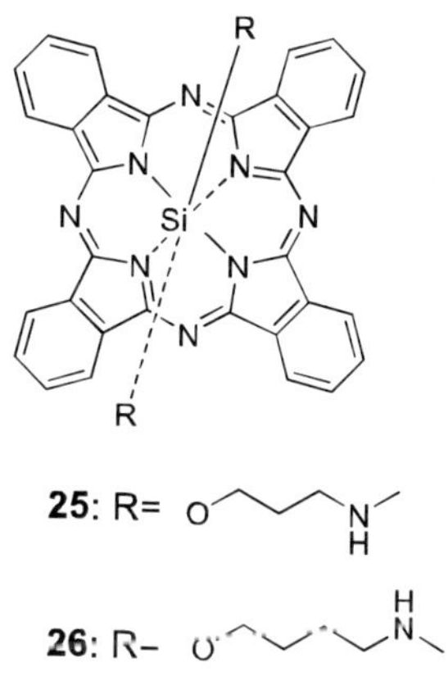

By Jiang XJ.

Figure 16. Chemical structures of phthalocyanines **24** - **26**.

Bi-Yuan Zheng et al. reported a series of asymmetric silicon(IV) phthalocyanines axially substituted by tyramine and other oligomeric ethylene glycol moieties to improve the singlet oxygen generation and hydrophilicity of photosensitizer. The asymmetric compound SiPc5 (**27**) (Figure 17) could effectively kill human gastric cancer cells BGC-823 *in vitro* with IC_{50} value of 3 nM, which was attributed to its high cellular uptake. The mechanism studies revealed that **27** induced cell apoptosis with light irradiation by accumulating in mitochondria [52].

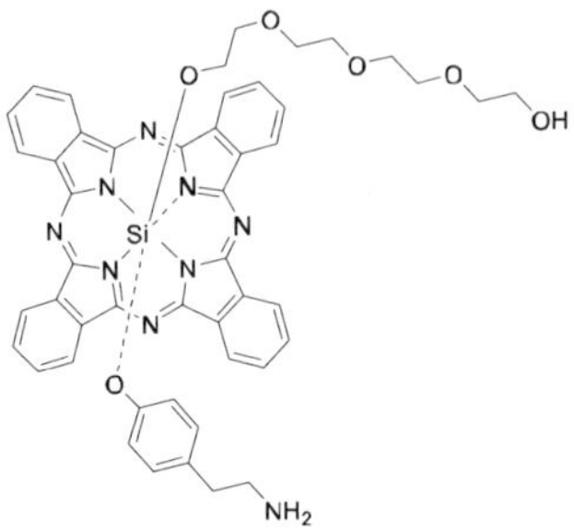

27 By Zheng BY

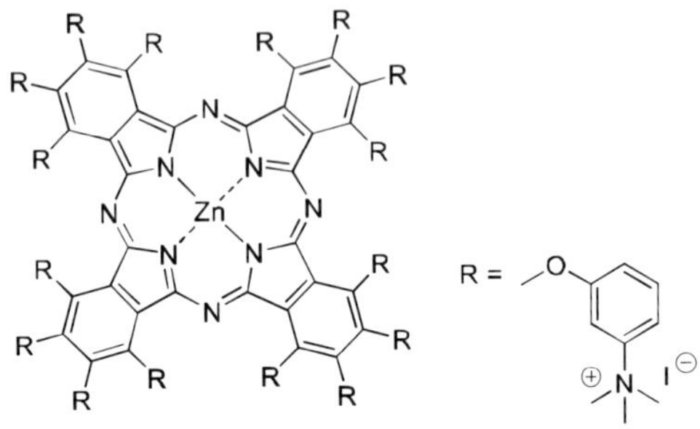

28 By Zhang BD

Figure 17. Chemical structures of phthalocyanines **27** and **28**.

A cationic zinc(II) phthalocyanine (Pc4, **28**) was prepared for dual application of anti-tumor and anti-bacterial therapy by quaternization of precursor Pc3 by Bing-De Zheng et al. This modification enhanced its singlet oxygen generation and prevented **28** from self-aggregation in aqueous solution. Compound **28** presented strong anti-tumor and anti-bacterial activities *in vivo* and *in vitro*. 98.7% tumor growth inhibition on H22 tumor bearing mice could be achieved under **28**-PDT and its antibacterial activity was proved to be 85-fold more potent than methylene blue (MB, a known antimicrobial photosensitizer) against *S. aureus in vitro* [53].

BODIPY

Xiao-Mei Dai et al. introduced a galactose conjugated BODIPY based compound for the therapy of antibiotic resistant bacteria. Compound P(ATA-C4)-r-GAL-I2 (**29**) (Figure 18) could selectively bind to the membrane of both *P. aeruginosa* and *S. aureus* to disrupt their membrane by generating ROS without rising resistance of bacteria. It was found that 70% of bacterial biofilm could be eradicated by P(ATA-C4)-r-GAL-I2 PDT under concentration of 22 μg/mL [54].

Zhi-Liang Yu et al. synthesized a series of multi-halogenated aza-BODIPY compounds and the mono-iodinated compound **30** with *p*-methoxyphenyl moiety was found to be the most potent NIR photosensitizer. Its absorption peak located at 678 nm has strong molar extinction coefficient of 9.7×10^4 M^{-1} cm^{-1}. Its IC_{50} values against HeLa was as low as 76 nM *in vitro* and furthermore, its *in vivo* efficacy was proved to be better than photosensitizers ADPM06 and Ce6. Compound possesses strong NIR absorption, significant anti-tumor activity, low dark toxicity and thermal stability [55].

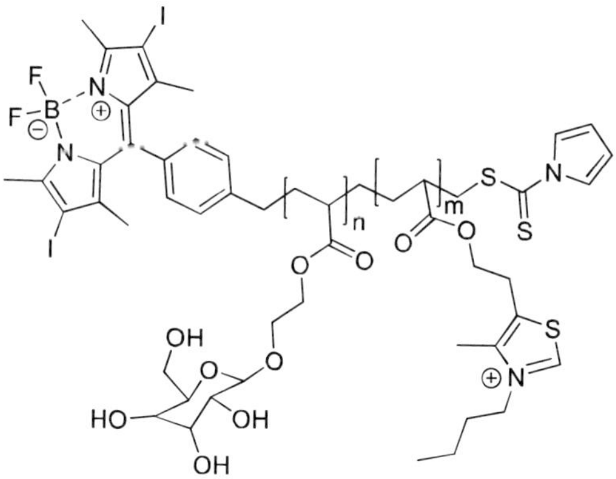

29 By Dai XM

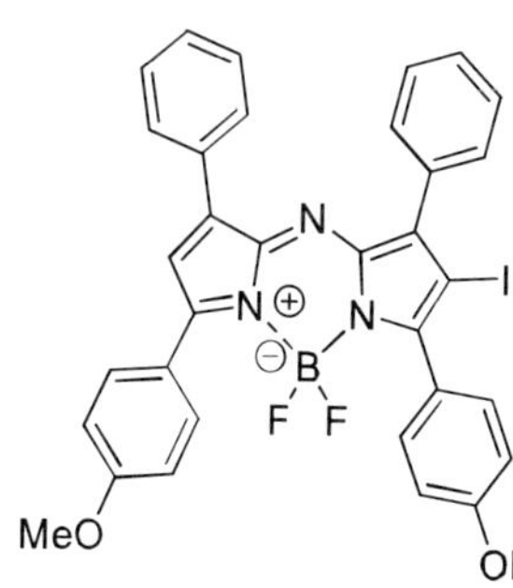

30 By Yu Z

Figure 18. Chemical structures of BODIPYs **29** and **30**.

Yong-Chao Liu et al. designed a pH activated BODIPY compound **31** (Figure 19) for photodynamic and photothermal therapy. Under acidic condition, the protonated compound would disaggregate and thus initiate the phototherapies. BODIPY **31** with low dark-toxicity was found to mainly distribute in lysosome, and produced intracellular ROS and killed 4T1 tumor cells effectively *in vitro* under 660 nm laser irradiation. It could also inhibit tumor growth in 4T1 tumor-bearing mice [56].

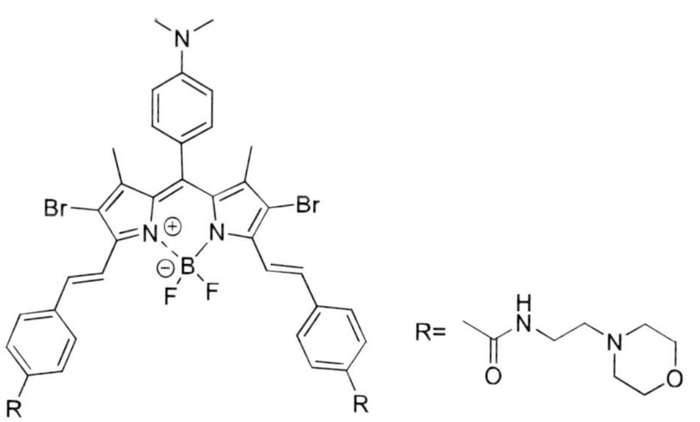

31 By Liu Y.

Figure 19. Chemical structure of BODIPY **31**.

The Applications PDT in China

At the end of 1970's, many Chinese researchers and organizations started to prepare new photosensitizers and laser equipment for PDT. In July 1981, Beijing Tongren Hospital treated a patient with basal cell carcinoma on left lower eyelid successfully, which was the first clinical application case in China [56]. PDT were developed quickly during the past four decades, a great number of patients with a large variety of diseases have been treated in many hospitals all over China. PDT protocols for treating nasopharyngeal carcinoma (NPC), liver carcinoma, port wine stains (PWS) and urethral condylomata acuminata were pioneered and established [57].

Cancer

Clinical research and application of PDT have been carried out against almost all types of histologic malignancies in China since domestically produced HiPorfin became available. A review of the literature from 1980 to 2012 showed that more than 5833 neoplastic patients were treated with PDT, including 1853 cases of esophageal cancer, 734 cases of gastric cancer, 111 cases of colorectal cancer, 70 cases of liver cancer, 488 cases of lung cancer, 800 cases of bladder cancer, 259 cases of nasopharyngeal cancer, 16 cases of breast cancer, 280 cases of brain tumor, 6 cases of cervical cancer, 915 cases of skin cancer, 119 cases of carcinoma of penis, vulva and oral cavity, and 182 other tumors. The efficacy of PDT in the treatment of intracavitary cancers and superficial cancers has been demonstrated [58, 59]. Compared with surgery, radiotherapy and chemotherapy, PDT not only has the characteristics of less trauma, less toxicity, higher specificity, better targeting and elimination of hidden cancer focus, but also it has unique advantages in the protection of appearance and important organ functions. As a new type of treatment for malignant tumors, it can achieve the goal of local radical cure for early localized malignant tumors and improve the quality of life of patients with advanced tumors. In 2010, the "Clinical Technical Operation Standard" of PDT was established and published by Ying Gu and other Chinese experts, which made remarkable contribution to the promotion of PDT in clinical application in China.

With the combination of PDT and imaging technology, the clinical application of PDT has been extended to deep-seated solid tumors such as liver cancer, pancreatic cancer, lung cancer and brain cancer. However, as the photosensitizers mainly used in PDT are HiPorfin, Photofrin and ALA, which have the problem of poor penetration and the occurrence of photosensitive skin reaction, the clinical application of PDT has slowed down to some extent in the last five years. The development of new, effective and safe photosensitive drugs is urgently needed.

Recommended Protocols

Nasopharyngeal Carcinoma are as Follows [60]

1. Ask about the medical history in details before treatment, and make a skin test to avoid allergic reactions to Hiporfin or Photofrin.
2. Set treatment parameters according to the patient's age, lesions and equipment conditions, which are generally 3.5 - 5 mg/kg for photosensitizer dose, 100 - 200 mW/cm^2 for light dose and 120 - 240 J/cm^2 for energy density, and record each treatment as reference of the next treatment.
3. Normal skin outside of the treatment area is covered with double layers of black fabrics. Then, 48 h after injection, PDT was performed under local anesthesia. The irradiating light was vertical with the treatment area, the distance between the fiber end and the tumor surface should not be less than 1 cm, and the spot area should completely cover the tumor surface and extend beyond 1 cm.
4. If the single spot cannot cover the surface of the tumor, multiple spots should be irradiated in sequence according to the dose requirement. The edge of each spot should overlap not less than 0.5 cm, and the total dose of each spot should be the sum of the total dose of each spot.
5. Within four weeks after treatment, the patients shall be protected from light as far as possible to minimize erythema, pruritus and other allergic reactions, which require symptomatic treatment.

Liver Carcinoma are as Follows [60]

1. The skin test should be performed to avoid allergic reactions to Hiporfin or Photofrin.
2. Interstitial PDT was performed under local anesthesia 48 h after administering 5 mg/kg photosensitizer Hiporfin or Photofrin. Then,

positioning probes (18G) were inserted into the tumor through percutaneous punctures under ultrasound guidance. Light delivery fibers (400 μm core diameter, 1- cm diffuser tip) were then placed into the tumor. An argon laser pumped dye laser (630 nm) was coupled into three diffuser fibers for simultaneously irradiation of three treatment spots. A light dose of 220 J/cm^2 was delivered at fluence rates of 300 - 350 mW/cm^2 for each spot [58].

3. The patients shall be protected from light within four weeks after treatment.

Port Wine Stains

In 1991, Ying Gu et al. from Chinese PLA General Hospital applied PDT to treat PWS with HiPorfin for the first time in China, and achieved a good efficacy, particularly for large-area and thick lesions [61]. The results showed that photodynamic therapy could effectively eliminate the red lesions without scarring or skin damage. In order to avoid prolonged skin photosensitization associated with HiPorfin, HMME (Hemoporfin) as a new photosensitizer with a rapid clearance rate was explored by De-Yu Xu. Since then, thousands of cases underwent a single or multiple HMME-PDT treatments and 94.8% lesions showed fair-to-excellent responses with the advantages of few acute reactions, fast healing and short skin photosensitive duration [58]. Besides, no recurrence was reported during 2 months to 9 years of follow-up observation of 58% patients. On December 3, 2016, new photosensitizer HMME was launched into the market and applied in clinic for the treatment of PWS by NMPA.

Recommended Protocols for PWS are as Follows [62, 63]

1. Ask about the medical history in details before treatment, and make a skin test to avoid allergic reactions to HMME.
2. Before treatment, adhesive tape is pasted along the edge of the treatment area, and normal skin outside of the treatment area are covered with double layers of black fabrics. People with a low tolerance and children could be given PDT under general anesthesia.
3. Set treatment parameters according to the patient's age, lesions and equipment conditions, which are generally 4 - 5 mg/kg for

photosensitizer dose and 90 - 100 mW/cm^2 for light dose, record each treatment as reference of the next treatment.

4. The treatment area was fully exposed in the same horizontal plane. The equipment was thoroughly debugged and maintained in a normal status. The irradiating light was vertical with the treatment area.
5. After irradiation, the patients may suffer from pain and edema at the local skin, which could be relieved with cold spraying or cold wet compressing for 3-5 days, and the small amount of exudation at the treated area could be gently dried with sterile cotton balls.
6. Within two weeks after treatment, the patients shall be protected to avoid light irradiation to minimize erythema, pruritus and other allergic reactions which require symptomatic treatment.

Condyloma Acuminatum

Condyloma acuminatum (CA) is a highly infectious disease caused by the human papillomavirus (HPV), with rapid growth of warts. Recurrence usually takes place after treatment, and results in significant economic and psychological burden to the patients [64]. As conventional treatments such as electrocautery, cryotherapy, and topical medications are mainly used to remove exogenous warts and cannot eliminate HPV infection, the recurrence rate of CA is high after treatment. Xiu-Li Wang et al. used firstly ALA-PDT in the treatment of male urethral CA in 2000 [65], and since then, a large amount of clinical data in China have shown that ALA-PDT has advantages in treating CA, such as minimal invasiveness, improved safety, and high patient tolerance. In addition, compared to traditional treatments, ALA-PDT can eliminate subclinical infections, reduce recurrence, and reduce cancer risk. Thus, topical ALA-PDT has been approved for the treatment of urethral CA in China in 2007, and now it is the first choice for the treatment of CA in urethral orifice and internal cavity.

Recommended Protocols

Intracavitary CA are as Follows [66]

1. If possible, endoscopy (urethroscopy, colposcopy, and anoscopy) should be used to assist diagnosis and localization of lesions.

2. A 10 - 20% ALA solution or gel should be applied to the surface of the lesions and its surrounding area within 1 cm. For the treatment of posterior urethral lesions, a 20% ALA solution can be directly injected through a specific catheter, and the mouth of the cavity can be sealed to avoid exposure to light for 3 h.
3. After the application of ALA, red light with an energy density of 100 - 150 J/cm^2 and a power density of 60 - 100 mW/cm^2 is recommended for irradiation, especially focusing on the warts.
4. Review the patient 1 week after treatment, and the treatment can be repeated if the lesion is not completely resolved. If the lesions resolve by < 50% after three ALA-PDT sessions, it is recommended that ALA-PDT is combined with other treatments or a different treatment can be adopted.

External Genitalia and Perianal CA are as Follows [65]

1. Before treatment, enhanced pretreatment programs should be used for lesion preparation. For example, after cleaning the lesion, physical methods such as a CO_2 laser can be used to remove keratinized thickened lesions.
2. The 10 - 20% ALA cream, gel, or solution should be applied externally on the surface of the lesion and its periphery within at least 1 cm and incubated for 3 - 6 h without light.
3. The rest of the treatment protocol is the same as that for intracavitary condyloma acuminate.

Age-Related Macular Degeneration

Since 2000 Visudyne has been approved by FDA for the treatment of CNV (choroidal neovascular disease) caused by age-related macular degeneration (AMD) using PDT, the clinical application of PDT has started in China almost at the same time. A multi-center clinical study held by the Peking Union Medical College Hospital, West China Hospital of Sichuan University, and the National Sun Yat-sen University confirmed the safety and efficacy of Visudyne-PDT in treatment of subfoveal classic CNV secondary to AMD from 32 cases, which was similar to the results of earlier studies of patients in Japan, Europe, and North America. The long-term efficacy of Visudyne-PDT in treatment of AMD has been further confirmed by many years follow-up data in China, such as improving or stabilizing the visual function of patients with CNV and controlling the progression of the disease. In particular, these studies also showed that the effect of delaying vision loss or improving vision of Chinese patients after PDT treatment was better than that of patients in Europe, and North America, which indicated that Asian people with more retinal pigment might be sensitive to PDT [59]. At present, Visudyne-PDT has become the first-line treatment of wet AMD in China.

Actinic Keratosis

Actinic keratosis (AK) is a common chronic precancerous skin lesion with high risk of malignant transformation, predominantly due to excessive ultraviolet exposure in sensitive individuals. Since Kennedy et al. [67] first validated the efficacy of 5-ALA-PDT in patients with AK in 1990, the application of PDT for AK treatment has been used in clinics worldwide. Compared to conventional regimens such as cryotherapy and medicine based on numerous clinical trials, PDT has a higher rate of clearance, superior cosmetic outcomes, and fewer side effects in the treatment of AK.

The clinical application of PDT for AK was developed well in the last decade. Hong Cai et al. reported the therapeutic response of Chinese patients with AK lesions to PDT [68]. 42 patients with a total of 56 AK lesions on the face underwent topical 20% ALA-PDT and 85.71% of those were cured following a single treatment and the remaining patients were cured following two or three treatment sessions at one month after therapy. Xiu-Li Wang et al. performed topical 20% ALA-PDT on 10 patients with AK on the nose. The results were founded that 12 of 13 lesions were completely relieved (92.3%) after 3 sessions of photodynamic therapy, and one lesion was partially relieved (7.7%). There was no recurrence within one year [69]. Numerous other studies also showed that ALA-PDT is an effective and safe method for the treatment of AK.

As ALA-PDT using narrow-band red led as light source has pain and other side effects, which can reduce patients' compliance and thus fail to achieve the desired effect, daylight photodynamic therapy (DL-PDT) using sunlight as a source of excitation was proposed and proved to be as effective as ALA-PDT by Hans Christian Wulf. In China, Wang XL et al. compared firstly the efficacy and safety of DL-PDT and ALA-PDT on 60 AK patients. The results showed that the effect of DL-PDT on Grade I and Grade II AK skin lesions was similar to that of red-light PDT. However, Grade III AK lesions still need to be strengthened with red light PDT for better efficacy [70]. At present, DL-PDT related research is still in the initial stage in China.

Bacteria-Resulted Diseases

Antibiotic resistance is expanding day by day, thereby increase the chances of more infections by resistant bacteria. Antimicrobial Photodynamic Therapy (aPDT) is a new and promising approach to kill microorganisms. It is a combination of light, photosensitizer and oxygen to produce oxidative damage to microorganisms. Since the active radicals can damage the charge-rich groups of various biologically active substances in the microorganisms such as hydroxyl group, sulfhydryl group, hydrazine, pyrimidine, they can effectively prevent the microorganism to develop drug resistance. aPDT is considered an alternative treatment of skin and mucosal lesions and infections.

In China, the clinical study of aPDT used for the treatment of local infections, including refractory diseases caused by local infections of the mouth and skin has been performed. Lin-Hua Ge and Rong Shu first reported the efficacy of methylene blue (MB)-PDT assisted subgingival scaling and root planning (SRP) for chronic periodontitis in 2008 [71]. The results showed that PDT could kill effectively three periodontal pathogens, *Porphyromonas gingivalis*, *Tannerella forsythensis*, and *Actinobacillus actinomycetemcomitans*. Xiu-Ling Luan, et al. found that Toluidine blue-PDT has a very good sterilization function of gingival plaque [72]. At present, phenothiazine photosensitizers are the main type of molecules for the clinic study and treatment of periodontal disease in China.

Disinfection of Blood Products

Blood transfusion including the infusion of hemocyte composition components, plasma and plasma protein self-products is a kind of effective therapeutic measures in clinical applications. As one of the important measures for clinical treatments, blood transfusion plays a significant role in the rescues and therapies of various diseases, such as, severe traumas, massive haemorrhages, hemopathies and so on. A variety of pathogens can be taken in blood, which brings about the risks of disease transmission. Therefore, it is greatly important to wipe out microbes and decrease the risks of blood diseases effectively for the security of blood products.

Photodynamic inactivation of viruses in blood or blood components has been used in clinical practice. Methylene blue (MB)-PDT, with the advantages of better inactivation effect, mild reaction conditions and less influence on cell components and bioactive substances in blood, was first used by Pohi in 1966 and approved to inactivate the virus for fresh frozen plasma (FFP) in most European countries in 1993. In China, MB-PDT is the only photochemical method approved for clinical use of inactivation of virus in blood products and receives good clinical results.

Other Diseases

In addition to the above applications which have been approved and widely used, the exertion of PDT for a variety of other diseases also began at the end of last century in China, including skin diseases such as acne and photoaging, gynecological diseases such as vaginal or cervical intraepithelial neoplasia, and immune-related diseases such as rheumatoid arthritis and psoriasis [59, 66, 73]. Numerous animal studies and a handful of human studies demonstrated that PDT is promising, but it is mostly in its infancy. The application of PDT in these diseases has not yet been standardized, thus future research will focus on the following aspects: clinical trials with large number of patients to optimize the indications and parameters of PDT in these diseases; development of novel PSs, light sources and methods to improve local oxygen supply to enhance the efficiency and reduce adverse reactions to PDT; basic research on the working mechanisms of PDT in different diseases, especially those related to disease recurrence after PDT and relevant strategies to overcome it; and combinations of PDT with other treatment options. With growing insights into these fields, the performance of PDT will be improved and may become an effective approach to treat more diseases [73].

Conclusion

Intensive efforts on preparation and study of applications of novel molecules for photodynamic therapy produced notable results in treatment of various diseases. Novel compounds with enhanced PDT activity are approved for clinical use and detailed protocols for their administration are well established.

References

[1] Dong, Z.X.; Ge, R.T.; Liu, L.F. Hiporfin. *Chin. J. New Drugs*, **2006**, 15, 1895-1896 (in Chinese).

[2] Zhang, H. *Studies on photochemical drugs for diagnosis and treatment of tumors: V: Study on semi-synthetic derivatives of hematoporphyrin*. Master thesis. Second Military Medical University, China, **1988**.

[3] Lu, P. *Studies on separation, identification and tumor-photobiological activities of the two positional isomers of the promising new drug for tumor-photodynamic therapy Herimether: Hematoporphyrin 3- and 8-monomethyl ether*. Master thesis. Second Military Medical University, China, **1996**.

[4] Pu, Y.; Chen, W.H.; Yu, Z.W. Research progress of Hemoporfin - Part one: Preclinical study. *Photodiagn. Photodyn.*, **2012**, 9, 180-185.

[5] Yu, L.; Li, X.Y. Advances the second generation of photosensitizer Hematoporphyrin Monomethyl Ether and its clinical application. *BME & Clin. Med.*, **2005**, 9, 365-368 (in Chinese).

[6] Chen, W.H.; Yu, J.X.; Yao, J.Z.; Shen, W.D.; Liu, J.F.; Xu, D.Y. Pharmacokinetic studies on Hematoporphyrin Monomethyl Ether : A new promising drug for photodynamic therapy of tumors. *Chin. J. Laser Med. Surg.*, **2000**, 9, 105-108 (in Chinese).

[7] Huang, J.L.; Chen, N.S.; Huang, J.D.; Liu, E.S.; Xue, J.P.; Yang, S.L.; Huang, Z.Q.; Sun, J.C. Metal phthalocyanine as photosensitizer for photodynamic therapy (PDT). *Sci. China Ser. B.*, **2001**, 44, 113-122.

[8] Xie, B.G.; Huang, H.F.; Huang, J.D.; Chen, N.S.; Huang, J.L. Determination and pharmacokinetics of photosensitizer $ZnPcS_2P_2$ uptake by K562 Cell. *Spectrosc. Spect. Anal.*, **2004**, 24, 1415-1418 (in Chinese).

[9] Chen, D.; Song, M.R.; Huang, J.L.; Chen, N.S.; Xue, J.P.; Huang, M.D. Photocyanine: A novel and effective phthalocyanine-based photosensitizer for cancer treatment. *J. Innov. Opt. Health Sci.*, **2020**, 13, 2030009 (13 pages).

[10] Xu, D.Y.; Yin, X.S.; Chen, W.H.; Zhang, H. Studies on the new tumor-photolocalizing and photochemotherapeutic agent Photocarcinorin (PsD-007). *Proc. SPIE 1616, International Conference on Photodynamic Therapy and Laser Medicine*, Beijing, China, **1991**, Vol. 1616.

[11] Fang, Q.C. Photodynamic therapy for cancer treatment and the new antitumor photosensitizer Sinoporphyrin Sodium. *Chin. J. New Drugs*, **2014**, 23, 1540-1545 (in Chinese).

[12] Rui, S.; Chao, L.; Jiang, Z.H.; Li, W.F.; Wang, A.P.; Wei, J.F. Preclinical study of antineoplastic Sinoporphyrin Sodium PDT via *in vitro* and *in vivo* models. *Molecules*, **2017**, 22, 112.

[13] Hu, J.M.; Wang, X.B.; Liu, Q.H.; Zhang, K.; Xiong, W.L.; Xu, C.S.; Wang, P.; Leung, A. W.N. Antitumor effect of Sinoporphyrin Sodium-mediated photodynamic therapy on human esophageal cancer eca109 cells. *Photochem. Photobiol.*, **2014**, 90, 1404-1412.

[14] Wang, X.B.; Hu, J.M.; Wang, P.; Zhang, S.L.; Liu, Y.C.; Xiong, W.L.; Liu, Q.H. Analysis of the *in vivo* and *in vitro* effects of photodynamic therapy on breast cancer by using a sensitizer, Sinoporphyrin Sodium. *Theranostics*, **2015**, 5, 772-786.
[15] Xiong, W.L.; Wang, X.B.; Hu, J.M.; Liu, Y.C.; Liu, Q.H.; Wang, P. Comparative study of two kinds of repeated photodynamic therapy strategies in breast cancer by using a sensitizer, Sinoporphyrin Sodium. *J. Photochem. Photobiol. B.*, **2016**, 160, 299-305.
[16] Wu, L.J.; Wang, X.B.; Liu, Q.H.; Leung, A. W.N.; Wang, P.; Xu, C.S. Sinoporphyrin Sodium mediated photodynamic therapy inhibits the migration associated with collapse of F-actin filaments cytoskeleton in MDAMB-231 cells. *Photodiagn. Photodyn.*, **2016**, 13, 58-65.
[17] Yan, X.F.; Niu, G.; Lin, J.; Jin, A. J.; Hu, H.; Tang, Y.X.; Zhang, Y.J.; Wu, A.G.; Lu, J.; Zhang, S.L.; Huang, P.; Shen, B.Z.; Chen, X.Y. Enhanced fluorescence imaging guided photodynamic therapy of Sinoporphyrin Sodium loaded graphene oxide. *Biomaterials*, **2015**, 42, 94-102.
[18] Yan, X.F.; Hu, H.; Lin, J.; Jin, A. J.; Niu, G.; Zhang, S.L.; Huang, P.; Shen, B.Z.; Chen, X.Y. Optical and photoacoustic dual-modality imaging guided synergistic photodynamic/photothermal therapies. *Nanoscale*, **2015**, 7, 2520-2526.
[19] Chu, C.C.; Lin, H.R.; Liu, H.; Wang, X.Y.; Wang, J.Q.; Zhang, P.F.; Gao, H.Y.; Huang, C.; Zeng, Y.; Tan, Y.Z.; Liu, G.; Chen, X.Y. Tumor microenvironment-triggered supramolecular system as an *in situ* nanotheranostic generator for cancer phototherapy. *Adv. Mater.*, **2017**, 29, 1605928.
[20] Zhu, B.; Li, S.X.; Yu, L.; Hu, W.; Sheng, D.D.; Hou, J.; Zhao, N.P.; Hou, X.J.; Wu, Y.C.; Han, Z.P.; Wei, L.X.; Zhang, L. Inhibition of autophagy with chloroquine enhanced Sinoporphyrin Sodium mediated photodynamic therapy-induced apoptosis in human colorectal cancer cells. *Int. J. Biol. Sci.*, **2019**, 15, 12-23.
[21] Li, L.; Wang, H.Y.; Wang, H.P.; Li, L.J.; Wang, P.; Wang, X.B.; Liu, Q.H. Interaction and oxidative damage of DVDMS to BSA: A study on the mechanism of photodynamic therapy-induced cell death. *Sci. Rep.*, **2017**, 7, 43324.
[22] Huang, Y.S.; Zhang, L.; Liu, L.T.; Ma, Y.; Liu, X. Effects of DVDMS-PDT on proliferation and migration of human tongue squamous carcinoma cell line HSC-4. *J. Oral Sci. Res.*, **2018**, 34, 1208-1211 (in Chinese).
[23] Ma, Y.; Zhang, L.; Huang, Y.S.; Liu, X. The effects of DVDMS-PDT on tongue squamous cell carcinoma SCC-15 cells. *J. Pract. Stomat.*, **2019**, 35, 562-567 (in Chinese).
[24] Liu, X.; Zou, H.; Deng, X.F.; Zheng, Y.W.; Yan, S.C.; Hou, X.Y.; Xiong, L.; Li, Q.L. Preparation of liposomal Sinoporphyrin Sodium and its photodynamic effect against cholangiocarcinoma cell *in vitro*. *Chin. J. Gen. Surg.*, **2019**, 28, 159-165 (in Chinese).
[25] Kong, F.H.; Zou, H.; Liu, X.; He, J.; Zheng, Y.W.; Xiong, L.; Miao, X.Y. miR-7112-3p targets PERK to regulate the endoplasmic reticulum stress pathway and apoptosis induced by photodynamic therapy in colorectal cancer CX-1 cells. *Photodiagn. Photodyn.*, **2020**, 29, 101663.

[26] Li, T.T.; Lv, H.Y.; Yang, L.; Xie, J.; Liu, C.; Xu, P.L.; Li, W.Y.; Wang, S.Y.; Yang, D.; Wu, T.; Yan, J.H.; Luo, F.H. Pharmacokinetics and tissue distribution of DVDMS-2 in tumor-bearing mice. *Photochem. Photobiol.*, **2020**, 96, 148-155.

[27] Liu, H.Q.; An, Y.W.; Li, Z.W.; Li, W.X.; Yuan, B.; Wang, J.C.; Jin, H.T.; Wang, C. Sinoporphyrin Sodium, a novel sensitizer for photodynamic and sonodynamic therapy. *Open Chem.*, **2020**, 18, 691-701.

[28] Chen, J.J.; Gao, L.J.; Liu, T.J. Photodynamic therapy with a novel porphyrin-based photosensitizer against human gastric cancer. *Oncol. Lett.*, **2016**, 11, 775- 781.

[29] Chen, J.J.; Mao, L.N.; Liu, S.P.; Liang, Y.L.; Wang, S.C.; Wang, Y.Y.; Zhao, Q.; Zhang, X.J.; Che, Y.J.; Gao, L.J.; Liu, T.J. Effects of a novel porphyrin-based photosensitizer on sensitive and multidrug-resistant human gastric cancer cell lines. *J. Photochem. Photobiol. B.*, **2015**, 151, 186-193.

[30] Meng, S.; Xu, Z.P.; Hong, G.; Zhao, L.H.; Zhao, Z.J.; Guo, J.H.; Ji, H.Y.; Liu, T.J. Synthesis, characterization and *in vitro* photodynamic antimicrobial activity of basic amino acideporphyrin conjugates. *Eur. J. Med. Chem.*, **2015**, 92, 35-48.

[31] Xuan, W.J.; Huang, L.Y.; Wang, Y.G.; Hu, X.Q.; Szewczyk, G.; Huang, Y.Y.; El-Hussein, A.; Bommer, J. C.; Nelson, M. L.; Sarna, T.; Hamblin, M. R. Amphiphilic tetracationic porphyrins are exceptionally active antimicrobial photosensitizers: *In vitro* and *in vivo* studies with the free-base and Pd-chelate. *J. Biophotonics.*, **2019**, 12, e201800318.

[32] Li, M.Y.; Gao, Y.H.; Zhang, J.H.; Mi, L.; Zhu, X.X.; Wang, F.; Zhou, X.P.; Yan, Y.J.; Chen, Z.L. Synthesis and evaluation of novel fluorinated hematoporphyrin ether derivatives for photodynamic therapy. *Bioorg. Chem.*, **2021**, 107, 104528.

[33] Zhang, L.J.; Yan, Y.J.; Liao, P.Y.; Margetić, D.; Wang, L.; Chen, Z.L. Synthesis and antitumor activity evaluation of a novel porphyrin derivative for photodynamic therapy *in vitro* and i*n vivo*. *Tumor Biol.*, **2016**, 37, 6923-6933.

[34] Zhao, Y.T.; Zhang, Z.Q.; Lu, Z.N.; Wang, H.; Tang, Y.L. Enhanced energy transfer in a donor-acceptor photosensitizer triggers efficient photodynamic therapy. *ACS Appl. Mater. Inter.*, **2019**, 11, 38467-38474.

[35] Liao, P.Y.; Wang, X.R.; Zhang, X.H.; Hu, T.S.; Zheng, M.Z.; Gao, Y.H.; Yan, Y.J.; Chen, Z.L. Synthesis of 2-morpholinetetraphenylporphyrins and their photodynamic activities. *Bioorg. Chem.*, **2017**, 71, 299-304.

[36] Liao, P.Y.; Gao, Y.H.; Wang, X.R.; Bao, L.L.; Bian, J.; Hu, T.S.; Zheng, M.Z.; Yan, Y.J.; Chen, Z.L. Tetraphenylporphyrin derivatives possessing piperidine group as potential agents for photodynamic therapy. *J. Photochem. Photobiol. B.*, **2016**, 165, 213-219.

[37] Liao, P.Y.; Wang, X.R.; Gao, Y.H.; Zhang, X.H.; Zhang, L.J.; Song, C.H.; Zhang, D.P.; Yan, Y.J.; Chen, Z.L. Synthesis, photophysical properties and biological evaluation of alkylaminoporphyrin for photodynamic therapy. *Bioorg. Med. Chem.*, **2016**, 24, 6040-6047.

[38] Feng, X.L.; Shi, Y.; Xie, L.F.; Zhang, K.; Wang, X.B.; Liu, Q.H.; Wang, P. Synthesis, characterization, and biological evaluation of a porphyrin-based photosensitizer and its isomer for effective photodynamic therapy against breast cancer. *J. Med. Chem.*, **2018**, 61, 7189-7201.

[39] Meng, Z.; Yu, B.; Han, G.Y.; Liu, M.H.; Shan, B.; Dong, G.Q.; Miao, Z.Y.; Jia, N.Y.; Tan, Z.; Li, B.H.; Zhang, W.N.; Zhu, H.Y.; Sheng, C.Q.; Yao, J.Z. Chlorin p_6-based water-soluble amino acid derivatives as potent photosensitizers for photodynamic therapy. *J. Med. Chem.*, **2016**, 59, 4999-5010.

[40] Zhang, X.J.; Meng, Z.; Ma, Z.Q.; Liu, J.H.; Han, G.Y.; Ma, F.J.; Jia, N.Y.; Miao, Z.Y.; Zhang, W.N.; Sheng, C.Q.; Yao, J.Z. Design and synthesis of novel water-soluble amino acid derivatives of chlorin p_6 ethers as photosensitizer. *Chin. Chem. Lett.*, **2019**, 30, 247-249.

[41] Zhang, X.J.; Han, G.Y.; Guo, C.Y.; Ma, Z.Q.; Lin, M.Y.; Wang, Y.; Miao, Z.Y.; Zhang, W.N.; Sheng, C.Q.; Yao, J.Z. Design, synthesis and biological evaluation of novel 3^1-hexyloxy chlorin e_6-based 15^2- or 13^1-amino acid derivatives as potent photosensitizers for photodynamic therapy. *Eur. J. Med. Chem.*, **2020**, 207, 112715.

[42] Gao, Y.H.; Zhu, X.X.; Zhu, W.; Wu, D.; Chen, D.Y.; Yan, Y.J.; Wu, X.F.; O'Shea, D. F.; Chen, Z.L. Synthesis and evaluation of novel chlorophyll a derivatives as potent photosensitizers for photodynamic therapy. *Eur. J. Med. Chem.*, **2020**, 187, 111959.

[43] Yan, Y.J.; Zheng, M.Z.; Chen, Z.L.; Yu, X.H.; Yang, X.X; Ding, Z.L.; Xu, L. Studies on preparation and photodynamic mechanism of chlorin P6-13,15-N-(cyclohexyl)cycloimide (Chlorin-H) and its antitumor effect for photodynamic therapy *in vitro* and *in vivo*. *Bioorg. Med. Chem.*, **2010**, 18, 6282-6291.

[44] Zhang, C.Y.; Zhang, L.J.; Li, J.W.; Li, J.H.; Wu, Z.M.; Zhang, L.X.; Chen, N.; Yan, Y.J.; Chen, Z.L. *In vitro* and *in vivo* antitumor activity of a novel chlorin derivative for photodynamic therapy. *Neoplasma*, **2016**, 63, 37-43.

[45] Zhang, L.J.; Bian, J.; Bao, L.L.; Chen, H.F.; Yan, Y.J.; Wang, L.; Chen, Z.L. Photosensitizing effectiveness of a novel chlorin-based photosensitizer for photodynamic therapy *in vitro* and *in vivo*. *J. Cancer Res. Clin.*, **2014**, 140, 1527-1536.

[46] Dong, Y.; Li, G.Z.; Wang, L.; Cao, L.; Li, Y.Q.; Zhao, W.J. Anti-tumor evaluation of a novel methoxyphenyl substituted chlorin photosensitizer for photodynamic therapy. *J. Photochem. Photobiol. B.*, **2020**, 211, 112015.

[47] Ke, X.S.; Tang, J.; Chen, J.J.; Zhou, Z.Y.; Zhang, J.L. Ionic conjugated chlorin-type photosensitizers based on porpholactone: Synthesis, photophysical properties, and photodynamic activity. *ChemPlusChem.*, **2015**, 80, 237-252.

[48] Zhang, L.J.; O'Shea, D.; Zhang, C.Y.; Yan, Y.J.; Wang, L.; Chen, Z.L. Evaluation of a bacteriochlorin-based photosensitizer's anti-tumor effect *in vitro* and *in vivo*. *J. Cancer Res. Clin.*, **2015**, 141, 1921-1930.

[49] Zhu, W.; Gao, Y.H.; Liao, P.Y.; Chen, D.Y.; Sun, N.N.; Nguyen Thi, P. A.; Yan, Y.J.; Wu, X.F.; Chen, Z.L. Comparison between porphin, chlorin and bacteriochlorin derivatives for photodynamic therapy: Synthesis, photophysical properties, and biological activity. *Eur. J. Med. Chem.*, **2018**, 160, 146-156.

[50] Jiang, X.J.; Lo, P.C.; Tsang, Y.M.; Yeung, S.L.; Fong, W.P.; Ng, D. K. P. Phthalocyanine-polyamine conjugates as pH-controlled photosensitizers for photodynamic therapy. *Chem.- Eur. J.*, **2010**, 16, 4777-4783.

[51] Jiang X.J.; Yeung S.L.; Lo P.C.; Fong W.P.; Ng, D. K. P. Phthalocyanine-polyamine conjugates as highly efficient photosensitizers for photodynamic therapy. *J. Med. Chem.*, **2011**, 54, 320-330.

[52] Zheng, B.Y.; Wang, L.; Hu, Q.Y.; Shi, J.; Ke, M.R.; Huang, J.D. Novel unsymmetrical silicon(IV) phthalocyanines as highly potent anticancer photosensitizers. Synthesis, characterization, and *in vitro* photodynamic activities. *Dyes Pigments*, **2020**, 177, 108286.

[53] Zheng, B.D.; Li, S.L.; Huang, Z.L.; Zhang, L.; Liu, H.; Zheng, B.Y.; Ke, M.R.; Huang, J.D. A non-aggregated Zinc(II) phthalocyanine with hexadeca cations for antitumor and antibacterial photodynamic therapies. *J. Photochem. Photobiol. B.*, **2020**, 213, 112086.

[54] Dai, X.M.; Chen, X.L.; Zhao, Y.; Yu, Y.J.; Wei, X.S.; Zhang, X.G.; Li, C.X. A water-soluble galactose-decorated cationic photodynamic therapy agent based on BODIPY to selectively eliminate biofilm. *Biomacromol.*, **2018**, 19, 141-149.

[55] Yu, Z.L.; Zhou, J.L.; Ji, X.; Lin, G.Y.; Xu, S.; Dong, X.C.; Zhao, W.L. Discovery of a monoiodo Aza-BODIPY near-infrared photosensitizer: *In vitro* and *in vivo* evaluation for photodynamic therapy. *J. Med. Chem.*, **2020**, 63, 9950-9964.

[56] Liu, Y.C.; Xu, C.Y.; Teng, L.L.; Liu, H.W.; Ren, T.Bi.; Xu, S.; Lou, X.F.; Guo, H.W.; Yuan, L.; Zhang, X.B. pH stimulus-disaggregated BODIPY: An activated photodynamic/photothermal sensitizer applicable to tumor ablation. *Chem. Commun.*, **2020**, 56, 1956-1959.

[57] Liu, F.G.; Li, X.S. Photodynamic therapy in China. *Chin. J. Laser Med. Surg.*, **2005**, 14, 67 (in Chinese).

[58] Huang, Z. Photodynamic therapy in China: Over 25 years of unique clinical experience Part Two-Clinical experience. *Photodiagn. Photodyn.*, **2006**, 3, 71-84.

[59] Li, L.B.; Li, W.M.; Xiang, H.L.; Qing, Z.P.; Wang, L.W.; Lin, J.R.; Ma, Y.; Zhen H. Photodynamic therapy: Clinical research and application in China. *Chin. J. Laser Med. Surg.*, **2012**, 21, 278-307 (in Chinese).

[60] Gu Y. *Clinical technical operation standard: Laser medicine*. Beijing: People's Military Medical Press, **2010**, 129-133.

[61] Gu, Y.; Li, J.H.; Guo, Z.H. Photodynamic therapy in treatment of PWS. *Beijing Med. J.*, **1991**, 13, 317 (in Chinese).

[62] Zhang, Y.J.; Zou, X.B.; Chen, H.X.; Yang, Y.G.; Lin, H.; Guo, X.J. Clinical study on clinical operation and post-treatment reactions of HMME-PDT in treatment of PWS. *Photodiagn. Photodyn.*, **2017**, 20, 253-256.

[63] Zhang, Y.J.; Yang, Y.G.; Zhang, Z.; Yang, Y.J.; Qiu, M.; Chen, H.X.; Li, L.; Zou, X.B. Clinical study on Hemoporfin PDT for infant facial port-wine stains. *Photodiagn. Photodyn.*, **2019**, 25, 106-110.

[64] Yin, G.W.; Li, X.Y.; Wang, X.Y.; Zhang, H. Effect of 5-aminolevulinic acid photodynamic therapy on the expression of apoptosis inhibitors Bcl-2 and survivin in keratinocytes of condyloma acuminatum. *Photodiagn. Photodyn.*, **2019**, 28, 53-57.

[65] Wang, X.L.; Xu, S.Z.; Zhang, C.R.; Zhu, X.J.; Xia, W. Aminolevulinic acid photodynamic therapy for male urethra condyloma acuminatum. *Chin. J. Derm. Venereol.*, **2000**, 14, 103-104 (in Chinese).

[66] Shi, L.; Wang, H.W.; Chen, K.; Yan, J.; Yu, B.; Wang, S.C.; Yin, R.; Nong, X.; Zou, X.B.; Chen, Z.; Li, C.X.; Chen, L.Q.; Zhang, C.L.; Zhang, F.R.; Zheng, H.Y.; Zheng, M.; Tu, P.; Xu, J.H.; Tao, J.; Kang, X.J.; Zeng, K.; Lu, Y.; Yu, N.; Lei, X.; Pan, M.; Ju, Q.; Gu, H.; Wang, X.L. Chinese guidelines on the clinical application of 5-aminolevulinic acid-based photodynamic therapy in dermatology (2021 edition). *Photodiagn. Photodyn.*, **2021**, 35, 102340.

[67] Kennedy, J. C.; Pottier, R. H.; Pross, D. C. Photodynamic therapy with endogenous protoporphyrin: IX: Basic principles and present clinical experience. *J. Photochem. Photobiol. B.*, **1990**, 6, 143-148.

[68] Cai, H.; Wang, Y.X.; Sun, P.; Yang, Z.Y.; Tian, R.; Liu, X.Y.; Li, Q.; Qiao, L.; Liu, W. Photodynamic therapy for facial actinic keratosis: A clinical and histological study in Chinese patients. *Photodiagn. Photodyn.*, **2013**, 10, 260-265.

[69] Han, J.T.; Wang, P.R.; Wang, X.L. Clinical effect and safety of 5-aminolevulinic acid photodynamic therapy in the treatment of nose actinic keratosis. *Chin. J. Ophthalmol. Otorhinolaryngol.*, **2018**, 18, 175-177 (in Chinese).

[70] Zhu, L.D.; Wang, P.R.; Zhang, G.L.; Zhang, L.L.; Liu, X.J.; Hu, C.; Yang, X.Q.; Sroka, R.; Zhou, Z.X.; Wang, X.L. Conventional versus daylight photodynamic therapy for actinic keratosis: A randomized and prospective study in China. *Photodiagn. Photodyn.*, **2018**, 24, 366-371.

[71] Ge, L.H.; Shu, R. Effects of photodynamic therapy on periodontal pathogens in subgingival plaque of chronic periodontitis. *J. Clin. Stomatol.*, **2008**, 24, 740-743 (in Chinese).

[72] Luan, X.L.; Qin, Y.L.; Bai, X.F.; Bi, L.J.; Zhang, Z.G.; Yu, S.L. Study of the effect of toluidine blue(TB)-mediated photodynamic inactivation to the periodontal pathogens from supragingival plaques. *Chin. J. Conserv. Dent.*, **2008**, 18, 94-97 (in Chinese).

[73] Yang, D.; Lei, S.X.; Pan, K.R.; Chen, T.; Lin, J.; Ni, G.C.; Liu, J.X.; Zeng, X.; Chen, Q.M.; Dan, H.X. Application of photodynamic therapy in immune-related diseases. *Photodiagn. Photodyn.*, **2021**, 34, 102318.

Chapter 8

Porphyrin-Based Photosensitizers for Dye-Sensitized Solar Cell Applications

Kala Ramakrishnan* **and Athira Pallikkara**
Department of Applied Chemistry, Cochin University of Science and Technology, Kochi, Kerala, India

Abstract

In the era of dye sensitized solar cells (DSSC), maximum efficiency reaches up to 14.2%, achieved by using porphyrin based photosensitizers. Prior to the birth of DSSC itself, there existed porphyrins as light harvesting molecules in nature. Inspired from such systems, scientists have been keenly trying to mimic and elaborate it. Light harvesting ability of these molecules finds applications in various fields such as dye sensitized solar cells, photodynamic therapy, bio-imaging and in therapeutics. In this chapter the application of porphyrin based dyes for photovoltaics is highlighted. From evaluation of these dyes, it is evident that 12 peripheral positions of the porphine core available for substitution makes it superior to other sensitizers. This chapter discusses how the structural features of the porphyrin dyes are being exploited in various electron transfer processes of natural systems. Also, the progress in the structural modification of bio-inspired porphyrin based systems through introducing donor-acceptor moieties, metallation, nanoarchitecture designs, cosensitization strategies etc. is critically analyzed which will be helpful to find a pathway in fine tuning the structure of the photosensitizer to meet the requirements of photovoltaic applications.

* Corresponding Author's Email: kala@cusat.ac.in.

In: Photosensitizers and Their Applications
Editors: Davor Margetic and Renjith Thomas
ISBN: 978-1-68507-880-5

Keywords: porphyrins, metallation, photosensitizers, dye sensitized solar cells

Introduction

Escalation of research in the area of solar cells is due to the increased energy demand that cannot be completely satisfied by the non-renewable energy sources, fossil fuels mainly, in a feasible way. Statistical analysis of energy consumption reveals that there is a sudden decrease in primary energy consumption in 2020 due to the COVID-19 pandemic. In spite of that, renewable energy has an augmented growth with solar energy recording its largest increase according to the bp Statistical Review of World Energy 2021 [1]. Currently solar power covers 27% of renewable energy and 3.2% of total power consumption, whereas in 2019 it was only 0.37% of the total power consumption [2]. Harnessing the solar power efficiently is the best option for meeting the energy demand in a cleaner manner. In this context, the benefits of environmental and economic friendly solar cells have to be remembered. The main benefits among them are:

- Zero energy production cost
- Cost effectiveness
- No environmental pollution (air, water and noise)
- Lower maintenance
- Long term energy source

There are three main types of solar cells belonging to first, second and third generations. First generation solar cells are the conventional ones consisting of silicon, while the second generation includes thin film solar cells and the third generation, also known as emerging photovoltaics, includes dye sensitized solar cells (DSSCs), perovskite solar cells, organic solar cells, quantum dot solar cells etc. Currently silicon based solar cells are pervasively used, but they face the problem of high cost on wholesale production. Amidst the three generations of solar cells, the 3rd generation is the leading one on account of the low cost of production, renewable nature and eco-friendliness, among which DSSCs have received much attention. A typical DSSC consists of four chief components: photoanode, photosensitizer, redox mediator and a cathode (Figure 1). Most commonly used materials in each component are

TiO_2, N719 dye, I^-/I_3^- and platinum respectively. Upon light irradiation the photosensitizer gets populated in the excited state and rapidly injects electrons to the photoanode. At the same time the oxidized dye absorbs electrons from the redox system, which undergoes redox reactions in the presence of cathode, ultimately resulting in the regeneration of dye [2]. Looking at the design of DSSCs, it is visible that the photosensitizers are of prime importance since the cycle begins with them. They harness sunlight and cause adequate charge separation between the anode and electrolyte, which hinders the recombination.

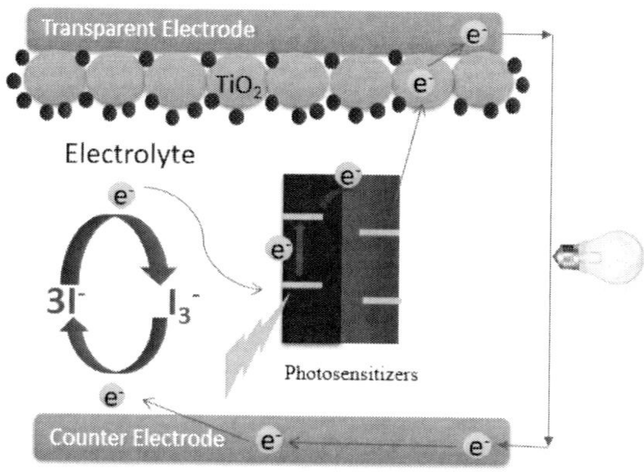

Figure 1. Schematic representation of DSSC design.

For the efficacious use of these dyes on DSSCs, a number of criteria have to be satisfied [2, 3, 4]. The dyes should:

- Satisfy the energetic criteria of cells, i.e., HOMO of the dyes should lie below the electrolyte potential and LUMO of the dye above the conduction band of photoanode [5].
- Have wide range of absorption in the solar spectrum.
- Have high molar extinction coefficient.
- Have structural features for hindering the aggregation of dyes on semiconductor.
- Have the capability of forming a monolayer on the semiconductor surface to avoid the electron recombination between semiconductor and electrolyte.
- Have the ability to bind effectively with the surface of photoanodes.
- Be non-toxic and highly stable.

Some inhibiting reactions like recombination can occur in DSSCs that reduce efficiency. The decay of excited electrons to ground state competes with electron injection from dye to TiO_2 and the dye regeneration competes with electron recombination between injected electrons in TiO_2 with dye and electrolyte [6]. Therefore, an efficient dye is the one which satisfies all the above requirements and overcomes the inhibiting processes.

A variety of photosensitizers has been utilized in the world of DSSCs, amidst of which ruthenium based organometallic dyes are the conventional photosensitizers. Research efforts in this path show serious defects of these dyes such as uneconomical nature, bio-incompatibility, phototoxicity and so on which are parallel to its enormous advantages [6]. In this context, porphyrin moieties seized our attention by virtue of their low cost, rigidity, stability, core metallation and easy synthesis [8]. Structure of porphyrin includes four pyrrole rings connected through methine bridges (Figure 2). It has 20 peripheral carbon atoms and four inner nitrogen atoms, in which meso and β positions are at hand for substitutions. Numerous structural modifications are feasible for this system due to the accessibility of total 12 positions (4 meso and 8 β sites). The difference in their substitutions results in a variety of molecules with contrasting properties.

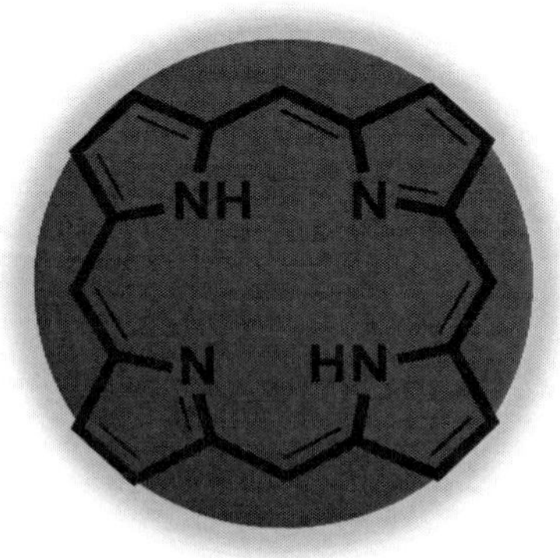

Figure 2. Structure of porphine.

Porphyrin molecules have their own assets such as long range absorption, structural rigidity, high extinction coefficients, feasibility for peripheral and core substitutions, high thermal and photostability [9, 10]. The optical properties of porphyrins manifest their photoactive behaviour. One Soret band around 400-450 nm and four Q bands above 500 nm and below 700 nm range are the characteristic features of the absorbance spectrum of porphyrin. Soret band also known as B band in the near UV is due to the strongly allowed electronic transition from ground S_0 (a_{1u}) to excited S_2 (e_g*) state, while Q bands in the visible region is the result of weakly allowed transitions from S_0 (a_{2u}) to S_1 (e_g*) states [11, 12]. Fluorescence emission spectrum of porphyrins generally shows dual emission, one band around 650 nm and another broader one at 720 nm for free base porphyrins [12]. Excited electrons in the S_2 state easily undergo internal conversion and occupy the S_1 state. The first emission band is Q*(0, 0) and the second one is called Q*(0, 1) as expressed in Figure 3. In spite of the above properties, the stumbling disadvantages of these dyes is that they have low intense absorptions above 500 nm and face the problem of aggregation.

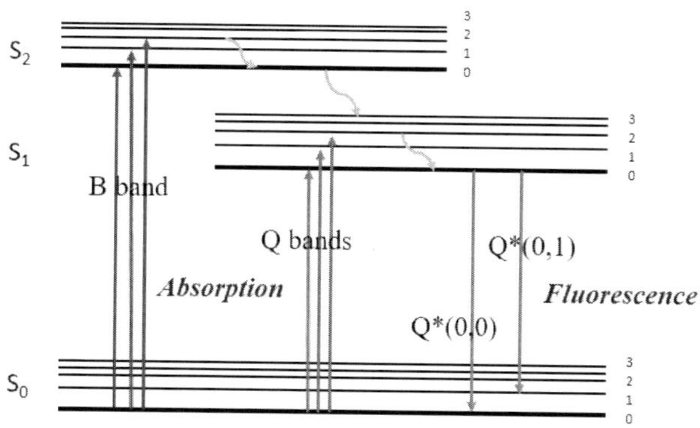

Figure 3. Schematic representation of the electronic transitions in the absorption and emission processes of porphyrin molecules.

The efficiency of DSSCs is greatly influenced by the mode of adsorption of porphyrin dye on TiO_2. The sensitization conditions such as the solvents used, time taken and the tilt angle of porphyrins with respect to TiO_2 are categorized under the modes of adsorption [13]. Electron transfer from the dye to TiO_2 can occur through bond or through space. Titled porphyrins demonstrate easy electron transfer through space, while the orthogonal porphyrins via bond. The difference in these angles, also affects their recombination rate [14]. As the space between dye and TiO_2 is reduced, the recombination becomes faster and it has an adverse effect on the efficiency of the solar cells. Considering the case of sensitization time, adsorption increases with increasing time. However, longer sensitization time causes a drop in their efficiency because of the porphyrin aggregation. The solvents involved in the adsorption have an important effect, protic solvents increase the photo conversion efficiency (PCE), whereas aprotic solvents lower the PCE due to the larger and smaller surface coverage of dyes on TiO_2 respectively [13]. The naturally occurring and artificially synthesized porphyrinic systems are assessed in the further text. Also, relation between the structure of porphyrin molecules and photovoltaic performance of DSSCs are evaluated.

Natural Dyes - Nature Creates Its Own Light Harvesting Molecules

Porphyrin can be regarded as the reason behind the life, because it is an inevitable part of most of the important biological molecules. Most common examples are chlorophylls, hemoglobin & myoglobin, cytochromes, catalases and peroxidases. Among the above, chlorophylls act similar to photosensitizers, whereas the residual ones behave as mediators in redox processes in the biological system. Considering the photosynthesis reaction, we can find analogy with the working principle of DSSCs. Green plants contain chlorophyll and other photoactive pigments that absorb sunlight in the red and blue region and reflect the green and near green areas of the spectrum. There are more than 5 types of chlorophylls, from which chlorophyll a and chlorophyll b are involved in the photosynthesis (Figure 4). Absorption spectra of chlorophyll a lies in the range of 400-450 nm and 650-700 nm, while chlorophyll b has absorption at a wavelength of 450-500 nm and at 600-650 nm. During the photosynthesis, chlorophylls present in the photoactive system get energized by light and these excited electrons are carried by the numerous electron carriers coming after chlorophyll a & b. Ample number of pigments structurally different from chlorophylls, such as carotenoids, phycoerythrin and phycocyanin are congregated in green plants and algae. They are together called photosystems. Presence of these pigments is boon to photosynthesis, because of their efficient light absorption in different parts of the electromagnetic spectrum [15]. As different from dye sensitized solar cells the light energy is converted to chemical energy in photosynthesis. The energized pigments can return to ground state either by fluorescence or conversion of excited energy into chemical energy by the transfer of electrons to the adjacent pigments [16]. This is depicted in Scheme 1, where D is the donor and A_1, A_2, A_3 etc. are the electron acceptors.

$$DA_1A_2A_3\ldots \xrightarrow{h\nu} D^*A_1A_2A_3\ldots \longrightarrow D^+A_1^-A_2A_3\ldots \longrightarrow$$

$$D^+A_1A_2^-A_3\ldots \longrightarrow D^+A_1A_2A_3^-\ldots$$

Scheme 1. Schematic representation of electron transfer in pigments involved in photosynthesis.

The electrons flow through numerous pigments that are involved in the production of carbohydrates and return back to these chlorophylls, meanwhile resulting in the splitting of water and producing oxygen. Hence the process is a cyclic one like the operation of dye sensitized solar cells.

A number of studies were executed on the direct use of natural porphyrin pigments as the dyes in solar cells. Living cells which contain chlorophylls are normally green in color, since they absorb red and blue. In the research the chlorophyll dyes are extracted directly from the green plants, bacteria, bryophytes or algae [3, 17, 18]. There are variations in the rate of adsorption of chlorophyll derivatives over TiO_2 surface. Chlorophyll c gets adsorbed easier than chlorophyll a due to the presence of carboxylic acid group in it. Presence of the keto and ester groups in chlorophyll a hinders the binding with TiO_2 [19]. Co-sensitization of chlorophyll a with chlorophyll b or c amplified the light harvesting and electron injection in the semiconductor-dye interface and achieved 5.4% efficiency [18]. Arof et al. reviewed chlorophyll dyes from various sources and stated a highest efficiency of 4.6% with chlorophyll c_2 from *Undaria pinnatifida* [3]. These chlorophyll dyes are found to have higher efficiency and lifetime than anthocyanins and carotenoids [20].

Despite the low cost, wide range of absorption, non-toxicity, bio-degradability and easy availability of chlorophyll based solar cells (CSCs), its commercialization faces a lot of problems [3, 19]. Compared to solar cells based on conventional ruthenium dyes, CSCs have low energy conversion efficiency and lower stability. The electron injection results in a cation core in ruthenium based dyes while in chlorophyll the entire molecule is oxidized, which may cause other reactions instead of regeneration. Opposite to chlorophylls in living cells that are reproduced in spring, dyes in DSSCs have to be sustained for an appreciable period in order to ensure the long life of solar cell [19].

The operation of other heme proteins such as hemoglobin, myoglobin, cytochromes etc. does not involve light. Also, it is interesting to note that all natural porphyrins are unsymmetrical in nature and all of them hold metals in their core. Investigations show that there is a great relation between the structures of porphyrin dyes and the performance of DSSCs. Because of that the structural study of porphyrin photosensitizers gained wide attention.

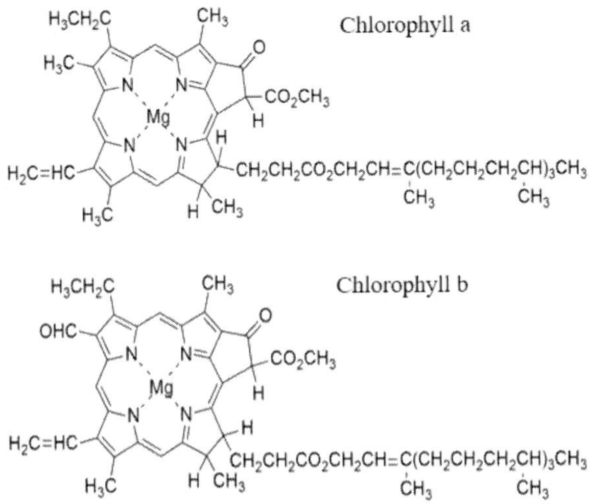

Figure 4. Structure of Chlorophyll a and Chlorophyll b.

Synthetic Dyes - Mimicking the Nature

Resemblance of the photosynthesis and dye sensitized solar cells introduced the chlorophylls as dyes in DSSCs, from which the porphyrin based dyes emerged. In 1935 Rothemund et al. reported the first laboratory synthesis of porphyrins from pyrroles and aldehydes [21, 22]. It became the basis for other new methods of synthesis such as those of Alder and Longo [23]. As described above, DSSCs emulate photosynthesis and because of that scientists all over the world try to mimic chlorophyll in search of new dyes for DSSC. The basic core was maintained as such and variations in their periphery were created. The primary modification of porphine involves the phenyl ring substitutions in the four meso positions [24], commonly called as tetraphenylporphyrin (TPP). Aryl ring substitution disrupts the planarity of the core and occupies the position perpendicular to it, which causes a reduction in the conjugation [25]. Insertion of metals, peripheral modifications with bulkier alkyl substituents, formation of nanohybrids or co-sensitization with other photoactive materials can effectively retain the planarity and overcome the above problems. Each of these modifications are discussed in detail.

Peripheral Modifications/Donor-Acceptor Moieties

In the development of porphyrin photosensitizers, designing one with high efficiency leads the way to production of antenna like systems. Appropriate substituents in the peripheral positions positively influence the photovoltaic behaviour. The consequences of alterations in the moiety are different for each substituent. Sometimes it results in red shift due to the J-aggregation and in some cases, it shows blue shift because of the deprotonation of acid groups or the H-aggregation of porphyrin systems. There is also change in the area of spectrum covered, such as broadening of bands indicating the improved light absorption, while the narrowing represents the lower light harvesting nature [26]. These changes were shown in both Soret and Q bands. Drastic change in the Q bands of comparable intensity with Soret band and shift to above 800 nm, make them suitable for the dye sensitized solar cell applications. Besides these spectral changes, the modifications also alter their energy levels. Analysis of the energy levels reveals that the presence of strong electron donors raised the HOMO energy levels, thereby reducing the bandgap of the molecules. In some situations, the elevation in the HOMO energy level hinders the regeneration process due to non-feasible electron transfer from electrolyte to the dye. For instance, bis(4-methoxyphenyl)amino groups in the core moiety raise HOMO energy level that suppresses the regeneration which adversely affects the photovoltaic behavior of DSSCs [26].

One of the most commonly investigated substituents is carboxylic acid group (-COOH) since this helps in anchoring of the dye on to the semiconductor surface. Each of the substituted positions either meso or β also have valuable importance [27]. There are three possible ways of binding between the carboxylic acid group and TiO_2 such as unidentate, bridging-bidentate and chelation as shown in Figure 5 [28]. Perpetually, the $-COO^-$ has a rigid bidentate binding with TiO_2. Depending on the position of this group, such as para-, meta- or ortho- there are deviations in the tilt angle, which results in the para- derivative being perpendicular to TiO_2, meta-derivative with some favourable angles in between 50-80° and ortho-derivative much closer to the surface of TiO_2 [29]. Ladomenou et al. reviewed the significance of some of the anchoring groups such as carboxylic acids, phosphonic & sulphonic acids and pyridines in the meso and β positions of porphine [27]. Among numerous anchoring groups, the presence of long alkyl chains, bulky substituents and large aromatic rings suppress the dye aggregation [13, 30].

Self-assembled porphyrin nanostructures are another class of modifications. Depending upon the size and shape of the self-assemblies, they are categorized into clovers, stars, nanoprisms, nanosheets, nanorods and nanospheres [31]. The robust nature of these porphyrins is a valuable quality in dye sensitized solar cells.

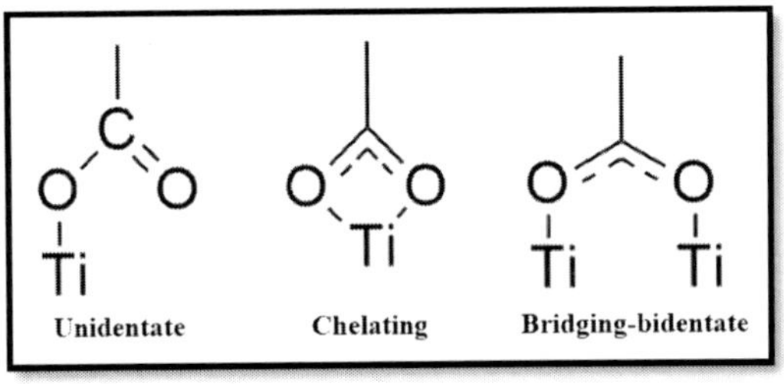

Figure 5. Binding interaction between -COOH groups of porphyrin with TiO_2.

Some of the modifications are specially named as push-pull or donor-π-acceptor (D-π-A) systems, owing to their characteristic structures. In both cases one or more donor and acceptor groups are present around the porphine core and the porphyrin moieties act as a π bridge between these groups. Push-pull molecular design in the asymmetric mode was found to be enhancing the cell performance. Highly asymmetric porphyrin with one carboxyl-phenylethynyl, two diarylamino and one mesityl moieties in the meso positions achieved an efficiency greater than 10% [25]. Another example of a D-π-A system is a diarylamino group as donor and an ethynylbenzoic acid as acceptor in which porphyrin itself acts as a π bridge [32]. A commonly used strong electron donating group is diphenylamino unit. In one of the reports, its derivative known as bis(4-methoxyphenyl)amino and a weak electron donating group, *p*-hexyloxyphenyl unit were incorporated with porphyrin dyes by varying their numbers. In addition to the electron donating nature, groups like *p*-hexyloxyphenyl hinder aggregation with the assistance of long alkyl chains [26]. Diphenylamino group as an electron donor and benzothiadiazole as an acceptor were used to design the push-pull porphyrins, wherein their mixture yielded a higher efficiency than their independent solutions [33].

Chances of getting higher efficiency are better for a cosensitized dye than a single dye. In contrast to that there are some limited examples of highly efficient single porphyrin dye. Firstly, very high efficiencies of 13% and 12% were achieved by two of the porphyrin dyes SM315 and SM371 respectively without co-sensitization [34]. Bulky amine donor known as bis(2',4'-bis(hexyloxy)-[1,1'-biphenyl]-4-yl)amine is present in both dyes but differs in the acceptor group by a benzothiadiazole ring in the SM315. The high rise efficiency noted for a single porphyrin dye without co-sensitization is 12.75% [35]. The introduction of benzene as a spacer between zinc porphyrin and the carboxylic group gave increased lifetime in comparison to the dye without benzene spacer and the rate of recombination was diminished.

From the above inferences, it is postulated that the carboxylic acid anchoring group has a higher efficiency than all other kinds of anchoring groups. Also, meso-diphenylamine derivatives as donors result in promising efficiencies. Nevertheless, synthesis of the asymmetric porphyrins with four different substituents is still challenging one.

Metallation

Porphine structure in Figure 2 has two pyrrole hydrogens in the core which are highly acidic in nature. Porphyrinato anions are obtained by the deprotonation of these NH protons. These anionic porphyrins have a remarkable affinity towards metal cations, which can occupy the central cavity. Insertion of metals to the interior of porphyrin molecule influences the electronic energy levels thereby altering the absorption of molecule and also the splitting of electron-hole pairs. Four to six coordinated complexes are produced by binding with almost all metal ions [36]. The spectral data of metallation commonly involves a bathochromic shift of Soret band and vanishing of one or more Q bands. That decreases spectral coverage as compared to free base porphyrins but seems to have a higher IPCE. The size, charge and spin multiplicity of the metals strongly influence its location in the core. They are categorized in two groups: in-plane (normal) and out-of-plane (OOP) complexes. If the ionic radius of the metal center is large (>90 pm), it occupies the position above the plane and shows special photophysical and chemical properties. The small size metal cations fit into the ligand core and the rate of formation of in-plane metalloporphyrins is lower due to the ligand rigidity [37]. For instance Ag^{+} (94 pm), Cd^{2+} (95 pm) and Bi^{3+} (103 pm) exhibit OOP characteristics, while Au^{3+} (85 pm) and Mn^{3+} (58 pm) form standard metalloporphyrins. Both of these groups of metalloporphyrins have significant roles in biological systems. For example, in deoxy-hemoglobin and myoglobin, high spin Fe^{2+} metal is detected out of plane and bacteriochlorophylls chelated with Mg^{2+} ions have a non-planar distorted structure [38].

The charge transfer transitions, either from ligand to metal (LMCT) or from porphyrin ring to metal are the reason behind the photophysical properties of metalloporphyrins. During the metal insertion in solution phase, first upon the porphyrin ring get contorted and solvated metal association occurs via outer sphere method (OOP complex), which is followed by slow process of the deprotonation of interior nitrogen atoms and results in a chelated ring [7].

Metal insertion alters the MO energy levels of sensitizers. Bathochromic shift in their spectrum indicates a lowering of the bandgap, and causes change in the HOMO-LUMO energy levels. Effective photovoltaic properties depend upon the matching energy levels of dye with respect to the photoanode towards an increased electron injection. Moreover, the planarity and symmetry of molecules are increased in presence of metals. Usually, the non-toxic and economic metals are used for the metallation of porphyrins [39]. Here, 4.8% efficiency was achieved by a zinc metallated porphyrin with four 3,5-xylyl groups in meso positions and an ethen-1-benzoic acid in one of the β positions [28]. Another postulation is that IPCE of diamagnetic zinc porphyrins is much greater than IPCE of the paramagnetic copper porphyrins with same substitutions.

Arkan et al. studied calcium, magnesium and zinc core modified porphyrins as photosensitizers and found that the Ca porphyrins have better photovoltaic properties due to their highest chemical potential and lowest electrophilicity index [39]. Some of the noble metals such as platinum, palladium, rhodium and ruthenium were inserted in porphyrins [40], wherein the peripheral modifications are done by propoxy and butoxy alkyl chains. These alkoxy chains avert aggregation, but the absence of conjugation badly affects the efficiency of the cell. Thus, these metallated porphyrins have a lower PCE than the standard N719 dye. Above the properties of these metallated porphyrins, their presence increases the cost of the sensitizers. Since the removal of metal lowers the cost of dyes and there by the price of DSSC, the free base porphyrins also have equal importance.

Nanohybrids/Co-Sensitization Strategies

Nanohybrids or nanocomposites are formed by the proper mixing of two or more photoactive materials that convert a short range absorption spectrum to panchromatic spectrum. This process of mixing is also known as co-sensitization. The absorption spectra of co-sensitized systems are similar to the spectra of photosynthesis pigments (Figure 5), which means the overlapping absorption spectra cover the entire UV-Visible spectral region. Similar to co-sensitization there are two more processes such as co-adsorbents and the addition of additives. All the three behave equally in DSSCs, like lowering the dye aggregation and charge recombination at the anode/dye/electrolyte junction.

For instance, deoxycholic acid (DCA) used as a co-adsorber of porphyrin derivative enhanced the efficiency from 2.90% to 4.22%. Presence of DCA cuts the porphyrin aggregates and improves the electron injection, which leads to the improvement of short circuit current (J_{sc}), open circuit voltage (V_{oc}) and power conversion efficiency (PCE) [41]. Chenodeoxycholic acid (CDCA) is similar to DCA which overcomes the aggregation of dyes due to their large π conjugated systems [33].

Even so, co-sensitization is widely studied and has better effect in dye based solar cells, than the remaining two. The best efficiency of DSSCs, 14.2% was achieved by the combination of an organic dye SGT-149 and a Zn metallated porphyrin based dye, SGT-021 through a cocktail method (the mixture of two or more dyes), along with cobalt electrolyte [42]. The absorption range of SGT-149 dye are at 350-400 nm and 500-630 nm, while SGT-021 has three bands at 360-400 nm, 400-500 nm and 630-750 nm. This clearly indicates the complementary absorption of two sensitizers which effectively covers the entire visible region of the solar spectrum.

While selecting dyes for co sensitization along with their spectral overlapping, the lifetime of excited electrons should be considered. Several research papers reveal that co-sensitized dyes have HOMO energy levels of higher potential than the electrolyte potential and their LUMO energy levels have more negative potential than the conduction band of the semiconductor, in NHE scale [43, 44]. The ways of sensitization also have significant roles in their performance. It can be sensitized either in stepwise manner or by using the mixed solution. The former one gives better results than the later. Similarly, in the combination of porphyrin and organic dyes, by stepwise sensitization in which porphyrin molecule is just above the TiO_2 and organic dye over it has higher efficacy [44]. As a result of this type of sensitization of one dye over the other the contact between semiconductor surface and electrolyte is reduced, thereby lowering the recombination. Also, the self-interaction within the dye molecules gets diminished that leads to the decreased aggregation and both of them promote better photovoltaic performance of DSSCs. Derivatives containing highly aromatic rings like anthracene are also used as co-sensitizers [26].

Co-sensitization of porphyrin molecules with nanomaterials are classified under nanohybrids, which are a special class of co-sensitizers. Carbon based nanomaterials like carbon nanotubes [45], carbon nanohorns [46], graphene oxides [47], and graphene quantum dots [48] belong to different categories of nanomaterials to which porphyrins are linked either covalently [45] or non-covalently [49]. Interactions involving a covalent bond between two or more components of nanohybrids is known as covalent interaction and they include the carboxylate bond, amide bond etc., while the non-covalent interactions comprise of electrostatic interactions, π-π stacking and hydrogen bonding [50, 51]. Similar to other co-sensitizing organic dyes, these materials also reduce recombination and aggregation and thereby increase the efficiency. In addition to that, the nanohybrids possess properties that make them advantageous in contrast to the conventional organic dyes, such as easy synthesis, availability, lower toxicity and eco-friendliness. These advantages give the nanomaterial incorporated porphyrin dyes a wide attention.

Consequently, the significance of cocktail sensitization is that they can improve the stability and efficiency of dyes, by overcoming the aggregation and recombination which are the main obstacles to higher efficiency.

Conclusion

This chapter is concentrated on the role of porphyrin based photosensitizers in dye sensitized solar cells. The improved photovoltaic properties of these solar cells depend on dyes, which are the inherent part of DSSCs. Among the numerous light harvesting materials, porphyrin molecules are the focus of attention, due to their high molar extinction coefficients, wide range of absorption, low cost and less toxicity. Also, various types of modifications are feasible for them that further enhance the power conversion efficiency of the cell. In this chapter, three main classes of strategies of modifications are discussed briefly, such as the push-pull mechanism, metallation and co-sensitization. Among the various substitutions, carboxylic acids and meso-diphenylamine derivatives in asymmetric porphyrins are found to be the best anchoring groups and electron donors respectively. In terms of photoconversion efficiency, metallated porphyrins outperform the non-metallated ones even though the former ones have a reduction in their spectral coverage area. Also, zinc metallated porphyrin are found to have higher PCE. Concerning co-sensitization, it is an advanced branch of modification and established as leading among the refurbishing methods. Cocktail photosensitizers make the absorption spectrum more panchromatic and thereby result in the higher efficiency of 14.2% up to date. Likewise, the efficiency inhibiting factors such as recombination and aggregation are reduced by the cocktail sensitizers. Another fascinating practice in porphyrin dyes development is that the combination of all the above modifications, a push-pull porphyrin with chelating metal, co-sensitization with another organic dye or inorganic material lead to an enhanced efficiency.

References

[1] Ritchie H, Roser M. Energy. 2015. *Most recent substantial revision in July 2018.* www.ourworldindata/energy.

[2] Pallikkara, A.; Ramakrishnan, K. Efficient Charge Collection of Photoanodes and Light Absorption of Photosensitizers: A Review. *Int. J. Energy Res.* 2021, *45* (2), 1425-1448.

[3] Arof, A. K.; Ping, T. L. Chlorophyll as Photosensitizer in Dye-Sensitized Solar Cells. *Chlorophyll; BoD-Books on Demand: Norderstedt, Germany* 2017, 105-121.

[4] Krishna, N. V.; Krishna, J. V. S.; Mrinalini, M.; Prasanthkumar, S.; Giribabu, L. Role of Co-Sensitizers in Dye-Sensitized Solar Cells. *Chem Sus Chem* 2017, *10* (23), 4668-4689.

[5] Chang, Y. C.; Wang, C. L.; Pan, T. Y.; Hong, S. H.; Lan, C. M.; Kuo, H. H.; Lo, C. F.; Hsu, H. Y.; Lin, C. Y.; Diau, E. W. G. A Strategy to Design Highly Efficient Porphyrin Sensitizers for Dye-Sensitized Solar Cells. *Chem. Commun.* 2011, *47 (31)*, 8910-8912.

[6] Listorti, A.; O'Regan, B.; Durrant, J. R. Electron Transfer Dynamics in Dye-Sensitized Solar Cells. *Chem. Mater.* 2011, *23 (15)*, 3381-3399.

[7] Sardar, S.; Kar, P.; Pal, S. K. Materials NanoScience The Impact of Central Metal Ions in Porphyrin Functionalized ZnO/TiO_2 for Enhanced Solar Energy Conversion. *J. Mater. Nanosci.* 2014, *1 (1)*, 12-30.

[8] Anghel, D.; Lascu, A.; Fratilescu, I.; Epuran, C.; Plesu, N. Review about Main Requirements for Porphyrin Derivatives as Components of Dye Sensitized Solar Cells. *J. Solar Energ. Res. Updat.* 2019, No. 24, 78-86.

[9] Bonnett, R.; Berenbaum, M. Porphyrins as Photosensitizers. *Photosensitizing compounds: their chemistry, biology and clinical use* 1989, *146* (2), 40-53.

[10] Al Mogren, M. M.; Ahmed, N. M.; Hasanein, A. A. Molecular Modeling and Photovoltaic Applications of Porphyrin-Based Dyes: A Review. *J. Saudi Chem. Soc.* 2020, *24 (3)*, 303-320.

[11] Dayer, M.; Moosavi-Movahedi, A.; Dayer, M. Band Assignment in Hemoglobin Porphyrin Ring Spectrum: Using Four-Orbital Model of Gouterman. *Protein Pept. Lett.* 2010, *17 (4)*, 473-479.

[12] Zakavi, S.; Hoseini, S. The Absorption and Fluorescence Emission Spectra of Meso-Tetra(Aryl)Porphyrin Dications with Weak and Strong Carboxylic Acids: A Comparative Study. *RSC Adv.* 2015, *5 (129)*, 106774-106786.

[13] Higashino, T.; Imahori, H. Porphyrins as Excellent Dyes for Dye-Sensitized Solar Cells: Recent Developments and Insights. *Dalt. Trans.* 2015, *44* (2), 448-463.

[14] Auwärter, W.; Écija, D.; Klappenberger, F.; Barth, J. V. Porphyrins at Interfaces. *Nat. Chem.* 2015, *7* (2), 105-120.

[15] https://www.api.simply.science/images/content/biology/cell_biology/photosynthesis/conceptmap/Photosynthetic_pigments.html.

[16] Gray, H. B.; Ellis, W. R. *Bioinorganic Chemistry;* Bertini, I.; Gray, H. B.; Lippard, S. J.; Valentine, J. S. Electron transfer Chapter 6; University Science Books Mill Valley, CA, 1994, 326-329.

[17] Calogero, G.; Citro, I.; Di Marco, G.; Armeli Minicante, S.; Morabito, M.; Genovese, G. Brown Seaweed Pigment as a Dye Source for Photoelectrochemical Solar Cells. *Spectrochim. Acta - Part A Mol. Biomol. Spectrosc.* 2014, *117* (October 2017), 702-706.

[18] Wang, X. F.; Koyama, Y.; Kitao, O.; Wada, Y.; Sasaki, S. ich; Tamiaki, H.; Zhou, H. Significant Enhancement in the Power-Conversion Efficiency of Chlorophyll Co-Sensitized Solar Cells by Mimicking the Principles of Natural Photosynthetic Light-Harvesting Complexes. *Biosens. Bioelectron.* 2010, *25* (8), 1970-1976.

[19] Kay, A.; Grätzel, M. Artificial Photosynthesis. 1. Photosensitization of TiO_{as} Solar Cells with Chlorophyll Derivatives and Related Natural Porphyrins. *J. Phys. Chem.* 1993, *97* (23), 6272-6277.

[20] Ammar, A. M.; Mohamed, H. S. H.; Yousef, M. M. K.; Abdel-Hafez, G. M.; Hassanien, A. S.; Khalil, A. S. G. Dye-Sensitized Solar Cells (DSSCs) Based on Extracted Natural Dyes. *J. Nanomater.* 2019, *2019*, 10 pages.

[21] Rothemund, P. Formation of porphyrins from pyrrole and aldehyde. *J. Am. Chem. Soc.* 1935, *57*, 2010-2011.

[22] Rothemund, P. A New Porphyrin Synthesis. *J. Am. Chem. Soc.* 1936, *157* (6), 625-627.

[23] Adler, A. D.; Longo, F. R.; Finarelli, J. D.; Goldmacher, J.; Assour, J.; Korsakoff, L. A Simplified Synthesis for meso-Tetraphenylporphine. *J. Org. Chem.* 1967, *32 (2)*, 476-476.

[24] Zeng, Z.; Zhang, B.; Li, C.; Peng, X.; Liu, X.; Meng, S.; Feng, Y. A Key Point of Porphyrin Structure Affect DSSCs Performance Based on Porphyrin Sensitizers. *Dyes Pigment.* 2014, *100* (1), 278-285.

[25] Kurotobi, K.; Toude, Y.; Kawamoto, K.; Fujimori, Y.; Ito, S.; Chabera, P.; Sundström, V.; Imahori, H. Highly Asymmetrical Porphyrins with Enhanced Push-Pull Character for Dye-Sensitized Solar Cells. *Chem. - A Eur. J.* 2013, *19* (50), 17075-17081.

[26] Chen, B.; Sun, L.; Xie, Y. S. Modulation of Photovoltaic Behavior of Dye-Sensitized Solar Cells by Electron Donors of Porphyrin Dyes and Cosensitization. *Chinese Chem. Lett.* 2015, *26* (7), 899-904.

[27] Ladomenou, K.; Kitsopoulos, T. N.; Sharma, G. D.; Coutsolelos, A. G. The Importance of Various Anchoring Groups Attached on Porphyrins as Potential Dyes for DSSC Applications. *RSC Adv.* 2014, *4* (41), 21379-21404.

[28] Nazeeruddin, M. K.; Humphry-Baker, R.; Officer, D. L.; Campbell, W. M.; Burrell, A. K.; Grätzel, M. Application of Metalloporphyrins in Nanocrystalline Dye-Sensitized Solar Cells for Conversion of Sunlight into Electricity. *Langmuir* 2004, *20* (15), 6514-6517.

[29] Hart, A. S.; Kc, C. B.; Gobeze, H. B.; Sequeira, L. R.; D'Souza, F. Porphyrin-Sensitized Solar Cells: Effect of Carboxyl Anchor Group Orientation on the Cell Performance. *ACS Appl. Mater. Interfaces* 2013, *5* (11), 5314-5323.

[30] Jia, H. L.; Ju, Z. M.; Sun, H. X.; Ju, X. H.; Zhang, M. D.; Zhou, X. F.; Zheng, H. G. Improvement of Photovoltaic Performance of DSSCs by Modifying Panchromatic Zinc Porphyrin Dyes with Heterocyclic Units. *J. Mater. Chem. A* 2014, *2* (48), 20841-20848.

[31] Medforth, C. J.; Wang, Z.; Martin, K. E.; Song, Y.; Jacobsen, J. L.; Shelnutt, J. A. Self-Assembled Porphyrin Nanostructures. *Chem. Commun.* 2009, *7345* (47), 7261-7277.

[32] Bessho, T.; Zakeeruddin, S. M.; Yeh, C. Y.; Diau, E. W. G.; Grätzel, M. Highly Efficient Mesoscopic Dye-Sensitized Solar Cells Based on Donor-Acceptor-Substituted Porphyrins. *Angew. Chem. - Int. Ed.* 2010, *49* (37), 6646-6649.

[33] Pan, J.; Song, H.; Lian, C.; Liu, H.; Xie, Y. Cocktail Co-Sensitization of Porphyrin Dyes with Additional Donors and Acceptors for Developing Efficient Dye-Sensitized Solar Cells. *Dyes Pigment.* 2017, *140*, 36-46.

[34] Mathew, S.; Yella, A.; Gao, P.; Humphry-Baker, R.; Curchod, B. F. E.; Ashari-Astani, N.; Tavernelli, I.; Rothlisberger, U.; Nazeeruddin, M. K.; Grätzel, M. Dye-Sensitized Solar Cells with 13% Efficiency Achieved through the Molecular Engineering of Porphyrin Sensitizers. *Nat. Chem.* 2014, *6* (3), 242-247.

[35] Yella, A.; Mai, C. L.; Zakeeruddin, S. M.; Chang, S. N.; Hsieh, C. H.; Yeh, C. Y.; Grätzel, M. Molecular Engineering of Push-Pull Porphyrin Dyes for Highly Efficient Dye-Sensitized Solar Cells: The Role of Benzene Spacers. *Angew. Chem. - Int. Ed.* 2014, *53* (11), 2973-2977.

[36] Shiragami, T.; Matsumoto, J.; Inoue, H.; Yasuda, M. Antimony Porphyrin Complexes as Visible-Light Driven Photocatalyst. *J. Photochem. Photobiol. C Photochem. Rev.* 2005, *6* (4), 227-248.

[37] Horváth, O.; Valicsek, Z.; Harrach, G.; Lendvay, G.; Fodor, M. A. Spectroscopic and Photochemical Properties of Water-Soluble Metalloporphyrins of Distorted Structure. *Coord. Chem. Rev.* 2012, *256* (15-16), 1531-1545.

[38] Scheer, H. An Overview of Chlorophylls and Bacteriochlorophylls: Biochemistry, Biophysics, Functions and Applications. *Chlorophylls and bacteriochlorophylls* 2006, 1-26.

[39] Arkan, F.; Izadyar, M. The Investigation of the Central Metal Effects on the Porphyrin-Based DSSCs Performance; Molecular Approach. *Mater. Chem. Phys.* 2017, *196*, 142-152.

[40] Stangel, C.; Daphnomili, D.; Lazarides, T.; Drev, M.; Krašovec, U. O.; Coutsolelos, A. G. Noble Metal Porphyrin Derivatives Bearing Carboxylic Groups: Synthesis, Characterization and Photophysical Study. *Polyhedron* 2013, *52*, 1016-1023.

[41] Mikroyannidis, J. A.; Charalambidis, G.; Coutsolelos, A. G.; Balraju, P.; Sharma, G. D. Novel zinc porphyrin with phenylenevinylene meso-substituents: Synthesis and application in dye-sensitized solar cells. *J. Power Sources* 2011, *196*, 6622-6628.

[42] Ji, J. M.; Zhou, H.; Eom, Y. K.; Kim, C. H.; Kim, H. K. 14.2% Efficiency Dye-Sensitized Solar Cells by Co-Sensitizing Novel Thieno[3,2-b]Indole-Based Organic Dyes with a Promising Porphyrin Sensitizer. *Adv. Energy Mater.* 2020, *10* (15), 1-12.

[43] Reddy, K. S. K.; Chen, Y. C.; Wu, C. C.; Hsu, C. W.; Chang, Y. C.; Chen, C. M.; Yeh, C. Y. Cosensitization of Structurally Simple Porphyrin and Anthracene-Based Dye for Dye-Sensitized Solar Cells. *ACS Appl. Mater. Interfaces* 2018, *10* (3), 2391-2399.

[44] Fan, S.; Lu, X.; Sun, H.; Zhou, G.; Chang, Y. J.; Wang, Z. S. Effect of the Co-Sensitization Sequence on the Performance of Dye-Sensitized Solar Cells with Porphyrin and Organic Dyes. *Phys. Chem. Chem. Phys.* 2015, *18* (2), 932-938.

[45] Baskaran, D.; Mays, J. W.; Zhang, X. P.; Bratcher, M. S. Carbon Nanotubes with Covalently Linked Porphyrin Antennae: Photoinduced Electron Transfer. *J. Am. Chem. Soc.* 2005, *127* (19), 6916-6917.

[46] Pagona, G.; Zervaki, G. E.; Sandanayaka, A. S. D.; Ito, O.; Charalambidis, G.; Hasobe, T.; Coutsolelos, A. G.; Tagmatarchis, N. Carbon Nanohorn-Porphyrin Dimer Hybrid Material for Enhancing Light-Energy Conversion. *J. Phys. Chem. C* 2012, *116* (17), 9439-9449.

[47] Ge, R.; Wang, X.; Zhang, C.; Kang, S. Z.; Qin, L.; Li, G.; Li, X. The Influence of Combination Mode on the Structure and Properties of Porphyrin-Graphene Oxide Composites. *Colloids Surfaces A Physicochem. Eng. Asp.* 2015, *483*, 45-52.

[48] Liu, Y.; Li, S.; Li, K.; Zheng, Y.; Zhang, M.; Cai, C.; Yu, C.; Zhou, Y.; Yan, D. A Srikaya-like Light-Harvesting Antenna Based on Graphene Quantum Dots and Porphyrin Unimolecular Micelles. *Chem. Commun.* 2016, *52* (60), 9394-9397.

[49] Zhu, M.; Li, Z.; Xiao, B.; Lu, Y.; Du, Y.; Yang, P.; Wang, X. Surfactant Assistance in Improvement of Photocatalytic Hydrogen Production with the Porphyrin Noncovalently Functionalized Graphene Nanocomposite. *ACS Appl. Mater. Interfaces* 2013, *5* (5), 1732-1740.

[50] Lu, Q.; Zhang, Y.; Liu, S. Graphene Quantum Dots Enhanced Photocatalytic Activity of Zinc Porphyrin toward the Degradation of Methylene Blue under Visible-Light Irradiation. *J. Mater. Chem. A* 2015, *3* (16), 8552-8558.

[51] Luo, X.; Zhang, Y.; Kim, S. H. Incorporation of Nanostructured Carbon Composite Materials into Counter Electrodes for Highly Efficient Dye-Sensitized Solar Cells. *Nanoscale Res. Lett.* 2018, *13*, 1-10.

Chapter 9

Recent Studies on the Performance of Sensitizers for Efficient Dye-Sensitized Solar Cells: From Dynamics to Kinetics of the Photovoltaic Processes

Faezeh Pakravesh and Mohammad Izadyar*

Research Center for Modeling and Computational Sciences, Department of Chemistry, Faculty of Science, Ferdowsi University of Mashhad, Mashhad, Iran

Abstract

Dye-sensitized solar cells (DSSCs) are the most promising emerging generation in photovoltaic technology in recent years due to their great efficiency, easy fabrication, and high capacity for producing abundant energy sources. A lot of efforts have been invested in analyzing different sensitizers such as metal-organic dyes, natural dyes, and metal-free organic dyes and their applications in DSSCs. Also, we applied various strategies such as the presence of the strong electron donor/acceptor and extended π-conjugated system, increasing the number of anchor groups and the effect of the auxiliary donor/acceptor, substituted groups, and different solvents to enhance the DSSC performance. This review highlights different aspects of the computational chemistry and photovoltaic parameters to calculate ground and excited states. Also, the evaluation of the quantum chemistry, and spectroscopic properties, and reasonable correlation between different parameters is important and provides basic knowledge for future studies through molecular approach for improvement of the physicochemical properties and efficiency of the DSSCs.

* Corresponding Author's Email: izadyar@um.ac.ir.

In: Photosensitizers and Their Applications
Editors: Davor Margetic and Renjith Thomas
ISBN: 978-1-68507-880-5

Keywords: dye-sensitized solar cell, sensitizer, power conversion efficiency, photovoltaic, electron injection

Acronyms

η_0	overall efficiency
η_e	electronic chemical hardness
APCE	absorbed photon to current conversion efficiency
a_x	exciton radius
AZO	aluminum zinc oxide
CCRD	2-(1,1-dicyanomethylene)rhodanine
COSMO	conductor like screening model
CPCM	conductor-like polarizable continuum model
DCM	methylene dichloride
D_{CT}	charge transfer distance
DFT	density functional theory
DMF	dimethylformamide
DSSCs	Dye-sensitized solar cells
EBE	exciton binding energy
EDOT	3,4-ethylenedioxythiophene
f	oscillator strength
FF	fill factor
FMO	frontier molecular orbitals
FTO	fluorine tin oxide
GGA	generalized gradient approximation
GZO	gallium zinc oxide
HOMO	highest occupied molecular orbital
HTMs	hole transfer materials
ICT	intermolecular charge transfer
IPCE	incident photon to current conversion efficiency
IPCM	isodensity polarizable continuum model
ITO	indium tin oxide
J_{sc}	short circuit current density
k_{inj}	electron injection rate
k_{rec}	electron recombination rate
LDA	local density approximation
LHE	light-harvesting efficiency

LUMO	lowest unoccupied molecular orbital
NBO	natural bond orbital
NLO	nonlinear optical
OMeTAD	(*N,N*-di-*p*-methoxyphenylamine) 9,9′-spirobifluorene
PANI	polyaniline
PCM	polarizable continuum model
Pcs	Phthalocyanines
P_{in}	incident solar power
PPy	polypyrrole
PT	polythiophene
q_{CT}	transferred charge
R_a	rate of the exciton formation
R_d	rate of the exciton dissociation
$r_{k,k'}$	electric dipole moment
S	overlap of the electron-hole distribution
t	scale of e-h separation
TCO	transparent conductive oxide
TD-DFT	time-dependent density functional theory
THF	tetrahydrofuran
V_{OC}	open-circuit voltage
WF	work function
XC	exchange-and-correlation
ΔG_{inj}	electron injection driving force
μ	electronic chemical potential
ω	electrophilicity
α	polarizability
β	hyperpolarizability
μ	dipole moment

Introduction

Global environmental concerns such as depletion of fossil fuel, deforestation, overpopulation, emission of acid gas, and formation of acid rain have created serious problems for the humans and environment [1-3]. On the other hand, the energy request has been enhanced significantly. So that, U.S. Energy Information Administration predicts that universal energy demand in 2040 will be 40% higher than the current rate [4]. Therefore, promoting renewable sources to overcome these problems has become a remarkable research issue in the 21st century [5, 6].

One of the safe renewable sources is solar energy which has many advantages such as availability, continuous, and less harmful environmental effects [7, 8]. The solar cell as an efficient and prominent photovoltaic technology has been developed to convert sunlight into electrical energy by photovoltaic effect [9, 10]. The photovoltaic effect was discovered by Becquerel in 1839 and thus he is called the pioneer of solar cell technology [11].

Solar cells are divided into three classes in terms of manufacturing technology. The first generation is the silicon solar cell [12], the second generation is the thin-film solar cell [13], and the third generation includes polymer [14], perovskite [15], quantum dot [16], and dye-sensitized solar cells (DSSCs) [17] (Scheme 1).

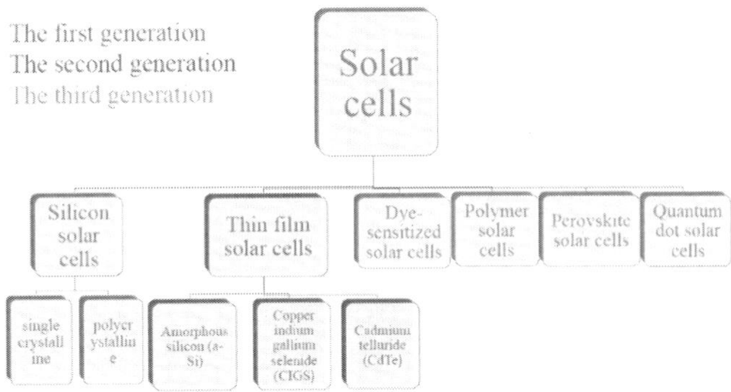

Scheme 1. Different generations of solar cells.

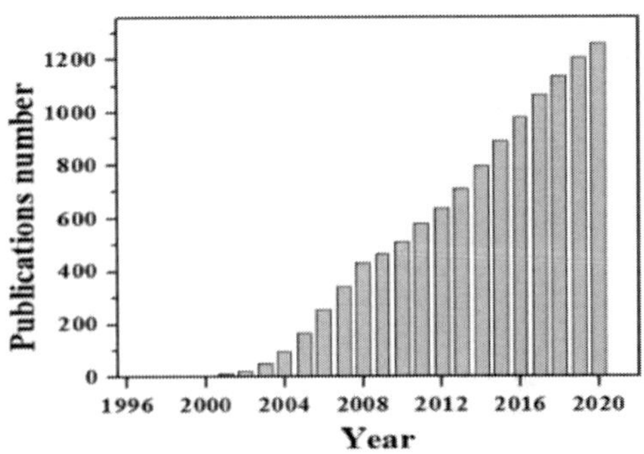

Figure 1. The histogram of the published articles in DSSCs with polymer-based electrolyte [18].

Dye-sensitized solar cells represent relatively good energy conversion efficiency and research on this kind of solar cell has made significant progress. Figure 1 shows the number of publications in the DSSCs with polymer-based electrolyte [18].

DSSCs emerged as prospective and effective candidates due to their unique features such as easy fabrication ability, low production cost, good stability, and less environmental pollution [19, 20]. DSSCs research was first introduced by O'Regan and Grätzel in 1991 [21]. The basic architecture of DSSCs depends on five key components including a transparent glass sheet, a mesoporous oxide layer, redox electrolyte, sensitizer, and counter electrode (Figure 2) [22, 23].

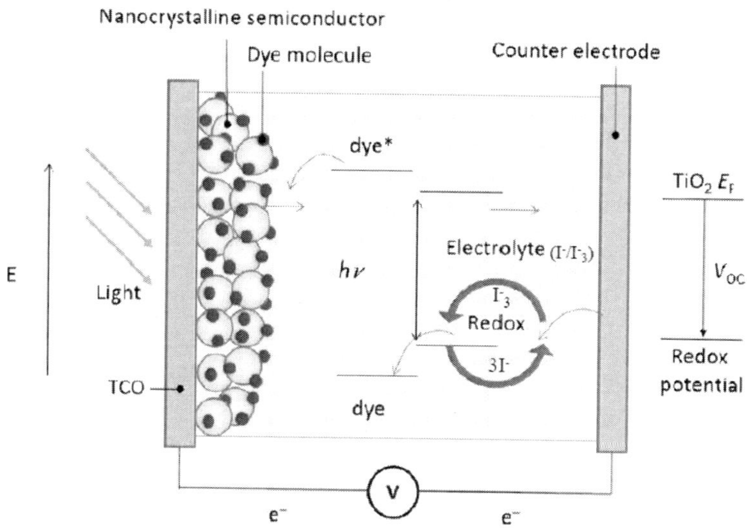

Figure 2. Schematic illustration of the DSSC [24].

The working principle of the dye-sensitized solar cells is based on the absorption of sunlight. First, dye molecules are excited from the highest occupied molecular orbital (HOMO) to the lowest unoccupied molecular orbital (LUMO) level by absorbing light. Excited dye molecules are injected into the conduction band of the semiconductor rapidly and then transferred to the counter electrode through the external circuit. The cathode inserts electrons into the electrolyte to generate oxidized dye and return it to the neutral state. Finally, triiodide reduces to iodide at the counter electrode and the circuit is completed [25-27].

During these processes, discordant reactions occur leading to charge recombination and this is disadvantageous for DSSCs. These reactions include electron recombination by donating electrons to the oxidized sensitizer and electron recombination by donating electrons to the oxidized redox couple at the TiO_2 surface [28].

Transparent conductive oxide (TCO), which is used as a substrate in the manufacturing of solar cells, provides good support for the photoanode and the counter electrode [29]. TCO must have high electrical conductivity, transparency, and flexibility leading to favorable solar cell performance. The most common transparent conductive material is indium tin oxide (ITO) [30]. The high cost and scarcity of indium have limited its use in DSSCs. Therefore, the application of cheaper and more flexible materials such as fluorine tin oxide (FTO) [31], gallium zinc oxide (GZO) [32], and aluminum zinc oxide (AZO) [33] can be an appropriate alternative to ITO. Although these materials are less transparent than ITO, they are more stable and have lower manufacturing costs.

In DSSCs, a counter electrode is applied to transfer electrons from the external circuit to the electrolyte, complete the circuit and generate electricity [34]. It also plays an important role as a catalyst for reducing triiodide ions. Low surface resistance, desirable catalytic properties, and resistant against electrolyte corrosion are the characteristics of an ideal counter electrode [35]. Platinum (Pt) is commonly utilized as the counter electrode in DSSC due to proper conductivity, high catalytic activity, and large surface area. Because of their scarcity, high cost, and instability in electrolyte and production of platinum iodide (PtI_4) salt, other materials as a suitable alternative to Pt, including carbon nanotube, graphene, graphite, and conductive polymers such as polypyrrole (PPy), polyaniline (PANI) and polythiophene (PT) are applied [36, 37].

Semiconductor provides an appropriate surface for the chemisorption of dye molecules and plays a crucial role in the transfer of the excited electrons from the dye to the counter electrode [38]. Basic criteria for selecting efficient semiconductors include excellent chemical stability under visible radiation, easy synthesis, biocompatibility, porosity, and high surface area for further adsorption of dye molecules [39]. Various nanoparticles such as titanium dioxide (TiO_2) [40], zinc oxide (ZnO) [41], cadmium selenide (CdSe) [42], lead sulfide (PbS) [43], tin oxide (SnO_2) [44], and niobium oxide (Nb_2O_5) [45] have been applied as semiconductors in solar cells. Among them, TiO_2 has attracted more attention due to the mentioned characteristics as well as its abundance, non-toxicity, and low cost.

Another main component of DSSCs is the electrolyte, which major function is to inject electrons from the counter electrode to the oxidized dye, as well as tuning the regeneration of dye molecules [46]. The high electrical conductivity, low viscosity, small dielectric constant, and great thermal stability are key factors of a superior electrolyte [47]. Mainly, three types of electrolytes have been used in DSSCs; liquid electrolyte, solid-state electrolyte, and quasi-solid electrolyte [48-50].

The liquid electrolyte consists of an additive, solvent, and a redox couple such as iodide/triiodide (I^-/I^-_3), bromide/tribromide (Br^-/Br^-_3), thiocyanate/ trithiocyanate ($SCN^-/(SCN)_3^-$, cobalt(II)/cobalt (III) (Co^{2+}/Co^{3+}), and ferrocene/ferrocenium (Fc/Fc^+). The I^-/I^-_3 redox couple is more advanced in comparison with other couples, due to the acceptable kinetic properties, fast dye regeneration, low recombination rate of electrons, easy preparation, and high conversion efficiency [51, 52].

Low chemical stability, leakage of solvent, and corrosion of the counter electrode are the liquid electrolyte disadvantages that solid-state electrolytes such as hole transfer materials (HTMs) including CuI, CuBr, CuSCN and 2,2′,7,7′-tetrakis (*N,N*-di-*p*-methoxyphenylamine) 9,9′-spirobifluorene (OMeTAD) have been adopted to overcome these problems. But inferior interfacial contact and less ionic conductivity lead to poor efficiency of solar cells that are not suitable for use in DSSCs [53].

Polymer gels such as polyethylene oxide, polyvinyl acetate, and polyvinylidene fluoride can be mentioned as quasi-solid electrolytes [54]. However, these types of electrolytes show lower performance of DSSCs than liquid electrolytes but they provide better long-term stability and interfacial contact, which are favorable for a proper DSSCs.

The sensitizer is the most prominent compound in DSSCs having a key role in solar light capture and power conversion efficiency [55]. The major properties associated with the remarkable sensitizer include great thermal stability, strong absorption in the visible region and on the TiO_2 surface, and good solubility in the solvent. Moreover, the alignment of LUMO and HOMO matches with the edge of the conduction band of TiO_2 and redox potential of electrolyte, respectively [56]. Therefore, selecting the efficient dye is an important issue.

Disparate architectures of the sensitizer including D-π-A [57], D-D-π-A [58], D-A-π-A [59], D-π_1-π_2-A [60], and D-π-A-π-A [61] have been expanded to improve intermolecular charge transfer (ICT) properties and performances of this generation of the solar cells. D as the electron donor such as triphenylamine [62], carbazole [63] and phenothiazine [64], π as the conjugated bridge such as benzene [65], thiophene [66] and furan [67], and A as the electron acceptor such as cyanoacetic acid [68], cyanoacrylic acid [69], and benzoic acid [70] were recommended. It is worth mentioning that for proper charge separation, HOMO and LUMO levels should be distributed at the donor and the acceptor groups, respectively.

Based on reports, a wide variety of sensitizers including metal-organic dyes such as ruthenium [71], porphyrins [72], and phthalocyanine [73] complexes, natural dyes [74], and metal-free organic dyes [75] have been developed to be used in DSSCs.

Ru(II) polypyridyl complexes (N3, N719, and N749) have been introduced as the best light absorption materials (Figure 3) [76, 77]. These compounds absorb in a wide range of visible light wavelengths and represent high quantum efficiency, which is originated from the presence of different ligands and π-conjugated systems [77]. But due to problems such as the high cost of Ru, long synthesis, and adverse environmental risks, they have been replaced by new compounds.

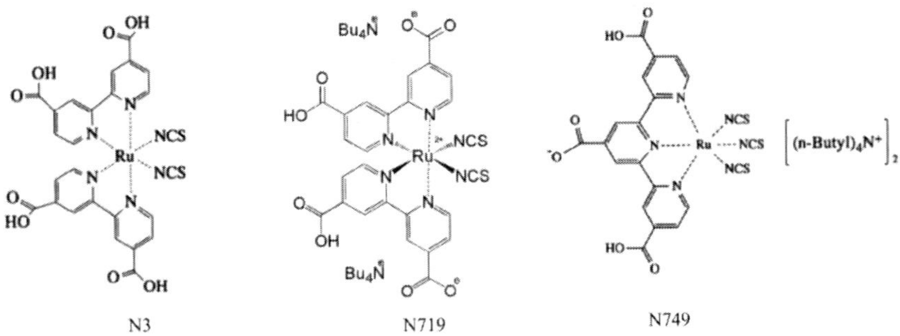

Figure 3. The structures of Ru-based photosensitizers of N3, N719, and N749 [76, 77].

Porphyrin compounds consist of four pyrrole rings and methine groups (=CH-) that exhibited more than 10% power conversion efficiency [78]. Porphyrins have absorption spectra in the Q-band (550-700) and Soret or B-band (350-500) area, which is more intense in B-band in comparison with Q-band [79]. Moreover, high molar absorption coefficient, structural flexibility, efficient electron transfer, and relatively low cost make them considerable sensitizers in DSSCs [80]. A series of these compounds are shown in Figure 4.

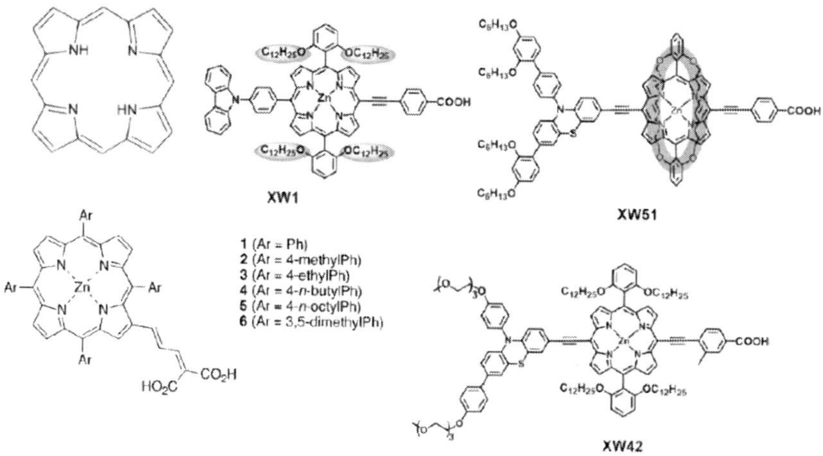

Figure 4. Molecular structures of porphyrin and derivatives [81, 82].

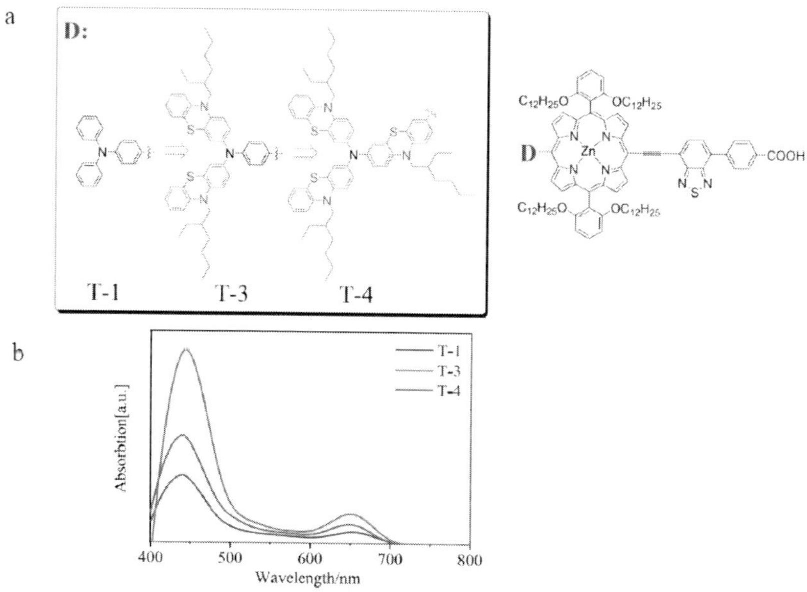

Figure 5. a) Chemical structures of the Zn-porphyrin dyes. b) The absorption spectra dyes in TiO_2 film [84].

For example, five porphyrin dyes derived from SM315 having a backbone of the D-π-A structure with different electron donors was evaluated for application in dye-sensitized solar cells [83]. All dyes showed a broad absorption band in both Soret and Q regions. Also, S101 dye exhibits an efficiency of 16.7%, which is greater than other dyes.

Shengzhong et al. examined the spectroscopy properties of phenylamine-based zinc-porphyrin dyes for highly efficient DSSCs [84]. The great molar absorption coefficient was observed, when phenothiazine units added to the triphenylamine donor of T-1 reference dye. Also, T-3 dye having three phenothiazine units represents the promising optical properties in comparison with T-2 dye having two phenothiazine. The chemical structures and absorption spectra of dyes are shown in Figure 5.

Phthalocyanines (Pcs) are composed of two-dimensional tetrapyrrolic macroheterocycles and 18 π-electrons [85]. Phthalocyanines (Figure 6) are similar to porphyrins in terms of spectroscopy properties (Soret and Q-bands). Also, good electrochemical stability, optical properties, and light-harvesting make them efficient candidates to be used in DSSCs [86].

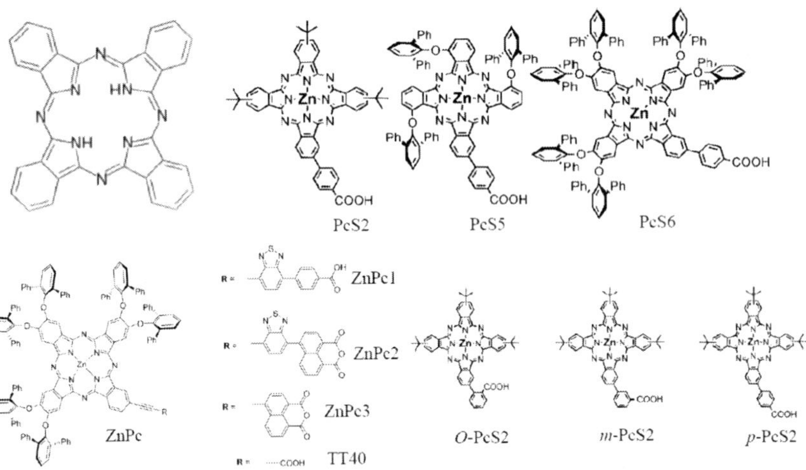

Figure 6. Molecular structures of phthalocyanine and derivatives [87-89].

In 2020, a comprehensive study was performed on the physicochemical properties of two zinc phthalocyanine sensitizers (PcS27 and PcS28) and the effect of cyclic alkoxy chains was investigated on the performance of DSSCs [90]. According to the results, PcS27 with 3-methylbutoxy chains exhibited 6.0% conversion efficiency and more favorable photovoltaic parameters such as open-circuit voltage (V_{OC}) and short circuit density (J_{sc}) in comparison with PcS28.

There is an important point that porphyrins and phthalocyanines aggregate in solutions leading to unfavorable charge recombination [91]. This behavior is originated from their planar configuration. Therefore, natural dyes were investigated.

Natural Dyes

The most popular natural dyes used in solar cells include chlorophylls, carotenoids, flavonoids, and anthocyanins [92]. These dyes unlike synthetic dyes are inexpensive, abundant, of high purity grade, environmentally friendly, and have strong absorption in the visible region, as well as they could be easily extracted from natural products such as fruits, flowers, and leaves [93, 94]. Some of the natural sensitizers applied in DSSCs within relative photovoltaic parameters are summarized in Table 1.

Table 1. Different natural dyes and their photovoltaic parameters

Plant source	Structural class	V_{OC}(V)	J_{SC} (mA.cm^{-2})	FF	η_0 (%)	Ref.
mangosteen pericarp	Anthocyanin	0.44	3.63	0.60	0.97	[95]
roselle	Anthocyanin	0.27	3.09	0.56	0.48	[95]
red cabbage	Anthocyanin	0.51	1.4	0.55	0.39	[95]
Thai berry	Anthocyanin	0.42	1.08	0.59	0.27	[95]
black rice	Anthocyanin	0.42	1.52	0.36	0.23	[95]
blue pea	Anthocyanin	0.48	0.61	0.43	0.13	[95]
Perilla	Chlorophyll	0.52	1.36	0.70	0.50	[96]
Petunia	Chlorophyll	0.61	0.85	0.61	0.32	[96]
China loropetal	Chlorophyll	0.51	0.84	0.63	0.27	[96]
Anethum graveolens	Chlorophyll	0.57	0.96	0.40	0.22	[97]
Arugula	Chlorophyll	0.59	0.78	0.42	0.20	[97]
Parsley	Chlorophyll	0.44	0.53	0.34	0.07	[97]
Capsicum	Carotenoid	0.41	0.23	0.63	-	[98]
Walnuts	Carotenoid	0.3	0.73	0.39	0.01	[99]
Achiote	Carotenoid	0.57	1.1	0.59	0.37	[99]
sour orange peel	Carotenoid	0.59	1.53	0.60	0.54	[100]
red grape	Anthocyanin	0.55	4.30	0.55	1.3	[101]
radish	Anthocyanin	0.49	5.10	0.59	1.47	[101]
red onion peel	Flavonoid	0.61	0.31	0.60	0.11	[101]
Cosmos sulphureus	Flavonoid	0.45	1.04	0.61	0.54	[102]
Botuje	Flavonoid	0.05	0.69	0.87	0.12	[103]
Punica granatum peel	Flavonoid	0.72	3.34	0.78	1.86	[104]

Chlorophylls as a green pigment consist of magnesium (Mg) as the metal center and four pyrrole rings [105]. Two main types of chlorophylls are chlorophylls 'a' and 'b', which absorb visible light [106]. The ability of light absorption in the visible region makes them ideal for DSSCs.

In 2020, a theoretical study in the gas phase and water was performed on the feasibility of chlorophyll-based natural dyes including Chl a, Chl b, Chl c2, Chl d, Phe a, Phe y, and Mg-Phe a [107]. The analysis of the LUMO and HOMO energy levels of the dyes indicated that the process of electron transfer and dye regeneration occurs efficiently in both phases and makes them favorable for DSSCs (Figure 7). The presence of the central Mg^{2+} and different functional groups on the chlorophyll structure has a key role in the energy of the frontier molecular orbitals and enhances the performance of the DSSCs.

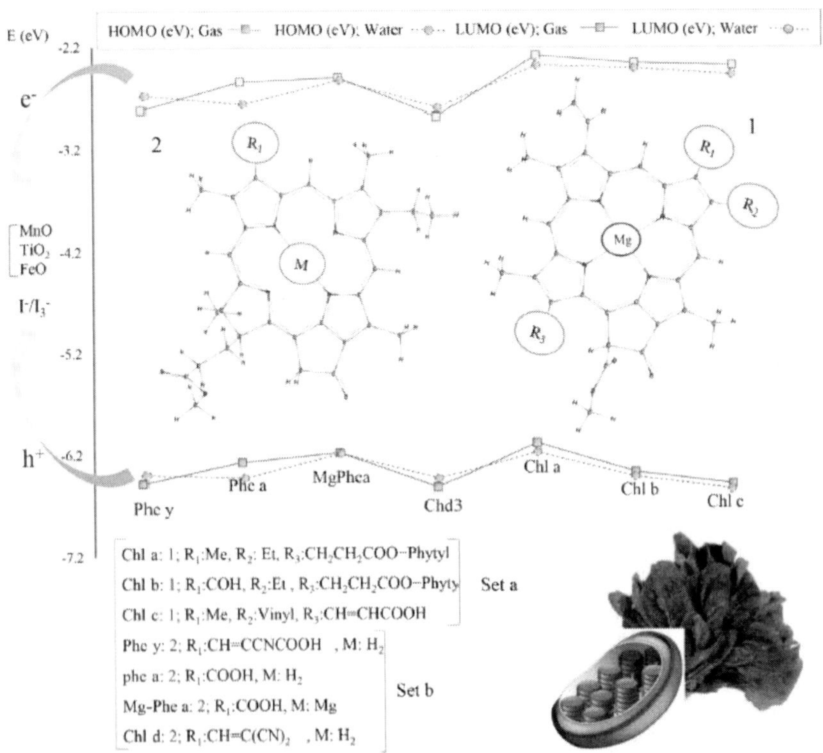

Figure 7. The energy diagram of the studied chlorophylls and the optimized structures within natural sources [107].

Another important group of natural dyes is carotenoid derivatives having C_{40} polyene backbone with 9-11 double bonds [108]. These dyes play an important role in light-harvesting and exhibit an efficiency of 2.6% [109]. Famous carotenoids are divided into carotene [110], and xanthophyll [111].

Yamazaki et al. examined photovoltaic and spectroscopic properties of carotenoid-based natural dyes including crocetin (8,80-diapocarotenedioic acid) and crocin (crocetin-di-gentiobioside) as photosensitizers in the DSSCs [112]. The results reveal that crocetin has the highest short circuit current density, efficiency (η_0), and molar absorption coefficient in comparison with crocin, which may be related to the presence of carboxylic groups.

Natural dye based on flavonoids have a fundamental C6-C3-C6 carbon skeleton with two phenyl rings joined together through three carbon bridges constituting a third ring [113]. The color of the flavonoids depends on the extent of C-ring oxidation. According to the reports, only some flavonoids can absorb light in the visible region [114, 115]. Therefore, the wavelengths of visible light that are absorbed by dye molecules have a key role to distinguish the different colors of the flavonoid.

In 2020, charge transfer descriptors of the flavonoid-based photosensitizers set a (luteolin, naringenin, genistein, apigenin, and chrysin) and set b (fisetin, morin, quercetin, isorhamnetin, and kaempferol) were investigated for use in the photovoltaic devices in the gas phase and water media by density functional theory (DFT) and time-dependent DFT (TD-DFT) approaches [116]. Also, different correlations between charge transfer parameters and intrinsic molecular descriptors were analyzed (Figure 8). A decrease in the chemical hardness (η_e) leads to an increase in the polarization and the excited-state lifetime of the flavonoids. Also, water as the medium improves the polarizability (α) and excitation lifetime of the dyes (τ) in comparison with the gas phase, due to the electrostatic interactions of water and natural dyes.

Anthocyanins are water-soluble pigments in plants similar to flavonoids in terms of the C6-C3-C6 carbon framework [117]. Cyanidin [118], delpinidin [119], malvidin [120], pelagonidin [121], peonidin [122], and petunidin [123] were introduced as best-known anthocyanins. One of the critical factors for determining the color of anthocyanins is pH. So that the acidic ($pH<7$) and alkaline ($pH>7$) conditions lead to red and blue-purple anthocyanins, respectively [124].

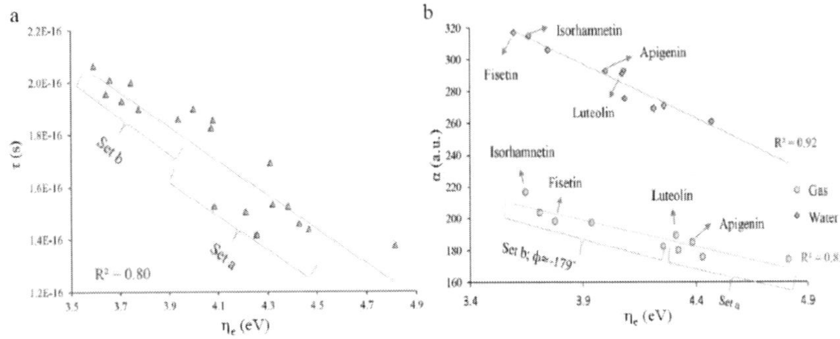

Figure 8. The linear correlation of the a) excited-state lifetime and b) polarizability of the dyes with the chemical hardness [116].

UV-Vis absorption spectra and performance of dye-sensitized solar cell based on crude and purified anthocyanins extracted from mangosteen pericarp, roselle, red cabbage, Thai berry, black rice, and blue pea were reported by Chaiamornnugool et al. [95]. Purified extracts of red cabbage, black rice, and blue pea shows a greater molar absorption coefficient than their crude extracts. Also, the obtained results indicate that the efficiency of purified dyes is higher in comparison with crude dyes except for mangosteen pericarp. Dyes extracted from roselle and red cabbage exhibited a greater efficiency (0.71%).

Metal-Free Organic Dyes

One of the pioneering compounds are metal-free organic dyes that are attractive for use in DSSCs due to their proper characterizes such as tunable optical properties, excellent performance, ease of synthesize, high structural flexibility, and molar absorption coefficient [125-127]. There are different metal-free organic dyes including triphenylamine [128], coumarin [129], carbazole [130], phenothiazine [131], quinoxaline [132], and indoline [133], which are of interest as considerable materials. Some metal-free organic dyes used in DSSCs within relative photovoltaic properties are represented in Table 2.

Triphenylamine-based organic dyes are widely used in DSSCs [139]. These compounds exhibit great electron-donating ability and hole transport properties, as well as, improved photovoltaic properties and high efficiency in DSSCs, because of the fine π-conjugated system [140, 141]. First, Yanagida et al. reported the efficiencies of 3.3% and 5.3% for two dyes based on triphenylamine [142].

Table 2. Different metal-free organic dyes and their photovoltaic parameters

Structure	**V_{OC} (V)**	**J_{SC} (mA.cm^{-2})**	**FF**	**η_0 (%)**	**Ref.**
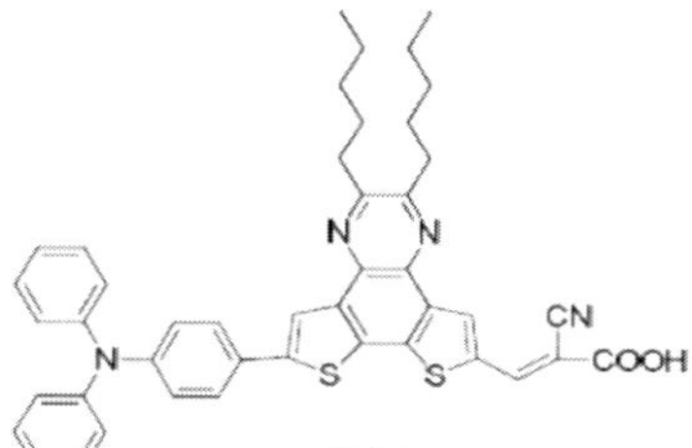 DQ1	0.74	13.11	0.65	6.37	[134]
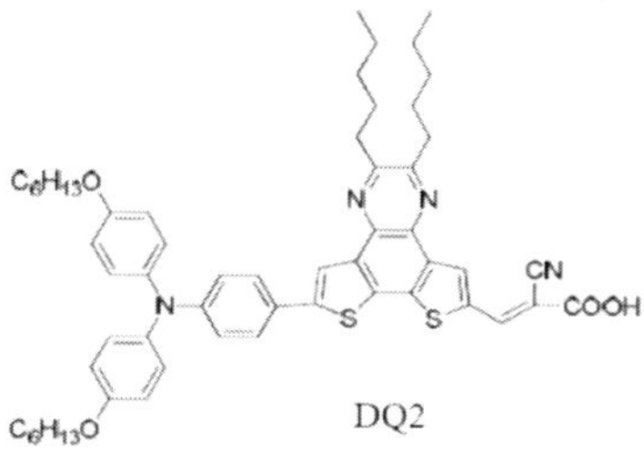 DQ2	0.66	9.31	0.65	4.03	[134]
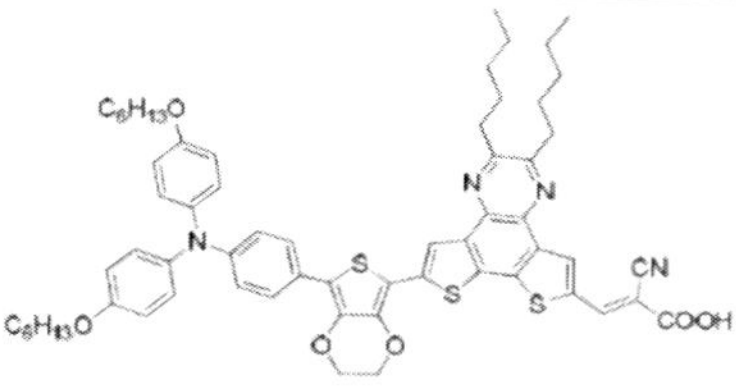 DQ3	0.65	9.01	0.68	4.01	[134]
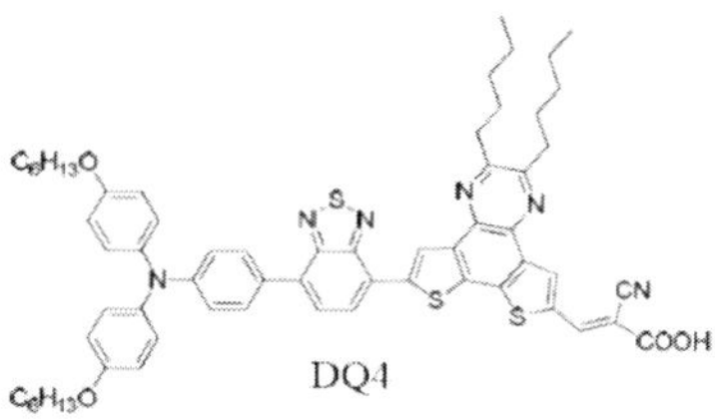 DQ4	0.72	14.51	0.65	6.78	[134]
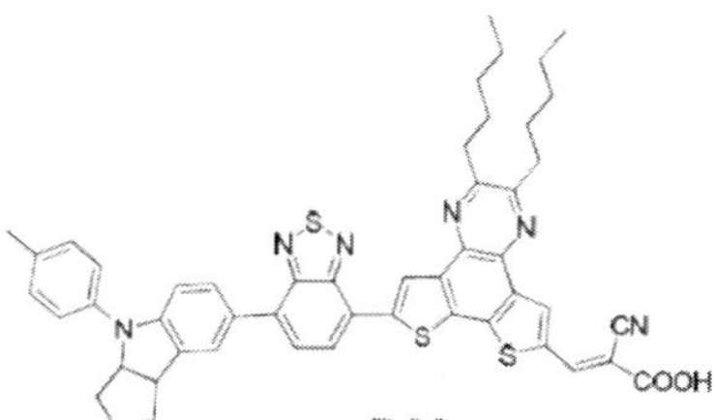 DQ5	0.68	17.61	0.59	7.12	[134]

Table 2. (Continued)

Structure	V_{OC} (V)	J_{SC} (mA.cm^{-2})	FF	η_0 (%)	Ref.
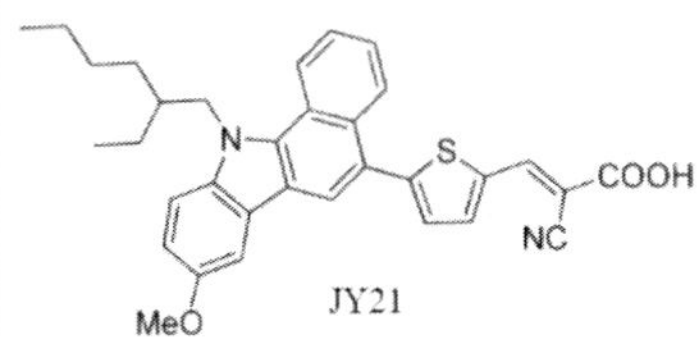 JY21	0.73	12.4	0.66	6.01	[135]
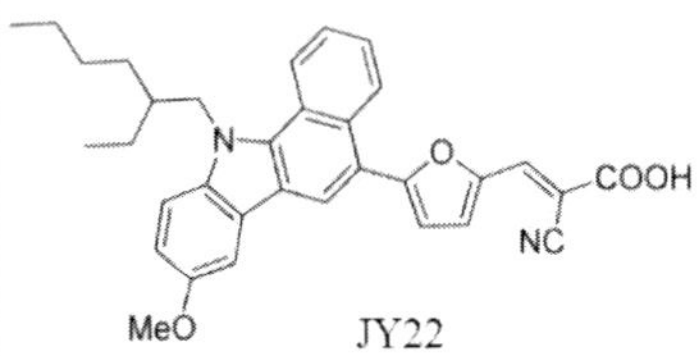 JY22	0.76	13.80	0.66	6.93	[135]
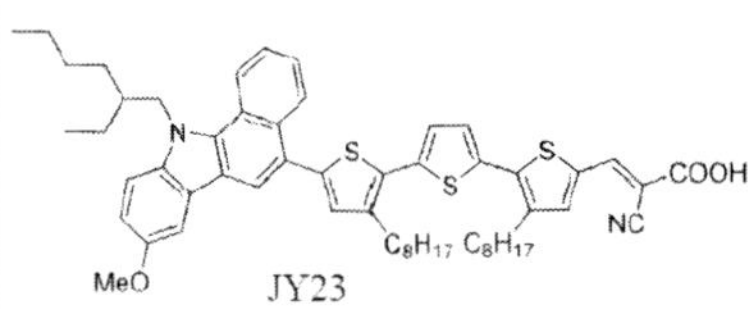 JY23	0.74	14.80	0.68	7.54	[135]
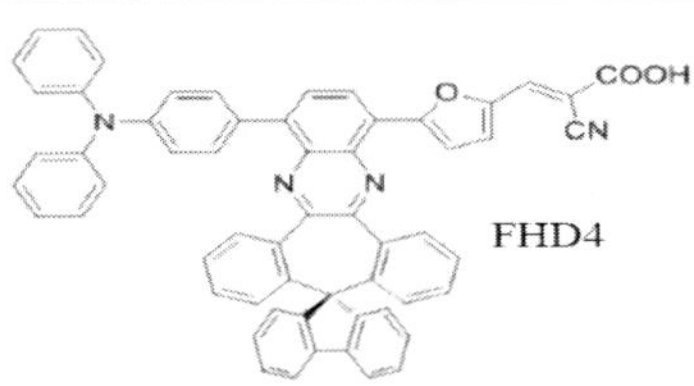 FHD4	0.69	9.33	0.71	4.61	[136]
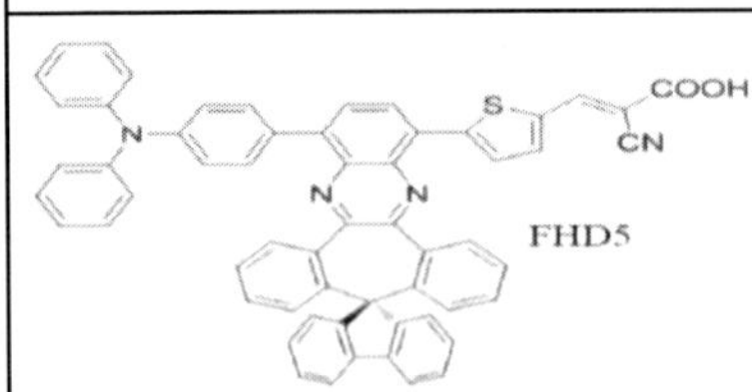 FHD5	0.65	7.46	0.65	3.18	[136]
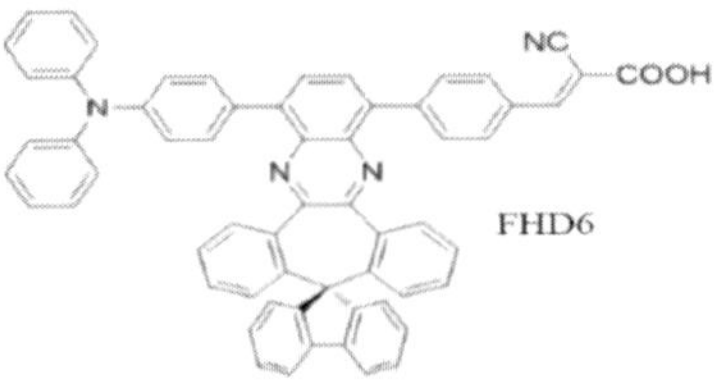 FHD6	0.71	6.97	0.73	3.59	[136]

Structure	V_{OC} (V)	J_{SC} (mA.cm^{-2})	FF	η_0 (%)	Ref.
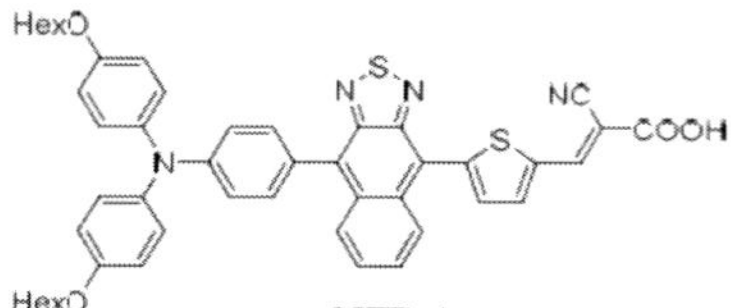 NTD-1	0.57	4.35	0.76	1.94	[137]
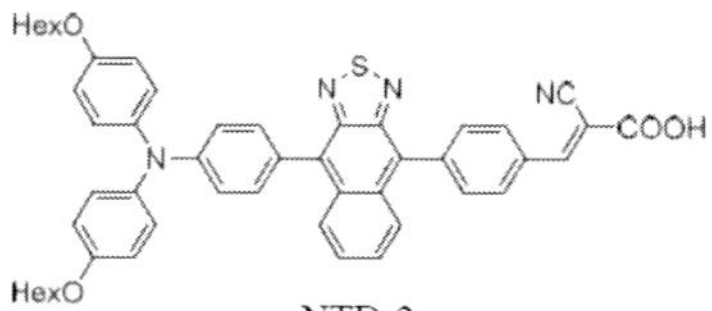 NTD-2	0.68	12.71	0.74	6.37	[137]
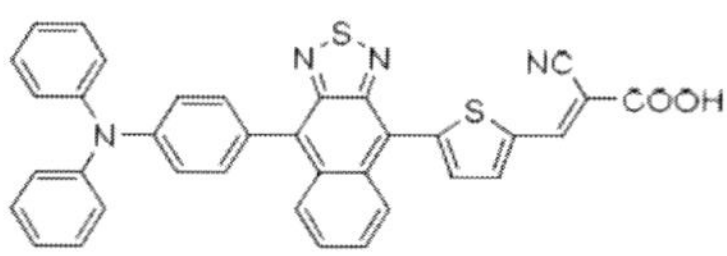 NTD-3	0.55	2.84	0.76	1.17	[137]
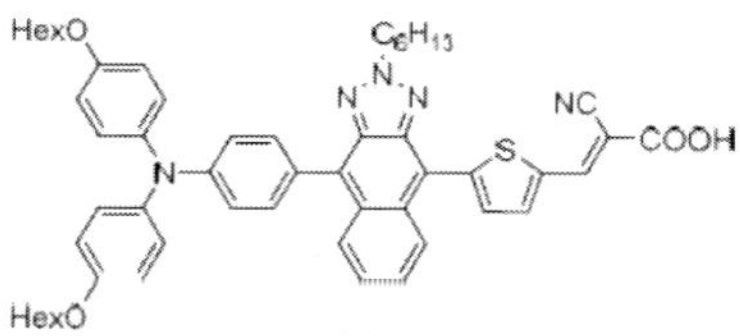 NTz-1	0.71	14.73	0.69	7.23	[137]
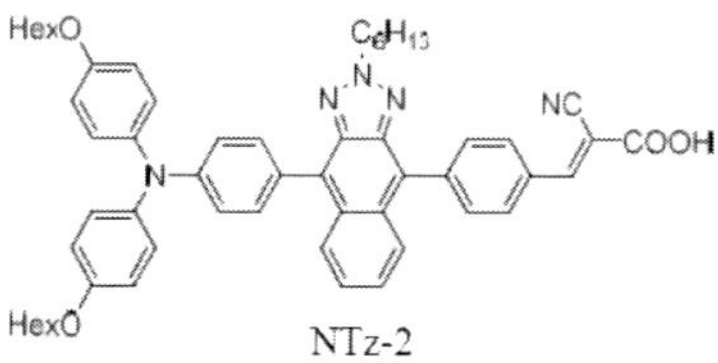 NTz-2	0.74	13.13	0.73	7.09	[137]
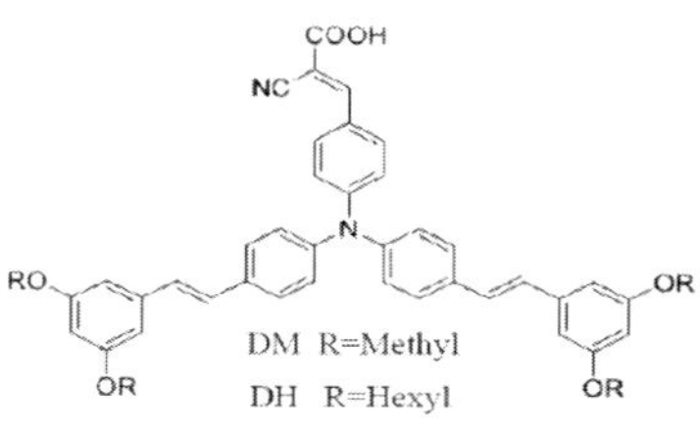 DM R=Methyl DH R=Hexyl	0.70 0.73	3.60 6.68	0.67 0.64	1.69 3.12	[138] [138]

Ferdowsi et al. evaluated the feasibility of using two metal-free organic dyes as the photosensitizers containing electron-rich triphenylamine, electron-deficient 4-(benzo[c] [1,2,5]thiadiazol-4-yl)benzoic acid, and various π-conjugated heteroaromatics such as thiophene (L156) and 4,8-bis(4-hexylphenyl)benzo[1,2-b:4,5-b']dithiophene (L224) [143]. Also, the effect of iodide/triiodide, $[Co(bpy)_3]^{3+/2+}$ (bpy=2,2'-bipyridine), and $[Cu(tmby)_2]^{2+/+}$ (tmby=4,4',6,6'-tetramethyl-2,2'-bipyridine) complexes as redox mediators and 18NR-T and 30NR-D transparent TiO_2 films on the efficiency of DSSCs was examined. Based on results, the highest power conversion efficiencies belong to L156-based DSCs with 30NR-D TiO_2 film and $[Cu(tmby)2]^{2+/+}$ complex as the electrolyte.

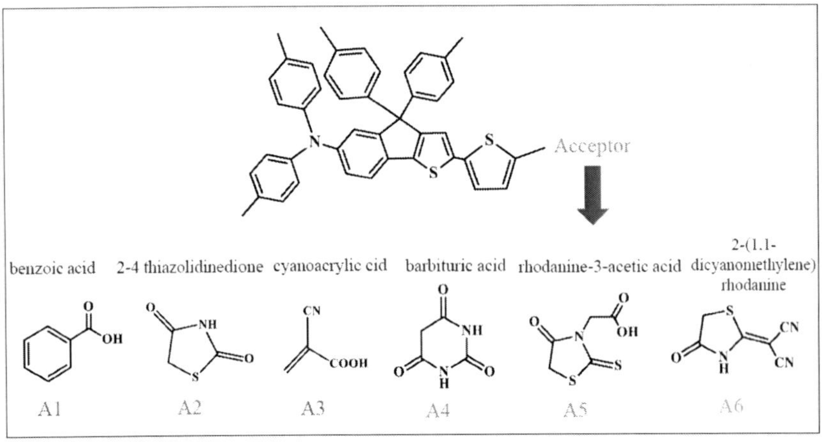

Figure 9. Molecular structures of the D-π-A dye molecules having various anchoring groups.

Slimi et al. reviewed the photovoltaic performance of six triphenylamine-based organic dyes with different electron acceptors (benzoic acid, 2,4 thiazolidinedione, cyanoacrylic acid, barbituric acid, rhodanine-3-acetic acid, and 2-(1,1-dicyanomethylene) rhodanine acid) by using DFT and TD-DFT methods (Figure 9) [144]. The results indicate that A6 dye having 2-(1,1-dicyanomethylene) rhodanine acid acceptor unit shows better results and performance in comparison with other dye, due to excellent light-harvesting efficiency (LHE), and lower electron injection driving force (ΔG_{inj}).

Another metal-free organic dye is carbazole that has advantages such as efficient thermal stability, charge carrier mobility, electron/donor group, and quantum yield. These features make them favorable to be used in DSSCs [145,

146]. Computational studies have been applied to estimate the optoelectronic properties of carbazole-based organic dyes with different π-spacers by Ennehary et al. [147]. They reported that the presence of pyrrole unit as the π-spacer enhances the open circuit-voltage and electron transfer performance in comparison with other π-spacers.

Also, in 2018, new carbazole-based metal-free organic dyes as photosensitizers with D-D-π-A architecture have been developed [148]. In this model, carbazole as the donor electron, 4-methoxyphenyl as the auxiliary donor, and cyanoacetic acid, 2,4-thiazolidinedione, and barbituric acid as acceptor groups have been used. Among these dyes, D_1 exhibits a proper power conversion efficiency (2.20%), fill factor (67.7%), and short circuit current density (5.22 $mAcm^{-2}$). Also, these compounds have been applied as co-sensitizers with NCSU-10 dye, in which D_3 dye represents better PCE (8.32%) when co-sensitized with NCSU-10.

Phenothiazine-based organic dyes are introduced as remarkable sensitizers in DSSC [149, 150]. Phenothiazine is a butterfly-shaped molecule that exhibits good optical and hole-transporting properties and high thermal/photochemical stability. Also, the attendance of sulfur and nitrogen heteroatom improves the electron-donating ability of this compound [151]. In 2007, the first phenothiazine-based DSSCs reported by Tian et al. exhibited the final efficiency of up to 5.5% [152].

For example, a theoretical study on some phenothiazine-based organic dyes with various π-spacers shows that the insertion of the thiophene-benzothiadiazole-thiophene unit as the π-spacer can improve light harvesting efficiency, dipole moment (μ), and broad absorption spectrum [153]. Also, in 2020, an accurate estimation has been performed on photovoltaic properties of a series of phenothiazine-based organic dyes (2a, 2b, and 2c) and compared with N719 reference dye [154]. The results reveal that the studied dyes have good photovoltaic characteristics including open-circuit voltage, fill factor, and power conversion efficiency in comparison with reference N719 dye.

Quinoxaline derivatives have been extensively employed in DSSCs [155, 156]. Their strong electron-deficient property, good thermal stability, availability, and red shift in the absorption spectra make them ideal for application.

A joint experimental and computational work was performed on the quinoxaline-based organic dyes (WS-82-84) to evaluate the effect of different photovoltaic and spectroscopy parameters [157]. These dyes exhibit proper FMO distribution lead to facilitate charge transfer from donor to acceptor.

Also, some parameters such as molar extinction coefficient, incident photon to current efficiency (Figure 10) and short-circuit current density are more favorable in the case of WS-84 having with long alkyl chains functionalized cyclopentyldithiophene unit.

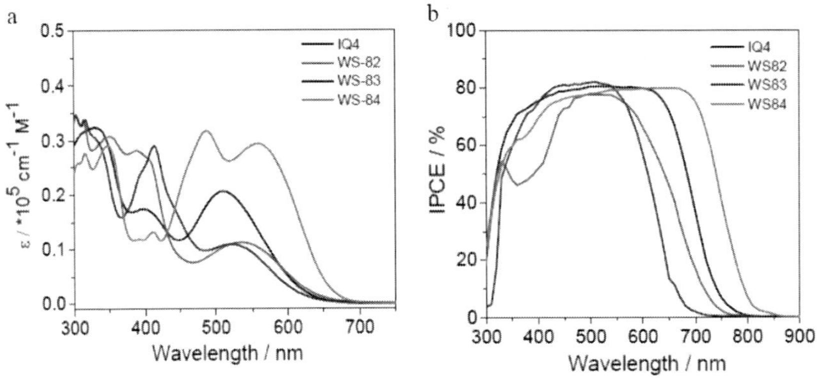

Figure 10. a) The absorption spectra of WS-82, WS-83, and WS-84 dyes. b) IPCE spectra of the designed dyes [157].

Also, in 2018, a theoretical insight has been applied on some quinoxaline-based organic dyes, which triphenylamine as the donor, indoloquinoxaline as a subordinate acceptor, thiophene/furan as the π-bridges, and COOH/DCRD as the anchoring groups were used [158]. An improvement is observed in the charge transfer dynamic and kinetic of these dyes when cyanoacrylic acid is utilized as the anchoring group and thiophene as the π-spacer.

Computational Chemistry

For better modeling of solar cells, access to molecular information, and improvement of efficiency over time, theoretical studies have become very important [159, 160]. For this purpose, different software such as Gaussian [161], AFORS-HET [162], SCAPS-1D [163], PC1D [164], SIESTA [165], MULTIWFN [166], MATLAB [167], and QUANTUM ESPRESSO [168] are used.

One of the most important applications is Gaussian for evaluating the solar cells properties, the optimal structure of molecules, bond energies, and absorption spectra by using the laws of quantum and classical mechanics [161].

AFORS-HET software is proper for simulating the solar cell characteristics such as internal and external quantum efficiency (IQE and EQE), current-voltage (V-J) diagrams, and local cell currents [162].

SCAPS-1D is a powerful program for simulating photovoltaic devices, especially thin-film solar cells such as structures of the CdTe, CIGS, and CulnSe2 family [169]. Also, this software is suitable for analyzing the charge recombination process and electrical properties of the solar cells [163].

Another high-speed program for solar cells simulation is PC1D, which plays a key role in obtaining voltage-current parameters and performance of the solar cells [164].

SIESTA is computational chemistry software applied to calculate electronic structure, atomic forces, electron density, simulation of solids, and study of the absorption of photosensitizers on the semiconductors [165].

MULTIWFN is a great tool for examining wave functions and charge transfer indices such as the charge transfer distance (D_{CT}), the overlap of the electron-hole distribution (S), and scale of e-h separation (t) [166].

MATLAB is a mathematical modeling software applied for the development and simulation of photovoltaic systems and estimation of the characteristics of solar cells [167]. Also, this software is used to study the electronic structure, voltage, and power conversion efficiency of photovoltaic processes.

To investigate the electronic structures, the density of states, structural optimization, spectroscopic properties, band structure, interatomic forces, and quantum simulation of materials, QUANTUM ESPRESSO is a superior tool [168]. Also, ADF [170], Tinker [171], Spartan [172], and GAMESS [173] are proper programs for prediction molecular structure optimization, vibration frequencies, and molecular properties such as dipole moment and polarizability.

According to accuracy and time spent on calculations, appropriate computational methods is chosen for studying solar cells. DFT is one of the efficient methods in quantum mechanics that is applied for optimizing geometries of photosensitizers and investigating the electronic structure in the ground state [174]. This method is based on Hohenberg-Kohn's theory that is a good way to determine chemical properties and calculate all the system characteristics in terms of electron density.

To have an insight into the photovoltaic processes of the DSSCs in the excited states, time dependent-density functional theory (TD-DFT) is recommended [175]. Also, this method has the main role for achieving molecular information of the photosensitizers such as excitation energy, oscillating strength, and simulation of the absorption spectra.

The natural bond orbital (NBO) method, proposed by Löwdin [176], is a comprehensive computational method for understanding intermolecular interactions, donor-acceptor interactions, bond order, energy levels of the FMO, delocalization of the electron density [177], and quantum chemistry reactivity descriptors such as the electronic chemical hardness (η_e), electronic chemical potential (μ), work function (WF), and electrophilicity (ω) (Eqs. 1-4) [178-181] where E_{HOMO} is the energy of the highest occupied molecular orbital and E_{LUMO} is the energy of the lowest unoccupied molecular orbital. Also, IE and EA are equal to ($-E_{HOMO}$) and ($-E_{LUMO}$), respectively.

$$\eta_e = E_{LUMO} - E_{HOMO} \quad (1)$$

$$\mu = \frac{E_{HOMO}+E_{LUMO}}{2} \quad (2)$$

$$WF = \frac{(IE+EA)}{2} \quad (3)$$

$$\omega = \frac{\mu^2}{2\eta_e} \quad (4)$$

Polarizable continuum model (PCM) [182], isodensity polarizable continuum model (IPCM) [183], conductor like screening model (COSMO) [184], and conductor-like polarizable continuum model (CPCM) [185] are essential solvation models for the description of solvent effects, of which CPCM is more favorable solvation model in comparison with other models. Moreover, there are two main models for examination of the solvent effects, including the implicit solvent model [186] and the explicit solvent model [187]. The implicit solvent model is computationally cheaper, of lower accuracy, and is using solvent as a continuous medium but in the explicit solvent model, solvent molecules are located around the solute. This model has high authenticity and precision but is computationally expensive model.

According to the literature, the utilization of the suitable functional and basic set is important for evaluating the solar cell properties. One of the successful functionals is the exchange-and-correlation (XC) functionals [188] that consist of three major categories:

Local density approximation (LDA) such as PW92 [189], VWN [190], and Xonly [191].

Generalized gradient approximation (GGA) such as PBE [192], BLYP [193], XLYP [194], BEEx [195], LB94 [196], PW91 [197], and BP86 [198] and meta-GGA such as TPSS [199], M06-L [200], MS0 [201], MS1 and MS2 [202].

Global hybrid functionals with a constant amount of Hartree-Fock exchange such as B3LYP [203], PBE0 [204], WB97XD [205], TPSSh [206], M06-2X [207], and BMK [208] and range-separated/long-range-corrected hybrid functionals with a 20-25% of the Hartree-Fock exchange such as CAM-B3LYP [209], and LC-WPBE [210].

In order to describe molecular orbitals, different basis sets are applied, such as 6-31++G(d,p), 6-311++G(2d,2p), and LANL2DZ [211, 212].

One of the most important benchmarks for determining the performance of DSSCs is sunlight-to-electricity conversion efficiency (η_0) (Eq. 5) [213], which is influenced by the open-circuit voltage (V_{OC}), short circuit current density (J_{SC}), fill factor (FF), and incident solar power (P_{in}).

$$\eta_0 = \frac{J_{SC} \cdot V_{OC} \cdot FF}{P_{in}} \tag{5}$$

FF can be defined as the maximum power expected for the solar cell (P_{max}) by the output of J_{sc} and V_{oc}, obtained from Eq. 6 [214].

$$FF = \frac{P_{max}}{J_{SC} V_{OC}} \tag{6}$$

V_{OC} is characterized as the potential measured when the current in the cell is zero and the external charge is not joined, which is calculated by Eq. 7 [215].

$$V_{OC} = E_{HOMO}^{Donor} - E_{LUMO}^{Acceptor} \tag{7}$$

J_{SC} is described by the interaction between the sensitizer and semiconductor, absorption coefficient, and molecular structure of sensitizer that can be estimated by Eq. 8 [216].

$$J_{SC} = \int_{\lambda 1}^{\lambda 2} \frac{SI}{hc/e\lambda} LHE(\lambda)\Phi_{inj}\eta_{coll} d\lambda \tag{8}$$

J_{SC} is significantly dependent on the light-harvesting efficiency (LHE). SI, h, e, c, Φ_{inj}, and η_{coll} are the solar radiation intensity, the Planck constant, the unit charge, the speed of light in vacuum, the electron injection efficiency, and the charge collection efficiency, respectively. Φ_{inj} and η_{coll} can be evaluated by Eqs. 9 and 10, respectively [217].

$$\Phi_{inj} = 1/\left(1 + \frac{\tau_{ing}}{\tau_{relax}}\right) \tag{9}$$

$$\eta_{coll} = 1/\left(1 + \frac{\tau_{trans}}{\tau_{rec}}\right) \tag{10}$$

where τ_{inj} is the electron injection lifetime from the excited dye to the TiO_2, τ_{relax} is the relaxation lifetime for excited state of the dye in solution, τ_{trans} is the electron transport time from the CB of TiO_2 to the electrode, and τ_{rec} is the electron-hole recombination lifetime. η_{coll} can be considered to be constant in the systems whose components are the same, while different only in sensitizers. According to Marcus' theory, τ_{inj} and τ_{rec} are equal to electron injection rate (k_{inj}) and electron recombination rate (k_{rec}), respectively. (Eq. 11) [218].

$$k_{inj/rec} = A\sqrt{\frac{\pi}{\hbar\lambda\, k_B T}}\exp(-\beta r)\exp\left[\frac{-(\Delta G^0 + \lambda)^2}{4\lambda\, k_B T}\right] \tag{11}$$

where A is constant (2.5×10^{-2}), β is the attenuation factor ($0.7 Å^{-1}$), r is the electron transfer distance, k_B is the Boltzmann constant, T_{cell} is the temperature of the solar cell, ΔG^0 is the driving force for the reaction, and λ is the reorganization energy.

As mentioned earlier, LHE is the ability of solar energy capture by a sensitizer that can be expressed by Eq. 12 [219].

$$LHE = 1\text{-}10^{-f} \tag{12}$$

where f is the oscillator strength corresponding to the special wavelength that is determined from the TD-DFT analysis, directly. According to quantum mechanical theory, for transition probability between the initial and final states, a transition moment is examined and oscillator strength corresponding to this transition is obtained from Eq. 13 [220].

$$f_{ij} = \frac{8\pi^2 mcv}{3e^2 h} GM_{ij}^2 \tag{13}$$

where m is the electron mass, v is the wavenumber, G is the degeneracy of the final state, and M_{ij} is the dipole moment operator. Another important parameter in the DSSCs is the incident photon to current conversion efficiency (IPCE) that can be calculated by Eq. 14 [221].

$$IPCE = 1240.\frac{J_{SC}(\lambda)}{\lambda\, \varphi_{\lambda}} \tag{14}$$

where λ is the wavelength of the light-harvesting material and Φ(λ) is the incident photon flux. One of the effective parameters in the quantum efficiency is the absorbed photon to current conversion efficiency (APCE), which is known as the ratio between IPCE and LHE (Eq. 15) [222].

$$APCE = \frac{IPCE}{LHE} \tag{15}$$

Upon illumination, the electrons in the dye molecules are excited from the ground state to the excited state, producing a hole in the HOMO level. As a result, an electron-hole pair, exciton, is generated by a strong Coulomb interaction between this pair. Therefore, the analysis of the exciton behavior in the DSSCs is a very important issue. Exciton binding energy (EBE) that shows the charge separation in the solar cells can be calculated theoretically by Eq. 16 [223].

$$EBE = E_{gap.elec} - E_{0\text{-}0} \tag{16}$$

where $E_{0\text{-}0}$ is the vertical excitation energy and E_{gap} is the band gap energy of the dye. The ratio of the Coulomb and exchange interactions between the electron and hole is known as the material-dependent constant (α), obtained from Eq. 17 [224].

$$EBE = \frac{(\alpha-1)^2\, \mu_x\, e^4\, k^2}{2\alpha^2\, \hbar^2\, \varepsilon^2} \tag{17}$$

where ε is the dielectric constant of the donor component, $\hbar$ is the reduced Planck's constant, $k = (4\pi\varepsilon_o)^{-1} = 9 \times 10^9\ N\ m^2\ C^{-2}$, ε_o is the vacuum permittivity, and μ_x is the reduced mass of the exciton. In organic materials, the effective masses of electrons and holes are adequate to free-electron mass ($m_e^*=m_h^*=m_e \rightarrow \mu_x = 0.5m_e$).

The rate of the photon absorption and exciton dissociation rate for a singlet excitation are estimated according to Eqs. 18 and 19, respectively [225, 226].

$$R_a^S = \frac{4\,k\,e^2\,(E_{LUMO}-E_{HOMO})^3 a_x{}^2}{3c^3\varepsilon^{1.5}\hbar^4} \tag{18}$$

$$R_d = \frac{8\,\pi^2}{3\,\hbar^3\varepsilon^2 E_B}\,[E_{LUMO}^{D} - E_{CB}^{A} - E_B]^2\,(\hbar\omega_\vartheta)\,\mu_x\,a_x{}^2 \tag{19}$$

in which ω_ϑ is the frequency of the incident phonon to the dyes and a_x is the radius of Bohr's exciton that can be determined by Eq. 20 [227].

$$a_x = \frac{\alpha^2 \mu\,\varepsilon}{(\alpha-1)^2\,\mu_x} a_0 \tag{20}$$

where μ is the reduced mass of the electron in a hydrogen atom and $a_0 = 5.29 \times 10^{-11}$ m is the Bohr radius.

For better understanding of photovoltaic behavior of the DSSCs and improvement of photocurrent response of sensitizers, the nonlinear optical properties (NLO) analysis such as isotropic polarizability (α), dipole moment (μ), and total first-order hyperpolarizability (β_{tot}) (Eqs. 21-23) is recommended [228-230].

$$\alpha = \frac{1}{3}\left(\alpha_{xx} + \alpha_{yy} + \alpha_{zz}\right) \tag{21}$$

$$\mu = \sqrt{\mu_x^2 + \mu_y^2 + \mu_z^2} \tag{22}$$

$$\beta_{tot} = \sqrt{\left(\beta_{xxx} + \beta_{xyy} + \beta_{xzz}\right)^2 + \left(\beta_{yyy} + \beta_{yxx} + \beta_{yzz}\right)^2 + \left(\beta_{zzz} + \beta_{zxx} + \beta_{zyy}\right)^2} \tag{23}$$

where α_{xx}, α_{yy}, and α_{zz} are tensor components of polarizability, μ_x, μ_y, and μ_z are vector components of dipole moment, and β_{xxx}, β_{xyy}, β_{xzz}, β_{yyy}, β_{yxx}, β_{yzz}, β_{zzz}, β_{zxx}, and β_{zyy} are the tensor components of hyperpolarizability.

The investigation of the intra-molecular charge transfer characteristics is beneficial to enhance the energy conversion of the DSSCs. An improvement in these parameters provides valuable insight into the efficient design of molecules. Quantum of charge transfer (q_{CT}) is one of the main factors that is calculated by Eq. 24 [231].

$$q_{CT} = \int \rho_+(r)dr = \int \rho_-(r)dr \quad (24)$$

where $\rho_+(r)$ and $\rho_-(r)$ are increment and depletion of the density. Another character for describing the CT is the length of charge transfer (D_{CT}) and is theoretically expressed by Eq. 25 [232].

$$D_{CT} = |r_+ - r_-| \quad (25)$$

where r_+ and r_- are the barycenters of the density distributions, derived from Eqs. 26 and 27, respectively [233].

$$r_+ = (x_+, y_+, z_+) = \frac{1}{q_{CT}} \int r\rho_+(r)dr \quad (26)$$

$$r_- = (x_-, y_-, z_-) = \frac{1}{q_{CT}} \int r\rho_-(r)dr \quad (27)$$

The overlap of the e-h distributions (S) is a useful charge transfer parameter that can be calculated through Eq. 28 [234].

$$S_{\pm} = \int \sqrt{C_+(r)/A_+}\sqrt{C_-(r)/A_-}dr \quad (28)$$

where A is the normalization factor and C^+ (r) and C^- (r) are equal to ρ^+ (r) and ρ^- (r) integrals. Also, t denotes the scale of e-h separation calculated from the difference between D_{CT} and H (Eq. 29) [235].

$$t = D_{CT} - H \quad (29)$$

where H index is introduced as the half of the sum of the centroids axis along the electron transfer direction that can be determined by Eq. 30 [236].

$$H = \frac{\sigma_{+x} + \sigma_{-x}}{2} \tag{30}$$

where $\sigma_{\pm x}$ is the root-mean-square deviation for the hole and electron distribution along the x-axis. One of the primary parameters of CT is the lifetime of the excited state that can be calculated according to Eq. 31 [237]. A longer lifetime in the excited state facilitates charge transfer processes, which increases efficiency.

$$\tau = \frac{c^3}{2\,(E_{flu})^2 f} \tag{31}$$

where E_{flu} is the fluorescent energy.

The Effective Factors on the Efficiency of DSSCs

The issues of efficiency improvement and increasing light absorption in the DSSCs are important. Therefore, many studies have been done to achieve better efficiency of the DSSCs. The presence of strong electron donor/acceptor and extended π-conjugated system, increasing the number of anchor groups and the effect of the auxiliary donor/acceptor, substituted groups, and different solvents are important ways to attain this purpose. In this section, we examine each of these factors.

The Strong Electron Donor/Acceptor Effect

To study the effect of different electron donors and acceptors, a set of organic dyes is usually designed. For example, Zhang et al. investigated the effects of various donor species on the photovoltaic performance of three dyes BP (carbazole), BF (dibenzofuran), and BT (dibenzothiophene) [238]. Figure 11 illustrates the chemical structures of these dyes. These donor compounds contain heteroatoms of nitrogen, sulfur, and oxygen. Because they are rich in electrons, their presence increases the ability of electron-donating. According to the results, BF dye having a dibenzofuran donor displays a marked impact on the open-circuit voltage, fill factor, performance, and dye loading.

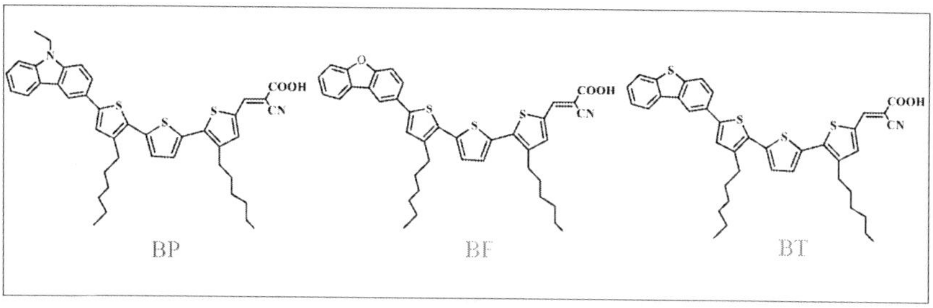

Figure 11. Molecular structure of BP, BF, and BT dyes.

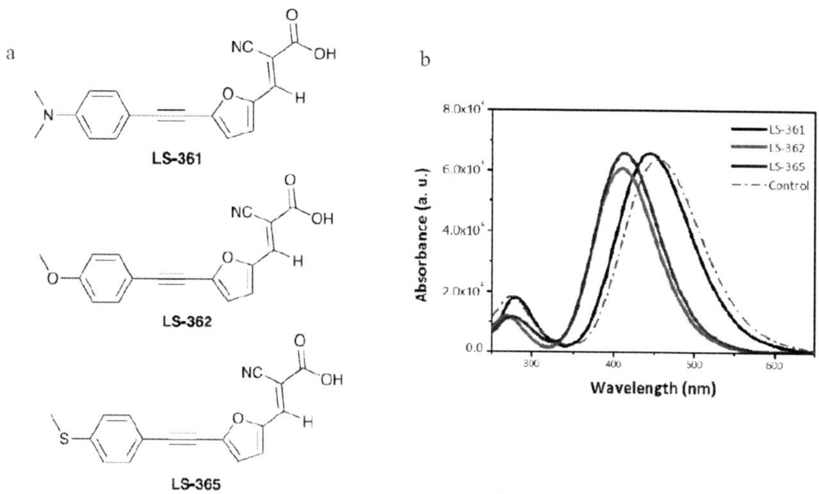

Figure 12. a) Chemical structures of the LS-361, LS-362, and LS-365 dyes. b) UV-Visible absorption spectrum of the dyes at CAM-B3LYP/6-311++G (d,p) level of the theory [239].

Serrano et al. introduced three metal-free organic sensitizers (LS-361, LS-362, and LS-365) with dimethylaniline (Me_2N-), 4-methoxyphenyl (MeO-), and 4-methylthiophenyl (MeS-) as electron donors attached to electron acceptor of cyanoacrylic acid through furylethynyl π-bridge (Figure 12a) [239]. The absorption spectra study of these sensitizers showed that the LS-361 with dimethylaniline electron donor has a red shift and higher molar absorption coefficient than LS-362 and LS-365 (Figure 12b). This behavior is due to the strong electron-donating of the dimethylaniline fragment. Also, in addition to the desired spectral properties, LS-361 sensitizer represents efficient photovoltaic processes in comparison with LS-362 and LS-365.

In 2020, the effects of varied arylamine donors (carbazole (TD1), phenothiazine (TD2), and triphenylamine (TD3)) for use in the solar cell were studied (Figure 13) [240]. Based on the obtained results, TD1 exhibits more interesting properties such as higher short circuit current density, final efficiency, and percentage of electron transitions between the molecular orbitals, longer excited-state lifetime, lower oxidation potential, and optical band gap energy. These great characteristics may originate from the higher electron enrichment of *N*-phenyl-carbazole in TD1 dye. Also, the co-sensitizations of TD1-TD3 with ruthenium complex N719 were investigated, yielding a power conversion efficiency up to 8.02%, higher in comparison with N719 (7.29%). In addition, incorporating TD-1 with N719 further enhances the efficiency of DSSCs (8.02%) which is higher than TD-2 (7.74%) and TD-3 (7.60%) dyes.

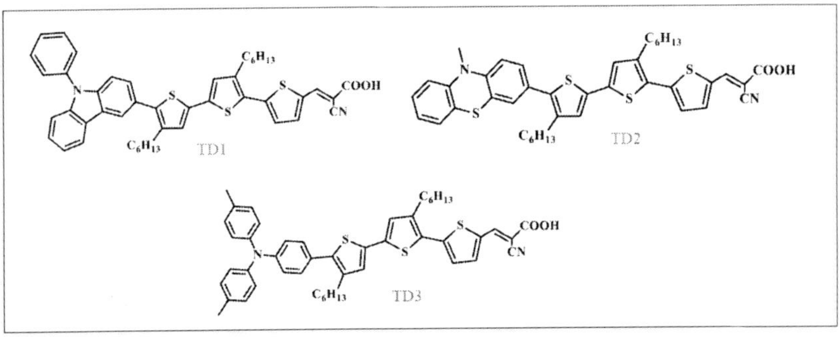

Figure 13. Molecular structures of TD1, TD2, and TD3.

In 2020, donating strength of donor fragment was evaluated in thirteen donor-π spacer (butadiene)-acceptor (cyanoacrylic acid) systems by DFT and TD-DFT approaches [241]. The analysis of linear correlations between electron-donating strength (ΔV_{ma}) and change in absorption maxima ($\Delta\lambda_{max}$) and eV_{oc} were performed. An enhancement in the donating strength improves the values of $\Delta\lambda_{max}$ and eV_{oc}. *N,N*-dialkylaniline, and julolidine are proposed as the most promising electron donor candidates be applied in the DSSCs due to the stronger electron-donating capacity of the donor moieties.

Yen et al. studied the optical properties and energy conversion efficiency of organic dyes based solar cell, An-1, An-2, An-3, and An-4 (Figure 14a) [242]. These dyes carry the electron rich of different amines are connected to cyanoacrylic acid group through to 2-(6-substituted-anthracen-2-yl)-thiophene π-conjugated bridge to model a D-π-A type structure. Considering the incident photo to current conversion efficiency and absorption spectra of the dyes in Figure 14b and c, An-3 represents a superior IPCE and molar absorption coefficient with respect to the other dyes, because of strong electron-donating ability of bis-diethylfluorenyl amino group. In addition, high short circuit current density, oscillator strength and low ground state oxidation potential introduced as other favorable properties of this dye (An-3).

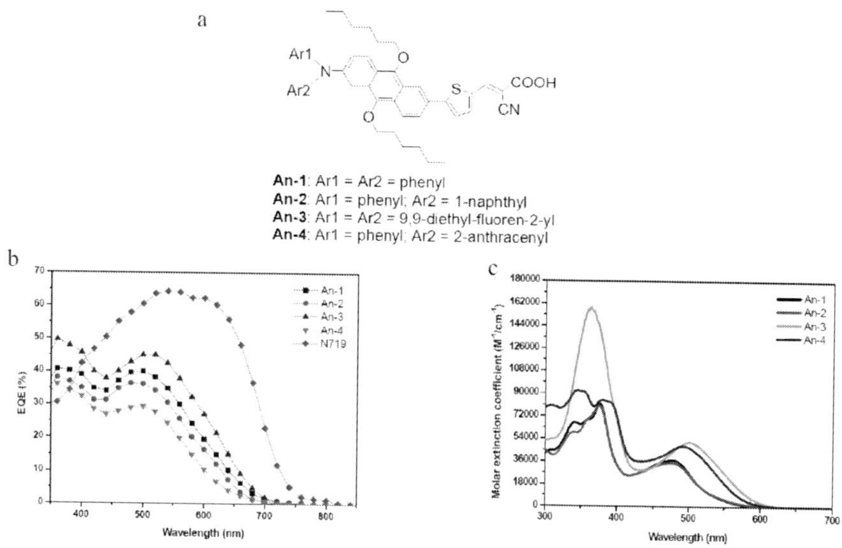

Figure 14. a) Molecular structures of the An-1, An-2, An-3, and An-4 dyes. b) IPCE spectra of dyes. c) Absorption spectra of the studied dyes in THF solutions [242].

Xiao et al. examined the application of different acceptors including thiophene cyanoacetic acid (D6), benzothiadiazole benzoic acid (D7), and benzothiadiazole thiophene cyanoacetic acid (D8) on the performance of dye-sensitized solar cells [243]. Based on data analysis, the employment of thiophene cyanoacetic acid represents more favorable photovoltaic properties such as open-circuit voltage, short circuit current density, efficiency, and light-harvesting ability. Also, the electron density of the HOMO is distributed on the indolo[3,2-b] carbazole, and LUMO is distributed on the electron acceptor. Figure 15 displays the optimized structures and frontier molecular orbitals distribution of these dyes.

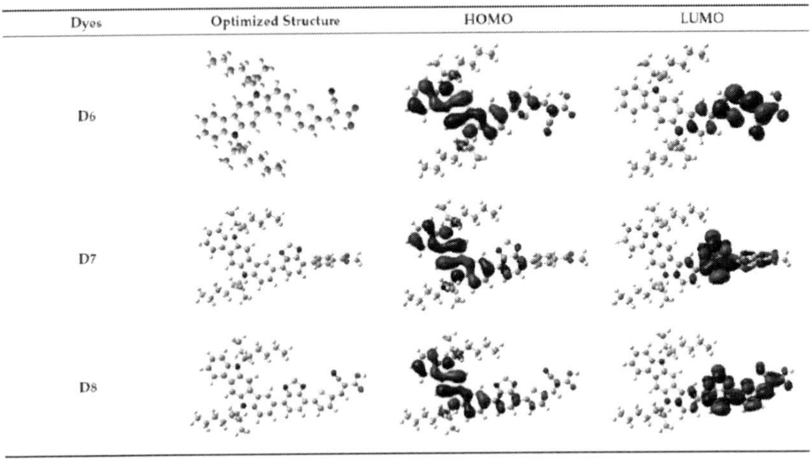

Figure 15. Optimized structures and electron distributions in the HOMOs and LUMOs of organic dyes [243].

In 2019, four diphenylamine-based organic dyes (DDC, DDR, DDH, and DDB) with different electron acceptors (cyanoacetic acid, rhodanine acetic acid, 4-hydrazinylbenzoic acid, and barbituric acid) were synthesized and designed by Kesavan et al. [244]. The experimental and theoretical results were in agreement and both results identified the cyanoacetic acid as the superior electron acceptor related to the strong electron-withdrawing ability of this acceptor. Moreover, the calculated electronic transition of these dyes shows that DDR having rhodanine acetic acid electron acceptor has a lower HOMO-LUMO gap in comparison with other electron acceptors (Figure 16) that lead to shift the wavelength to higher values. The predicted trends in the band gap energy, respectively, are as follow: DDH > DDC > DDB > DDR.

A theoretical investigation was performed on electronic features of organic dyes through the change in the anchoring groups (carboxylic acid, cyanoacrylic acid, pyridine-*N*-oxide, squaric acid, non-conjugated rhodanine-3-acetic acid, and conjugated rhodanine derivative) [245]. The insertion of squaric acid with an electron donor (triphenylamine) through π-bridge (3,4-ethylene dioxythiophene) improves charge transfer descriptors such as charge transfer distance, amount of charge transfer, the scale of e-h separation, and dipole moment. Also, this anchor group (squaric acid) improves J_{SC}, red-shift in absorption wavelength, light-harvesting efficiency, and sunlight irradiation.

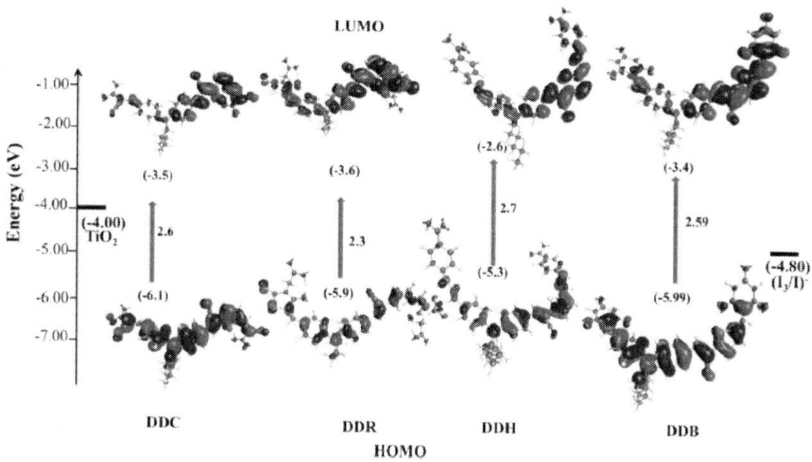

Figure 16. Calculated HOMO/LUMO levels and band gap energy of dyes DDC, DDR, DDH, and DDB [244].

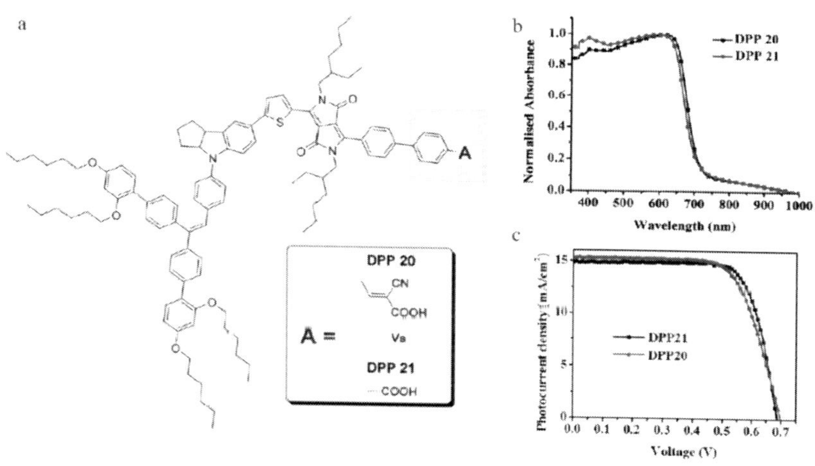

Figure 17. a) Chemical structures of DPP 20 and DPP 21 dyes. b) The absorption spectra of the dyes on the TiO_2 film. c) The curve of the voltage-current of studied dyes [246].

Prediction of spectroscopy properties and voltage-current behavior of diketopyrrolopyrrole sensitizers (DPP 20 and DPP 21) with various electron withdrawing groups (cyanoacrylic acid and carboxylic acid) was performed by Ganesan et al. (Figure 17a) [246]. According to the voltage-current plot in Figure 17b, the highest values of open-circuit voltage and short circuit current density belong to DPP 20 dye having cyanoacrylic acid electron acceptor.

Also, molar extinction coefficient of this dye (DPP 20) has increased slightly in comparison with DPP 21 having carboxylic acid electron acceptor (Figure 17c). These improved characteristics can be originated from strong electron withdrawing capacity of cyanoacrylic acid.

Quan et al. employed different electron donors and acceptors in dithienopyrrolobenzothiadiazole-based organic dyes (H1, H2, and H3) to investigate their effect on the performance of the DSSCs (Figure 18) [247]. Data analysis demonstrates that benzothiadiazole as the electron acceptor in H2 and H3 dyes enhances the light-harvesting efficiency and IPCE spectra, due to the high electron-withdrawing capacity of the benzothiadiazole. Also, according to the voltage-current curve of this dyes, dihexyloxy substituted triphenylamine as the electron donor improves the electrical properties of the solar cells based on H3 dye such as V_{OC} and J_{SC}.

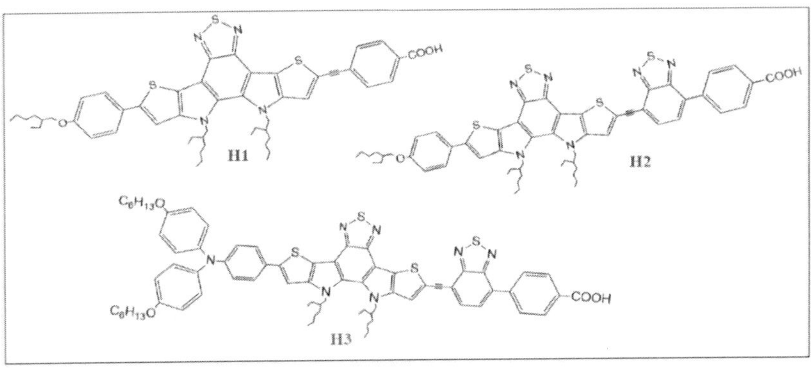

Figure 18. Molecular structures of H1, H2, and H3.

The Effect of the Extended π-Conjugated System

The extended π-conjugated system is a key factor for achieving the promising performance of the DSSCs and easy charge transfer from the donor to the acceptor. For example, a theoretical framework was established to investigate the π-spacers effect of the D-π-A phenothiazine-based dyes by El Mzioui et al. and compared to reference dye of JY56' having 4-isopropyl-4H-dithieno[3,2-b:2,3-d]pyrrole (DTP) π-spacer (Figure 19a) [248]. Alkoxyphenyl- and methyl-substituted phenothiazine (PTZ) moiety as an electron donor, cyanoacrylic acid as an electron acceptor, and cyclopentadithiophene (CDT), rhodanine (RD), and 2,7-dihydronaphtho[1,2-

d:5,6-d']diimidazole (NDI) as the π-spacers were utilized in this system. Based on the obtained results, the presence of 2,7-dihydronaphtho[1,2-d:5,6-d']diimidazole π-spacer in 3 dye leads to decrease driving force of electron injection (G_{inj}) and HOMO-LUMO gap. Also, this (3 dye) shows a broader absorption range and higher light-harvesting ability (Figure 19b and c) in comparison with other dyes. These favorable properties may originate from the extended π-conjugated system of NDI.

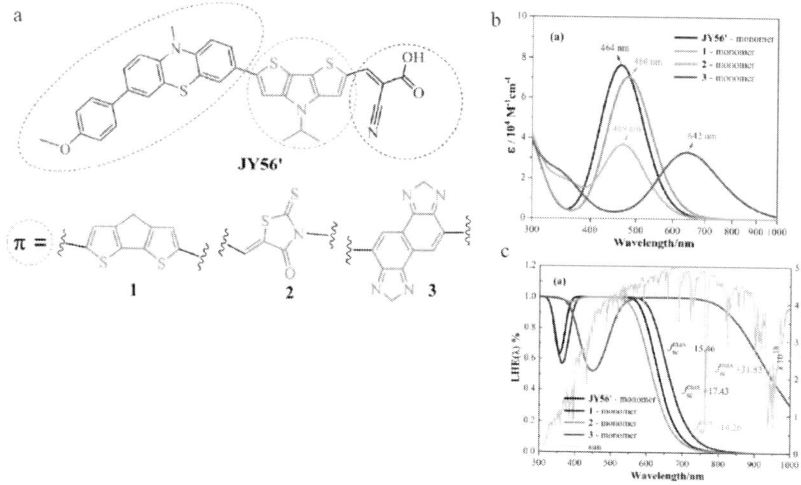

Figure 19. a) Chemical structures of studied dyes. b) UV-vis absorption spectra of dyes in dichloromethane solution. c) Simulated maximum short-circuit current density and light-harvesting efficiency curves of designed dyes [248].

Li et al. analyzed the effects of disparate π-spacers such as thiophene, furan, and terthiophene on the performance of pyrrolo[3,2,1-kl]phenothiazine-based D-π-A type organic dyes (JY40-JY42) for possible application in the DSSCs [249]. The obtained results indicate that terthiophene having a greater extended π-conjugated system in comparison with other π-spacers, results in proper characteristics such as red-shift in wavelengths, excellent short circuit current density, and power conversion efficiency in JY42 dye.

The study on the effect of different π-spacers of series of 3,3'-dithioalkyl-2,2'-bithiophene (SBT)-based organic was applied by alkyl chains and conjugated thiophene units and 3,3-dialkyl-2,2'-bithiophene (BT)-based dye was utilized as comparison (Figure 20a) [250]. The analysis of voltage-current and IPCE curves indicate that the favorable values of short circuit current density and incident photon to electron conversion efficiency is related to 1a dye having thiophene and thioalkyl groups (Figure 20b and c). Also, SBT-based dyes represent the superior characteristics such as higher J_{SC}, η_0, FF, λ_{max}, and lower electron-transport resistance (R_{ct2}) in comparison with BT-based dye, due to the great π-conjugated extension of SBT than BT.

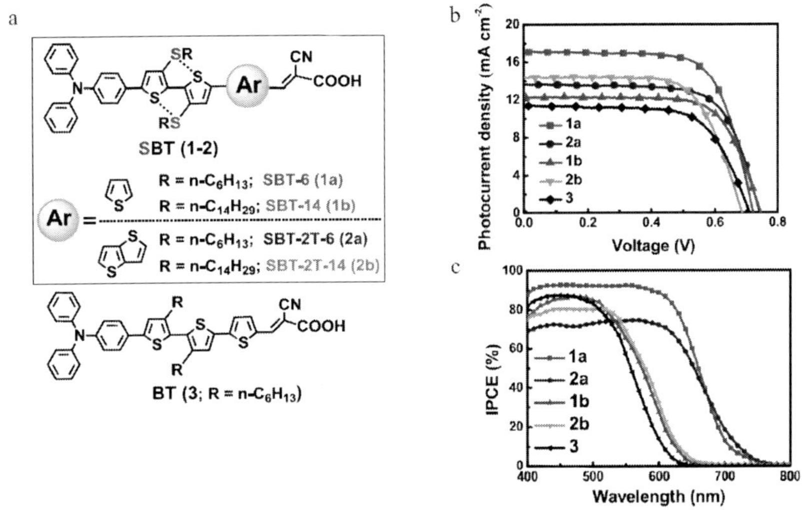

Figure 20. a) Chemical structures of designed dyes. b) The curve of the voltage-current of dyes. c) The curve of IPCE of studied dyes [250].

In 2020, molecular design of indacenodithiophene-based organic dyes was performed for tuning optoelectronic properties by changing the π-conjugated system [251]. These structures consist of diaryl amino as the donor linked to the acceptor unit of cyanoacrylic acid via different π-conjugated systems such as indacenodithiophene (IDT), diacenodithieno [3,2-b]thiophene (IDTT), and pyrroloindacenodithiophene (NIDT) unit (Figure 21). The computational results reveal that the extension of the conjugation length of IDTT provides broadened adsorption bands, lower band gap, and greater oscillator strength. Also, there is a linear relationship between the normal dipole moment (μ_{normal}), dye surface concentration (γ), and semiconductor conduction band shift (Δ_X) ($\Delta_X = -\frac{q\mu_{normal}\gamma}{\varepsilon_0\,\varepsilon}$). The examination of the dye adsorption on the $(TiO_2)_{46}$ showed that the normal dipole moments of all IDT dyes are negative, shifting towards the higher energy levels. Heteroatom substitution of the bridging atoms by electron-rich nitrogen atoms in NIDT dye leads to an enhanced dipole moment and Δ_X.

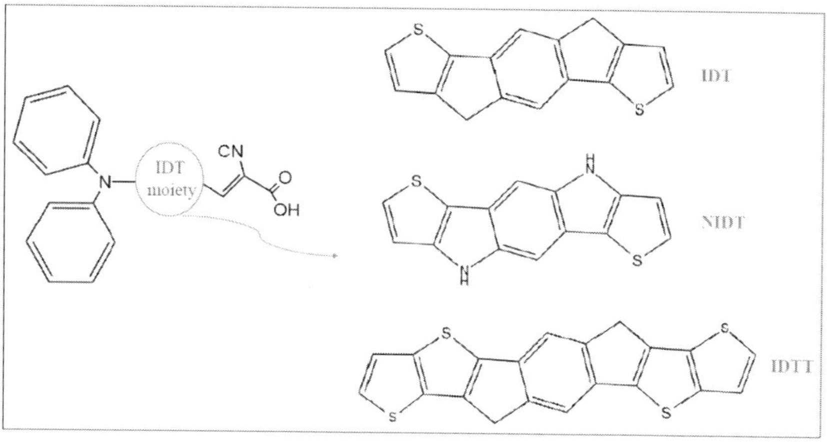

Figure 21. Molecular structures of the indacenodithiophene-based organic dyes.

Table 3. Exciton radius (a_x), exciton binding energy (EBE), the rate of the exciton formation (R_a), and the rate of the exciton dissociation (R_d) of the investigated dyes [252]

Dye	a_x (Å)	EBE (eV)	10^{-8} R_a (s^{-1})	10^{-49} R_d (s^{-1})
TC201	2.00	1.19	0.17	0.81
TC202	1.90	1.26	0.13	0.71
TC203	1.83	1.30	0.12	0.66
TC601	5.90	0.40	1.50	3.77
H-P	1.80	1.33	0.13	0.68
F-P	1.81	1.32	0.13	0.68
FF-P	1.78	1.34	0.14	0.69
T-F	1.65	1.44	0.10	0.57
P1B	1.47	1.62	1.21	0.84

In 2021, molecular engineering of triphenylamine-based metal-free organic dyes has been performed which comprises triphenylamine unit as the electron donor, cyanoacrylic acid and benzoic acid moiety as the electron acceptors, and different π-conjugated systems such as anthracene phenyl, anthracene vinyl phenyl, anthracene ethynyl phenyl, ethynyl anthracene phenyl, styryl phenyl, styryl-2-fluorophenyl, styryl-2,6-difluorophenyl, styryl furan, and styryl [252]. The study of the kinetic properties of these compounds and linear correlations between different kinetic parameters showed that TC601 having the ethynyl anthracene phenyl π-conjugated system has proper kinetic behavior such as maximum values of the exciton formation/

dissociation rate and exciton radius and the minimum exciton binding energy value (Table 3). The broader π-conjugated system increasing the resonance effect may be the source of this superior behavior. Also, the corresponding current-voltage (J-V) curve exhibits the highest values of the open-circuit voltage and short circuit current density for P1B having the benzoic acid as the electron acceptor and TC601 dyes, respectively (Figure 22).

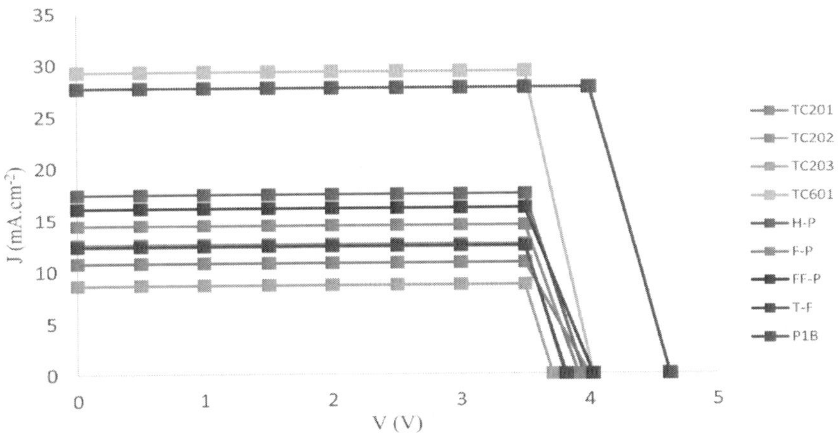

Figure 22. The simulated curve of the voltage-current dyes based on triphenylamine [252].

The Effect of the Number of Anchor Groups

On the basis of the literature reports, increase in the number of the anchor groups improves intra-molecular charge transport, photovoltaic parameters, and PCE of the solar cells. For example, the effect of mono- and di-anchoring organic dyes of Cz-D1 and Cz-D2 was studied by Raju et al. (Figure 23) [253]. In these sensitizers, carbazole was considered as the electron donor, *o,m*-difluoro substituted phenylene as the π-conjugated bridge, and cyanoacrylic acid as the electron acceptor. The obtained results shows that the enhancement of the π-spacer units and anchoring groups are in agreement with an increment in the open-circuit voltage, short-circuit density, and IPCE, which Cz-D2 dye having two π–spacers and anchoring groups is more efficient photosensitizer than Cz-D1 having a π-spacer and an anchoring group.

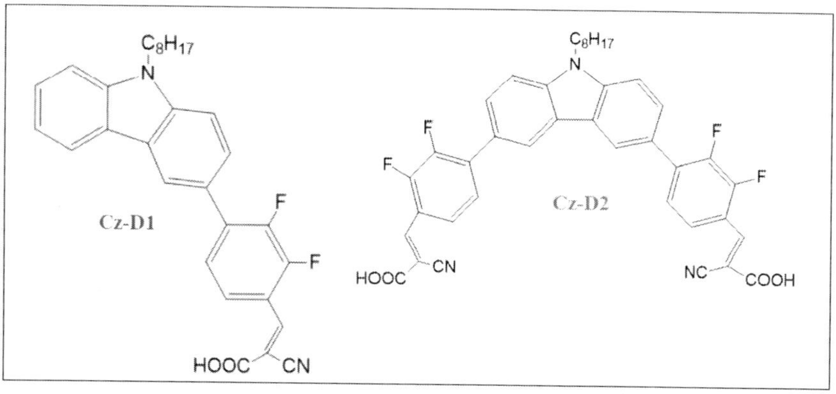

Figure 23. Chemical structures of Cz-D1 and Cz-D2.

Photovoltaic and spectroscopic studies of diphenylpyranylidene (DP) sensitized solar cells also demonstrate that D2 dye having dual cyanoacrylic acid anchoring groups is a promising dye in terms of greater open-circuit voltage, short-circuit density, power conversion efficiency, molar extinction coefficient, and light-harvesting capacity in comparison with D1 dye having mono cyanoacrylic acid anchoring group [254].

In 2019, the performance of twelve triphenylamine-based photosensitizers with D-π-A configuration was studied from the molecular engineering viewpoint [255]. Flavone, isoflavone, and triphenylamine as electron donors, cyanovinyl as π-bridge, and carboxylic acid as anchor group were used in these systems, in which some of them are composed of two anchor units. According to spectral analysis, dyes having two anchoring groups (FT3, FT6, FT9, and FT12) have longer wavelengths than mono anchoring dyes. Also, in the dyes, those having two anchoring groups and one flavone ring (FT3 and FT12) represent a higher red shift in the absorption band in comparison with the dyes having one anchoring group (FT1 and FT7), result in higher efficiency of the solar cell.

An experimental and theoretical study was applied on the impact of the number of anchoring groups to achieve higher DSSCs efficiency by Manoharan et al. [256]. In this research, a series of metal-free organic dyes was synthesized. These dyes are composed of diphenylamine or diethylamine as electron donors, acrylic acid or nitro vinyl as electron acceptors, and phenyl as π-spacer (Figure 24). Based on different analyses, di-anchoring dyes-based DSSCs (P2C and P2N) exhibit good optical properties, absorption maxima, and molar extension coefficient better than mono-anchoring dyes-based

DSSCs (EC, EN, PC, and PN). Also, P2C dye having carboxylic acid substituted di-anchoring has the most favorable characteristics such as open-circuit voltage, short circuit current density, excited-state lifetime, final efficiency, and incident photon to current conversion efficiency in comparison with nitro vinyl substituted di-anchoring P2N dye.

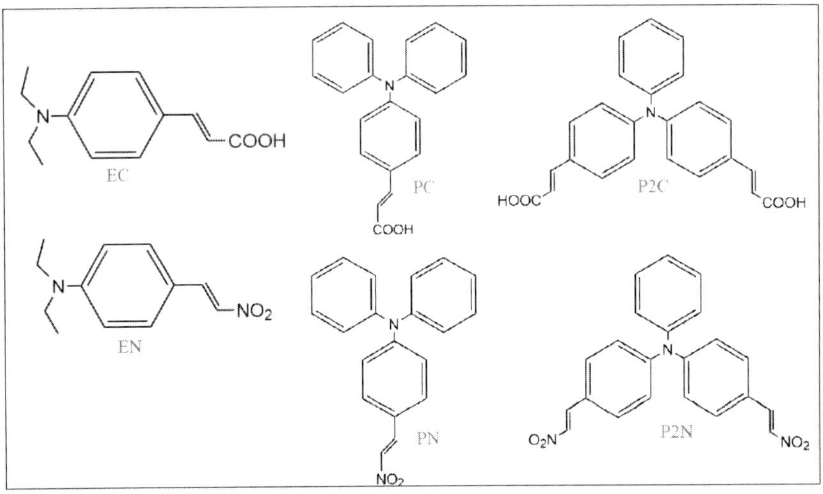

Figure 24. Molecular structure of the studied dyes.

Elsherbiny et al. compared different characteristics of three novel organic dyes by changing the number of the anchoring groups. DE-1, DE-2, and DE-3 are composed of one, two, and three anchoring groups (cyanoacrylic acid), respectively [257]. The obtained results reveal that increasing the number of anchoring groups elevates the dipole moment, polarizability, hyper-polarizability, oscillator strength, overall efficiency, open-circuit voltage, and short circuit current density. Therefore, DE-3 having the highest number of anchoring groups exhibits the eligible photovoltaic properties because an increase in the number of anchoring groups improves the electron-withdrawing ability, which in turn results in better photovoltaic properties.

A computational study on the optical features of boron dipyrromethene (BODIPY)-based sensitizers (P1-P6) (Figure 25a) reveal that increasing the number of anchoring groups plays a critical role in development of solar cells properties [258]. According to the results, a red-shift in the maximum absorption wavelength was observed in P2, P4, and P6 dyes having double anchoring groups in comparison with P1, P3, and p5 having single anchoring group (Figure 25b). Also, these sensitizers (P2, P4, and P6) exhibit the smaller band gap energy than P1, P3, and p5.

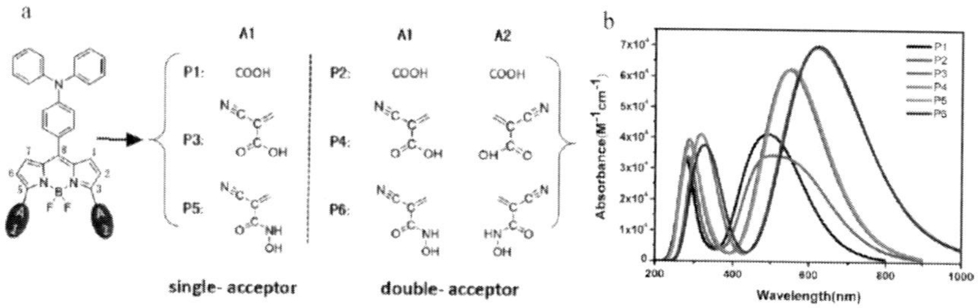

Figure 25. a) Molecular structures of the designed dyes. b) Calculated absorption spectra for studied dyes in CH_2Cl_2. [258].

The Effect of the Auxiliary Donor/Acceptor

The use of auxiliary donor and acceptor plays an important role in expanding the D-π-A structure, facilitating charge transfer, reduction of band gap, desirable spectral properties in the visible region, and increase of the efficiency of DSSCs. For example, Yang et al. synthesized and designed some phenothiazine dyes (PTZTB1-4) with the molecular motif $(D\text{-}\pi\text{-}A)_3L2$ based on reference R-SB dye having a mono D-π-A chain [259]. Phenothiazine was applied as electron donor, 4-methoxyphenyl and 4-octyloxyphenyl as auxiliary donors, thiophene as π-spacer, and cyanoacetic acid as electron acceptor. Moreover, the effect of the auxiliary donors and linkers of the *n*-hexyl or *n*-dodecyl group (L2) was investigated. According to the results, PTZTB1-4 dyes are the most favorable in terms of J_{SC} and PCE in comparison with the reference dye of R-SB. Moreover, PTZTB3 and PTZTB4 exhibit a broader IPCE response region and absorption intensity than other dyes. The greater electron-donating ability of the auxiliary donor of 4-octyloxyphenyl is the main reason for this behavior. Also, the incorporation of the longer carbon chain of the n-dodecyl linker in the PTZTB2 and PTZTB4 dyes leads to an improvement in the electron lifetime and open-circuit voltage.

In 2020, a computational study was performed on three anthracene-based organic dyes with various auxiliary electron donors such as diarylamine, carbazole, and indoline (Figure 26) [260]. Different properties of these compounds (DPA-BPT2, CAR-BPT2, and IND-BPT2) were compared with the reference dye of DPA-PHE without an auxiliary electron donor. The introduction of auxiliary electron donors in the studied dyes increases the oscillator strength, light-harvesting efficiency, and maximum wavelength and reduces the HOMO-LUMO gap in comparison with DPA-PHE. Also, among the dyes DPA-BPT2, CAR-BPT2, and IND-BPT2, DPA-BPT2 having diarylamine of auxiliary donor show a slight red-shift of 3-4 nm than other dyes. In the case of dye/$(TiO_2)_{46}$ systems, this behavior is not seen and IND-BPT2 dye-$(TiO_2)_{46}$ complex having indoline as the auxiliary donor represents a bathochromic shift of 80 nm than other dyes.

The substitution of the different auxiliary donors such as *N,N*-difluorenebenzenamine (DFBA), *N,N*-dihexylbenzenamine (DHBA), benzenehexyldithieno[3,2-b:2′,3′-d]pyrrole (BZ-HDTP) and benzene-substituted dithieno[3,2-b:2',3'-d]pyrrole (DTP) (BZDTP) were examined on the series of metal-free organic dyes (M43-M46) photovoltaic behavior by Dai et al. (Figure 27) [261]. The study of the addition electron donors effect indicates the introduction of these donors lead to promotion of dye-sensitized solar cells performance. Moreover, short circuit current density (15.3 mA/cm^{-2}), incident photon conversion efficiency (9.02%), and molar extinction coefficient (79600 $M^{-1}cm^{-1}$) have the maximum values in the case of M45 having BZ-HDTP auxiliary donor (Figures 28a-c).

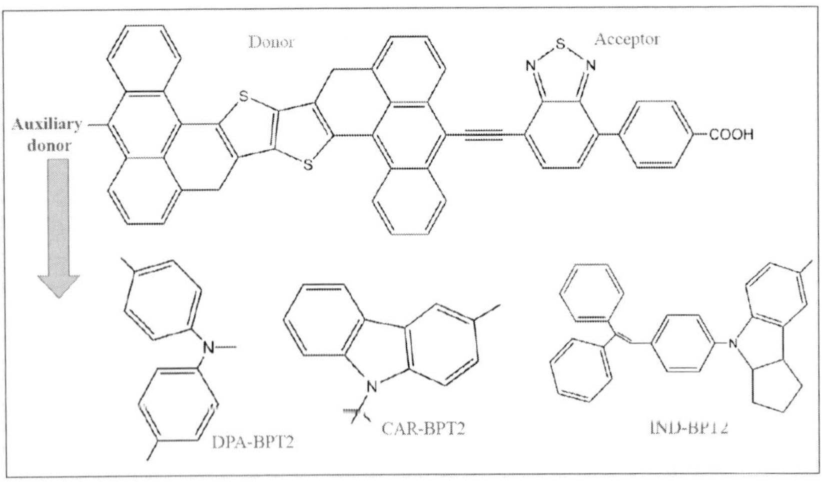

Figure 26. Molecular structures of the DPA-BPT2, CAR-BPT2, and IND-BPT2.

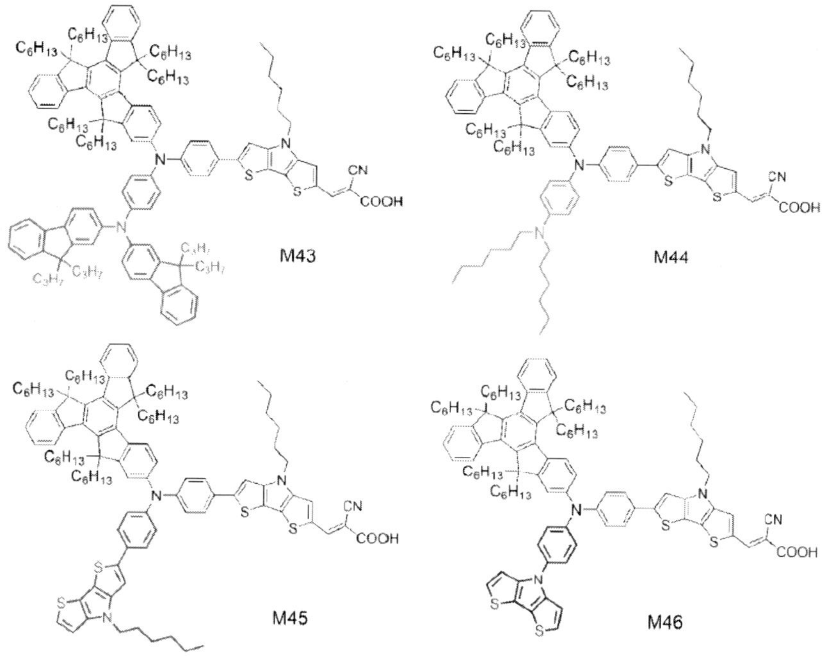

Figure 27. Molecular structure of M43, M44, M45, and M46 dyes [261].

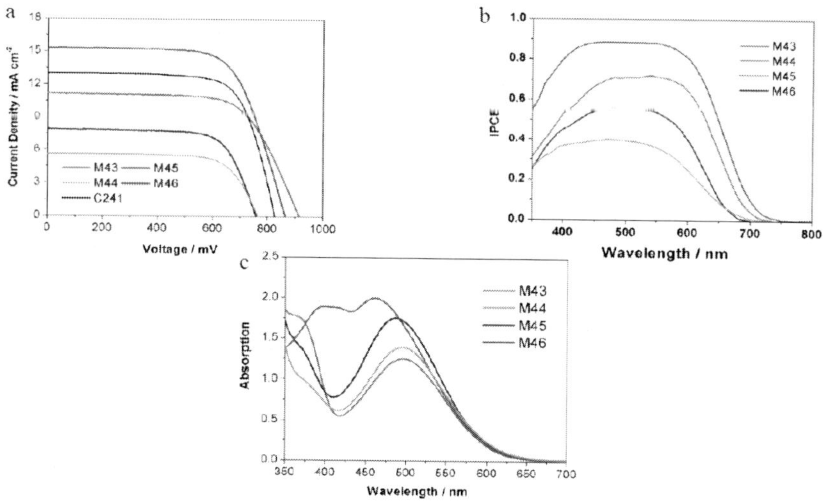

Figure 28. a) The curve of the voltage-current of M43, M44, M45, and M46 dyes. b) The curve of IPCE of studied dyes. c) Absorption spectra of the dyes on TiO_2 film [261].

Han et al. developed two phenothiazine-based metal-free organic dyes (WY5 and WY6) having D-D-π-A framework for highly efficient DSSCs (Figure 29) [262]. Different analyses were performed on these photosensitizers and compared with the reference dye of T0 having a D-π-A structure. The insertion of benzyl-based auxiliary electron donors in WY5 and WY6 increases short circuit current density, light-harvesting ability, red shift of the absorption region, and solar cell efficiency in comparison with T0 dye without auxiliary electron donor. Also, solar cell sensitized by WY6 having two methoxy groups attached to the phenyl is more efficient in terms of spectral properties, J_{SC}, and IPCE than that of WY5. This property is connected with the greater electron-donating ability of WY6 compared to WY5.

Mane et al. investigated the photovoltaic behavior of D-A-π-A organic dyes of ODn containing 1,3,4-oxadiazole as the auxiliary acceptor, triphenylamine as the electron donor, and different π-bridges and acceptors at the DFT/B3LYP/6-31G(d) level of theory (Figure 30) [263]. The presence of 1,3,4-oxadiazole as an auxiliary acceptor leads to advanced absorption properties in the visible region, charge transfer ability, and quantum yields, which makes them favorable to be used in DSSCs. Based on the results, OD3 dye is preferred, because of wider spectral coverage and red shift at maximum absorption wavelength, as well as better separation of the HOMO and LUMO electron density distributions on the dye structure.

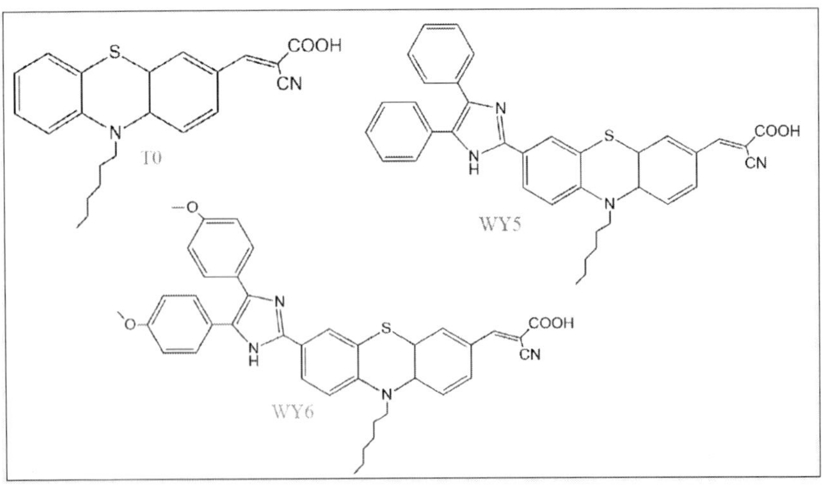

Figure 29. Chemical structure of T0, WY5, and WY6 dyes.

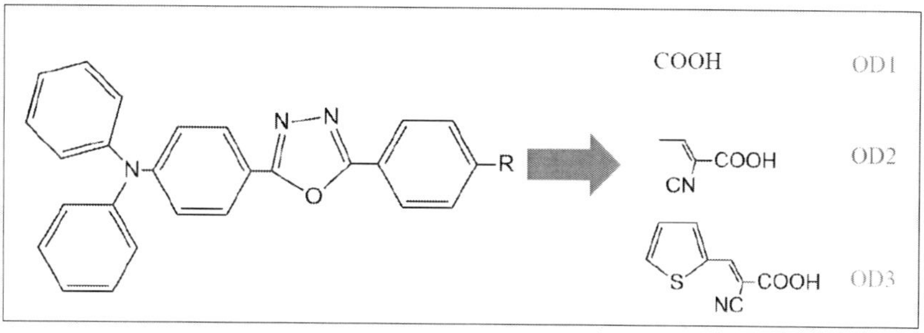

Figure 30. Molecular structures of the OD dyes.

The detailed investigation of some indoline-based sensitizers indicate that the better performance of dye-sensitized solar cells is strongly influenced through insertion auxiliary electron acceptors (Figure 31a) [264]. In this system, indoline derivative as the electron donor, thiophene as π-spacer, cyanoacrylic acid as electron acceptor, and benzothiadiazole, benzotriazole, benzopyrazine (quinoxaline), and benzoxadiazole as auxiliary electron acceptors were applied. Considering the incident photo to current conversion efficiency of the dyes in Figure 31b, all dyes having additional acceptor show greater efficiency in comparison with when there is not additional acceptor. Also, Due to the superior electron withdrawing ability of the benzoxadiazole additional acceptor unit, trend of the red-shift response in IPCE are according to: benzoxadiazole > benzothiadiazole > quinoxaline > benzotriazole.

Figure 31. Chemical structures of studied sensitizers. b) IPCE curve of dyes [264].

Spectral behavior, and the influence of the auxiliary acceptor and π-bridge, in metal-free organic dyes based on triarylamine (AFL7-AFL10) were examined by Xie et al. (Figure 32) [265]. In these compounds, triarylamine donors and different π-bridges with or without a benzothiadiazole auxiliary acceptor were applied. This study revealed that AFL9 and AFL10 dyes exhibit a red shift, lower band gap, proper intra-molecular charge transfer, and broader light absorption region than AFL7 and AFL8 dyes. The insertion of the benzothiadiazole auxiliary acceptor having proper electron-withdrawing ability can be considered for describing this superior behavior. Also, based on the study of the π-bridges effect, AFL7 dye having the bithiophene π-bridge has widespread attention due to remarkable features such as short circuit current density, overall power conversion efficiency, and oscillator strength.

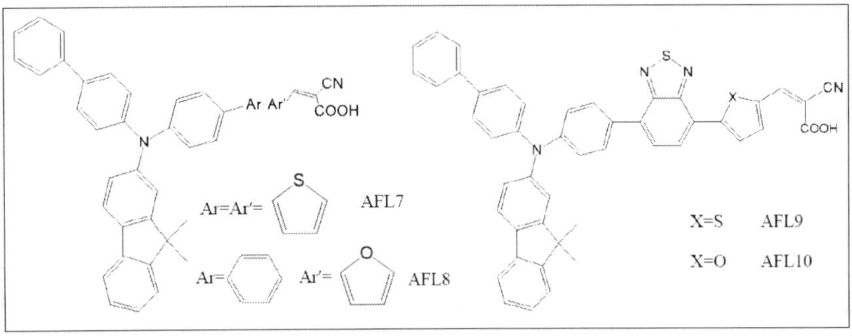

Figure 32. Molecular structures of the studied sensitizers.

Ji et al. published a review on some of dipentyldithieno[3,2-f:2',3'-h]quinoxaline-based dyes (IT, IPT, and IQT) possessing D-A-π-A backbone by modifying auxiliary acceptor (Figure 33) [266]. In the skeleton of these dyes, indoline moiety was utilized as an electron donor, dipentyldithieno[3,2-f:2',3'-h]quinoxaline (DPQ) unit as the π-bridge, cyanoacrylic acid unit as the electron acceptor, and 2,3-dipentylquinoxaline or 2,3-dipentylpyridopyrazine as the auxiliary acceptors. The study of the auxiliary acceptors effect showed that the introduction of additional acceptors broadens the absorption spectrum range and increases the molar extinction coefficient. Moreover, solar cells based on IQT dye have greater power conversion efficiency, fill factor, open-circuit voltage, and short circuit current density than that of other dyes. Although the electron-withdrawing ability of 2,3-dipentylpyridopyrazine in the IPT dye is predominant, it does not represent better properties due to the electron recombination and electron-trapping.

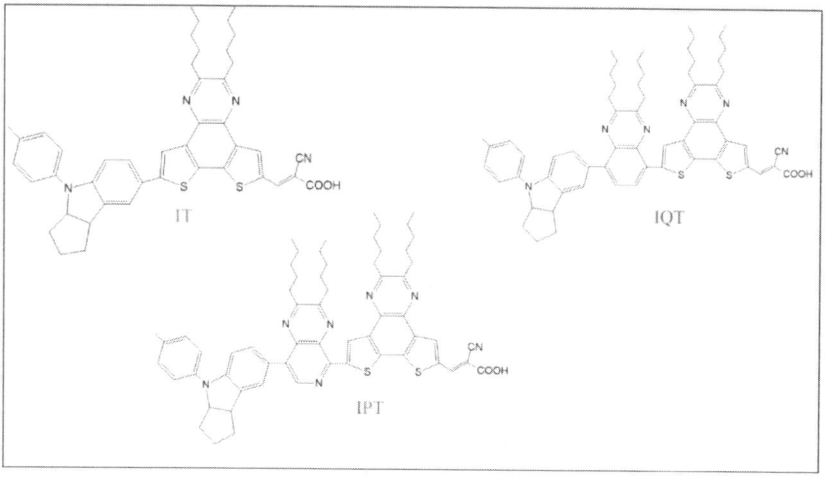

Figure 33. Chemical structures of the dyes IT, IPT, and IQT.

Wu et al. reported photovoltaic properties of metal-free organic dyes with di(1-benzothieno)[3,2-b:2′,3′-d]pyrrole (DBTP) as an auxiliary donor (WL5-10) having the molecular motif of D-D-π-A or D-D-A-π-A [267]. In this research, WL5,7-8 dyes containing triphenylamine or phenothiazine as the electron donor, and DBTP as an auxiliary donor with D-D-π-A architecture and WL9-10 dyes containing benzothiadiazole as an auxiliary acceptor with D-D-A-π-A architecture. The effects of the molecular structure and auxiliary donor/acceptor were examined. The obtained results reveal that DBTP-based dyes incorporated with an auxiliary donor having D-D-π-A configuration represent the most proper molar absorption coefficient, and photovoltaic performance in comparison with reference dye WL1a with D-π-A structure. Also, the study of the absorption spectra of the dyes illustrates a broader absorption band and red-shift in WL9-10 originated from the presence of an auxiliary acceptor of benzothiadiazole.

The Effect of Different Substituted Groups

To achieve promising photovoltaic properties, the investigation of different substituted groups such as halogen groups and central metal ion substitution revealed essential factors. For example, the effect of fluorine substitution on the photovoltaic behavior of some small molecules based on benzothiadiazole and phenothiazine was studied by Putri et al. (Figure 34) [268]. The incident photon conversion efficiency curve shows that BDP-FF dye having two fluorine atoms has the highest IPCE (0.78%) in comparison with BDP-F dye having one fluorine group (0.54%) and BDP without any fluorine group (0.49%). Also, the voltage-current curve of the studied dyes demonstrates that the highest values of the open-circuit voltage and short circuit current density are related to BDP-FF dye having two fluorine atoms. Therefore, the replacement of fluorine atoms can be considered as a useful strategy for improving the photovoltaic properties of solar cells.

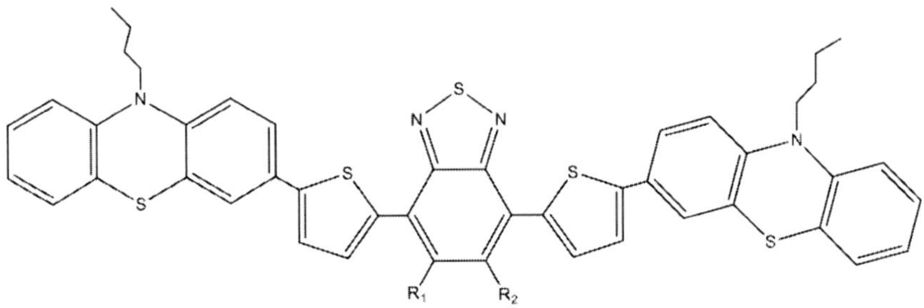

BDP; R_1=H, R_2=H
BDP-F; R_1=F, R_2=H
BDP-FF; R_1=F, R_2=F

Figure 34. Chemical structures of the studied dyes.

In 2020, Tingare et al. investigated various types of dyes based on oxindole as the electron acceptor, triphenylamine as the electron donor, and thiophene as the π-spacer (Figure 35a) [269]. In this work, the influence of substitution of different halogen groups (fluorine (F), bromine (Br), and iodine (I)) of the oxindole acceptor on the electrical and spectral properties of the studied dyes (TI111-TI116) was examined. TI111 is a halogen-free dye, while others are different in the position of the substituent with the anchoring group. According to the obtained results, the introduction of halogen groups improves the open-circuit voltage, short circuit current density, fill factor, solar cell efficiency, electron excited-state lifetime, maximum wavelength, and molar absorption coefficient of the corresponding dyes, when the halogens are in the para position with the anchoring group. Moreover, TI114 dye having Br substituent exhibits the most favorable electrical (V_{OC} and J_{SC}) and absorption properties in comparison with other dyes (Figures 35b and c).

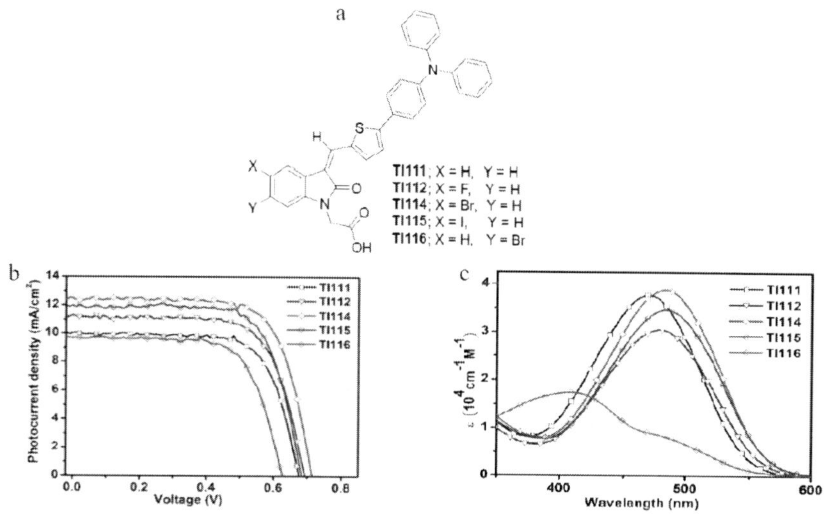

Figure 35. a) Molecular structures of oxindole dyes. b) The curve of the voltage-current of TI111-TI116 dyes. c) The absorption spectra of the oxindole dyes in CH_2Cl_2 [256].

Higashino et al. reported photovoltaic and electrical performance of the DSSCs based on the metallocorroles with para-carboxyphenyl groups (Au-A, Re-A, and Os-A) and meta-carboxyphenyl groups (Au-B, Re-B, and Os-B) (Figure 36a) [270]. According to the result, insertion Au metal in both para and meta positions improves the photovoltaic properties such as fill factor, open circuit voltage, short circuit current density, and power conversion efficiency (Figure 36b and c) in comparison with the other two metals. Also, investigation of spectroscopy properties of the compounds in THF reveal that presence of central metal has a remarkable role on improving the absorption spectrum.

In 2019, a comparative study was performed on the characterization of two series of metal-phthalocyanine dyes substituted with either (2-phenyl)phenoxy (M-PC1) and quinolineoxy (MPC2) for DSSCs application [271]. In this work, the structural modification of the dyes by the central metal atoms and the substitution affects the absorption spectra and photovoltaic performance of the compounds. According to the results, MPC2 dye substituted by quinoleinoxy shows considerable hyper-polarizability, light-harvesting efficiency, oscillator strength, open-circuit voltage, photo-current conversion efficiency, and Gibbs energy of the electron injection than MPC1 dye substituted by (2-phenyl)phenoxy. Also, the incorporation of Mo, Cr, Zn, and Cd displays a bathochromic shift and reduced excitation energy in both series of metal-phthalocyanine sensitizers.

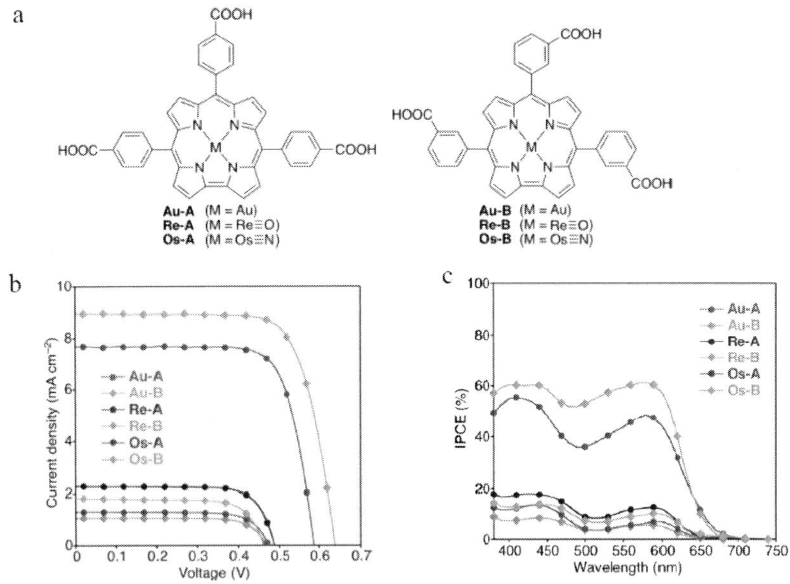

Figure 36. a) Molecular structures of metallocorroles with para-carboxyphenyl groups (Au-A, Re-A, and Os-A) and meta-carboxyphenyl groups (Au-B, Re-B, and Os-B).b) The curve of the voltage-current of compounds. c) IPCE spectra of compounds [270].

In 2021, the effect of the central metal ion on the photovoltaic properties of bacteriochlorin derivatives was studied [272]. Figure 37a shows the optimized structures of free-base (Fb) and metalbacteriochlorins of bisimide (MBC-I2; M= 2H, Be, Mg, Ca, Ni and Zn) within their electronic energies using the DFT method at the M06-2X/6-31 G level of theory. The replacement of free-base bacteriochlorin with different metals reduces quantum chemistry reactivity descriptors such as the electronic chemical hardness, absolute of electronic chemical potential, and electrophilicity, the ground and excited state oxidation potentials of the dye, Gibbs energy of the electron injection, and enhances the electron-driving force. Also, among the studied photosensitizers, CaBC-I2 was proposed as an excellent candidate to be used in DSSCs due to greater electron transfer rate constant, short circuit current density, final efficiency, and IPCE. The correlation of J_{SC} and η_0 versus IPCE illustrates that the dyes having a higher IPCE show a superior J_{SC} and η_0 (Figure 37b).

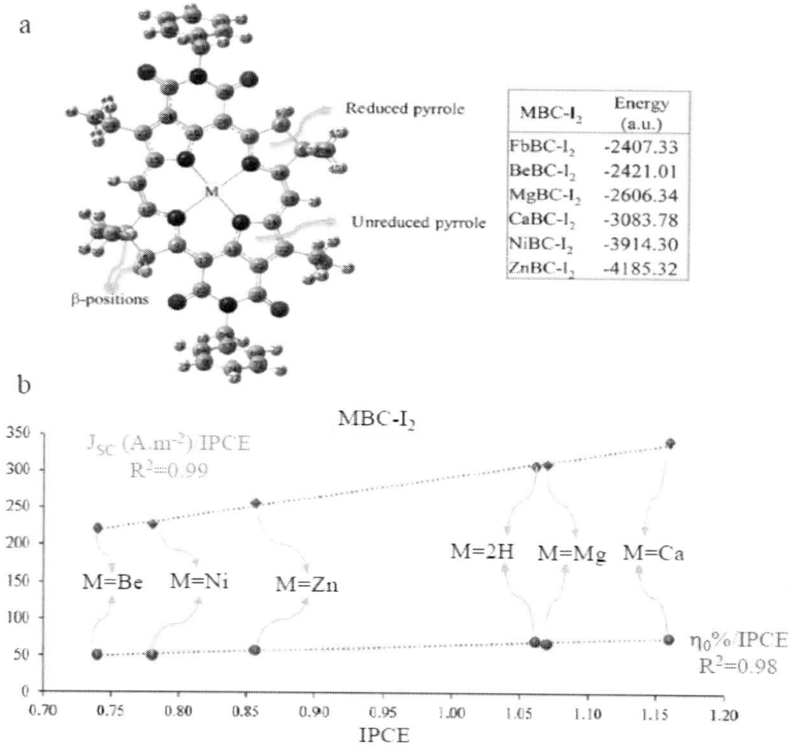

Figure 37. a) The optimized molecular structures of the dyes. b) The correlation of J_{SC} and η_0 versus IPCE [272].

Solvent Effects

On the basis of the latest reports, intermolecular interactions of the dye and the solvent can affect the absorption spectrum, charge transfer processes, photovoltaic behavior, and the efficiency of the solar cells. For example, in 2018, spectroscopic and physicochemical properties of the functionality of indoloquinoxaline-based dye-sensitized solar cells (QX22-QX25) were analyzed in different media such as gas, ethanol, toluene, THF, and methylene dichloride (DCM) using the quantum chemistry calculations [273]. In these structures, two acceptors of cyanoacrylic acid and 2-(1,1-dicyano-methylene)rhodanine (CCRD), a double triphenylamine as donor, an additional acceptor of indoloquinoxaline and two π-spacers furan and thiophene were applied. The presence of solvents leads to greater light-harvesting, oscillator strength, electric dipole moment ($r_{k,k'}$), red shift, and intensity in the absorption wavelength of the dyes in comparison with the gas phase. Figure 38 shows the LHE curve of the dyes in different media as the function of the maximum wavelength. Solvent effect analysis demonstrates that among the applied solvents, toluene is the favorable medium from dynamic and kinetic viewpoints.

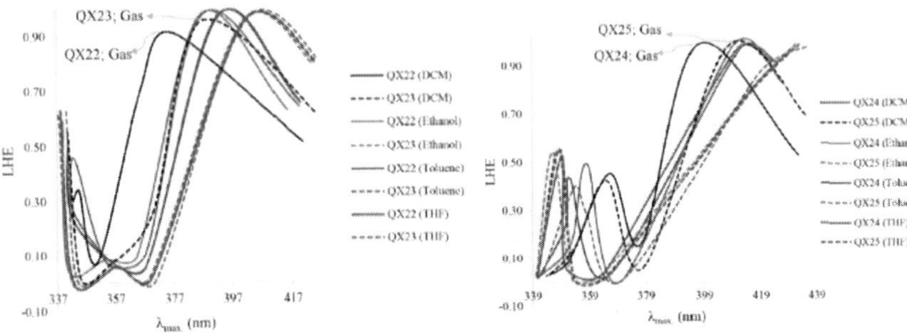

Figure 38. LHE of the dyes in different media as the function of the maximum wavelength predicted at M062X/6-311++G(2d,2p) level [273].

Chitumalla et al. reported electronic and photophysical properties of four thiocyanate-free cyclometalated ruthenium dyes in acetonitrile solvent and vacuum by the TD-DFT technique [274]. The energy levels of the sensitizers in a vacuum and acetonitrile show that the energy alignment of the LUMO levels of dyes is higher than the CB edge of TiO_2 in the solvent. This leads to easy electron transfer from the dye excited state to the semiconductor. But the LUMO energies are below the conduction band of TiO_2 in a vacuum. Also, the HOMO energy levels of the dyes are lower than the redox potential of the electrolyte in the solvent and vacuum results in appropriate dye regeneration. Therefore, the use of solvent is important for the description of electronic structures. Moreover, according to the simulated optical absorption spectra of the sensitizers, the presence of acetonitrile has a small effect on the absorption spectra, in which a red shift of about 5 nm is observed for M1 to M3 dyes and a blue shift of 1 nm for M4 dye.

Gebremariam et al. analyzed the absorption spectra properties of series of fluorenyl-decorative metal-free organic dyes coded GZ116, GZ126, GZ129, MA1116, and MA1118 using WB97X-D functional and 6-31G(d) basis set in dimethylformamide (DMF) (Figure 39a) [275]. The absorption spectra of the dyes in DMF represent two different absorption peaks. The shorter peaks region (335-600 nm) can be attributed to the π-π* transitions and longer peaks region (420-600 nm) can be assigned to the intramolecular charge-transfer transitions from triphenylamine donor to the cyanoacrylic acid anchoring moiety (Figure 39b).

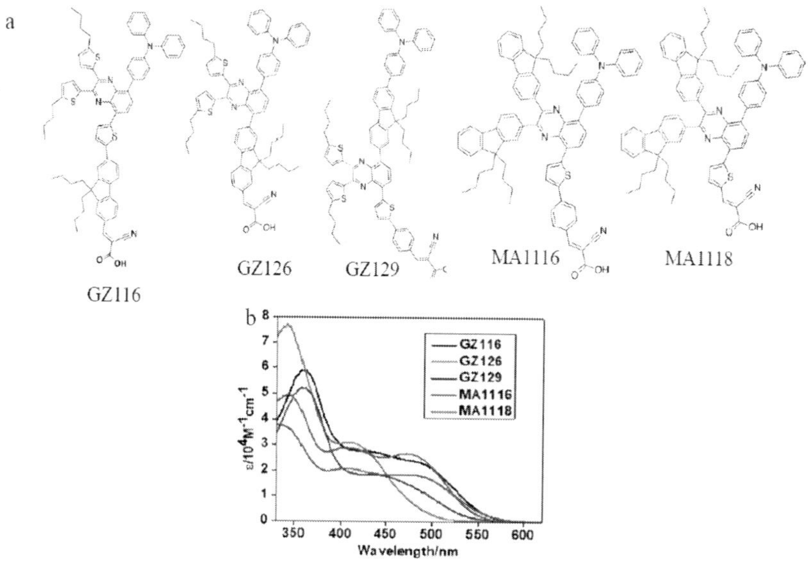

Figure 39. a) Molecular structures of GZ116, GZ126, GZ129, MA1116, and MA118 dyes. b) UV-visible spectra of the dyes in DMF [275].

Conclusion

In recent years, dye-sensitized solar cells have received great interest as noteworthy and versatile devices due to their features such as easy processibility, high potential, low cost, and environmentally friendly. To obtain detailed information on the photosensitizers application and development of the solar cells, disparate sensitizers such as metal-organic dyes, natural dyes, and metal-free organic dyes were applied in determining the photovoltaic performance. Also, computational methods as the pioneering procedures were examined to study different properties of the solar cells. This review enlightens promising methods to improve the DSSC performance such as the presence of the strong electron donor/acceptor and extended π-conjugated system, increasing the number of anchor groups and the effect of the auxiliary donor/acceptor, substituted groups, and different solvents from the theoretical and experimental points of view. Moreover, other appropriate strategies such as acceptable linear correlations between different photovoltaic properties and molecular descriptors, voltage-current curves, dynamic, kinetic, and spectroscopy properties help to describe and analyze the processes in the DSSCs.

References

[1] Solarin, S. A. An environmental impact assessment of fossil fuel subsidies in emerging and developing economies. *Environ. Impact Assess. Rev.* 2020, 85, 106443-106451.

[2] Saidi, K.; Omri, A. The impact of renewable energy on carbon emissions and economic growth in 15 major renewable energy-consuming countries. *Environ. Res.* 2020, 186, 109567-109591.

[3] Gielen, D.; Boshell, F.; Saygin, D.; Bazilian, M. D.; Wagner, N.; Gorini, R. The role of renewable energy in the global energy transformation. *Energy Strategy Rev.* 2019, 24, 38-50.

[4] EIA, U. Energy Information Administration: Annual Energy Outlook 2015 with projections to 2040. *US Energy Information Administration, Office of Integrated and International Energy Analysis, US Department of Energy, Washington, DC,* 2015, 20585.

[5] Babaei, M.; Azizi, E.; Beheshti, M. T.; Hadian, M. Data-Driven load management of stand-alone residential buildings including renewable resources, energy storage system, and electric vehicle. *J. Energy Storage* 2020, 28, 101221-101236.

[6] Elavarasan, R. M.; Shafiullah, G. M.; Padmanaban, S.; Kumar, N. M.; Annam, A.; Vetrichelvan, A. M.; Mihet-Popa, L.; Holm-Nielsen, J. B. A comprehensive review on renewable energy development, challenges, and policies of leading Indian states with an international perspective. *IEEE Access* 2020, 8, 74432-74457.

[7] Kabir, E.; Kumar, P.; Kumar, S.; Adelodun, A. A.; Kim, K. H. Solar energy: Potential and future prospects. *Renew. Sustain. Energy Rev.* 2018, 82, 894-900.

[8] Rabaia, M. K. H., Abdelkareem, M. A., Sayed, E. T., Elsaid, K., Chae, K. J., Wilberforce, T., Olabi, A. G. Environmental impacts of solar energy systems: A review. *Sci. Total Environ.* 2021, 754, 141989-142007.

[9] Nayak, P. K.; Mahesh, S.; Snaith, H. J.; Cahen, D. Photovoltaic solar cell technologies: analysing the state of the art. *Nat. Rev. Mater.* 2019, 4, 269-285.

[10] Sharma, D.; Mehra, R.; Raj, B. Comparative analysis of photovoltaic technologies for high efficiency solar cell design. *Superlattices Microstruct.* 2021, 153, 106861.

[11] Becquerel, M. E. Mémoire sur les effets électriques produits sous l'influence des rayons solaires. *C. R. Acad. Sci.* 1839, 9, 561-567.

[12] Harb, A. E. M. A.; Radwan, A.; Elsayed, K., Sedrak, M.; Ahmed, M. Influence of varying the Ethylene-Vinyl Acetate layer thicknesses on the performance of a

polycrystalline silicon solar cell integrated with a microchannel heat sink. *Sol. Energy* 2020, 195, 592-609.

[13] Liu, B.; Guo, J.; Hao, R.; Wang, L.; Gu, K.; Sun, S.; Aierken, A. Effect of Na doping on the performance and the band alignment of CZTS/CdS thin film solar cell. *Sol. Energy* 2020, 201, 219-226.

[14] Jia, T.; Zhang, J.; Zhong, W.; Liang, Y.; Zhang, K.; Dong, S.; Ying, L.; Liu, F.; Wang, X.; Huang, F.; Cao, Y. 14.4% efficiency all-polymer solar cell with broad absorption and low energy loss enabled by a novel polymer acceptor. *Nano Energy*, 2020, 72, 104718-104733.

[15] Arkan, F.; Izadyar, M. Computational modeling of the photovoltaic activities in EABX3 (EA= ethylammonium, B= Pb, Sn, Ge, X= Cl, Br, I) perovskite solar cells. *Comput. Mater. Sci.* 2018, 152, 324-330.

[16] Pandey, R.; Khanna, A.; Singh, K.; Patel, S. K.; Singh, H.; Madan, J. Device simulations: Toward the design of> 13% efficient PbS colloidal quantum dot solar cell. *Sol. Energy* 2020, 207, 893-902.

[17] Zhang, W.; Wang, L.; Mao, L.; Jiang, J.; Ren, H.; Heng, P.; Ågren, H.; Zhang, J. Computational protocol for precise prediction of dye-sensitized solar cell performance. *J. Phys. Chem. C* 2020, 124, 3980-3987.

[18] Hasan, M. M.; Islam, M. D.; Rashid, T. U. Biopolymer-based electrolytes for dye-sensitized solar cells: A critical review. *Energy Fuels* 2020, *34*, 15634-15671.

[19] Devadiga, D.; Selvakumar, M.; Shetty, P.; Santosh, M. S. Recent progress in dye sensitized solar cell materials and photo-supercapacitors: A review. *J. Power Sources* 2021, 493, 229698.

[20] Yahya, M.; Bouziani, A.; Ocak, C.; Seferoğlu, Z.; Sillanpää, M. Organic/Metal-organic photosensitizers for dye-sensitized solar cells (DSSC): Recent developments, new trends, and future perceptions. *Dyes Pigm.* 2021, 192, 109227.

[21] O'Regan, B.; Grätzel, M. A low-cost, high-efficiency solar cell based on dye-sensitized colloidal TiO_2 films. *Nature* 1991, 353, 737-740.

[22] Gong, J.; Sumathy, K.; Qiao, Q.; Zhou, Z. Review on dye-sensitized solar cells (DSSCs): Advanced techniques and research trends. *Renew. Sustain. Energy Rev.* 2017, 68, 234-246.

[23] Yang, G.; Sui, G.; Liang, Y.; Xue, X.; Feng, Y.; Zhang, B. Effects of interfacial adsorption configurations on dye-sensitized solar cell performance at the stoichiometric and defective TiO_2 anatase (101) surfaces: a theoretical investigation. *Phys. Chem. Chem. Phys.* 2020, 22, 4508-4515.

[24] Cole, J. M.; Pepe, G.; Al Bahri, O. K.; Cooper, C. B. Cosensitization in dye-sensitized solar cells. *Chem. Rev.* 2019, 119, 7279-7327.

[25] Al-Busaidi, I. J.; Haque, A.; Al Rasbi, N. K.; Khan, M. S. Phenothiazine-based derivatives for optoelectronic applications: A review. *Synth. Met.* 2019, 257, 116189-116204.

[26] Karim, N. A.; Mehmood, U.; Zahid, H. F.; Asif, T. Nanostructured photoanode and counter electrode materials for efficient dye-sensitized solar cells (DSSCs). *Sol. Energy* 2019, 185, 165-188.

[27] Arkan, F.; Izadyar, M. Recent theoretical progress in the organic/metal-organic sensitizers as the free dyes, dye/TiO_2 and dye/electrolyte systems; Structural

modifications and solvent effects on their performance. *Renew. Sustain. Energy Rev.* 2018, 94, 609-655.

[28] Ondersma, J. W.; Hamann, T. W. Recombination and redox couples in dye-sensitized solar cells. *Coord. Chem. Rev.* 2013, 257, 1533-1543.

[29] Koçyiğit, A.; Kerimli, G. The characterization of various TCOs and metal oxide layers for dye sensitized solar cells. *Mater. Today: Proc.* 2021, 46, 6947-6953.

[30] Wei, T. C.; Wan, C. C.; Wang, Y. Y. Poly (N-vinyl-2-pyrrolidone)-capped platinum nanoclusters on indium-tin oxide glass as counterelectrode for dye-sensitized solar cells. *Appl. Phys. Lett.* 2006, 88, 103122-103124.

[31] Chih-Hung, T.; Sui-Ying, H.; Tsung-Wei, H.; Yu-Tang, T.; Yan-Fang, C.; Jhang, Y. H.; Hsieh, L.; Chung-Chih, W.; Yen-Shan, C.; Chieh-Wei, C.; Chung-Chun, L. Influences of textures in fluorine-doped tin oxide on characteristics of dye-sensitized solar cells. *Org. Electron.* 2011, 12, 2003-2011.

[32] Pham, D. P.; Nguyen, H. T.; Phan, B. T.; Maenosono, S.;Tran, C. V. Influence of addition of indium and of post-annealing on structural, electrical and optical properties of gallium-doped zinc oxide thin films deposited by direct-current magnetron sputtering. *Thin Solid Films*, 2015, 583, 201-204.

[33] Ekmekci, M.; Ela, C.; Erten-Ela, S. Morphological characterization of aluminum-doped zinc oxide nanomaterials (AZO) for dye-sensitized solar cells. *Int. J. Appl. Ceram. Technol.* 2019, 16, 727-734.

[34] Wu, J.; Lan, Z.; Lin, J.; Huang, M.; Huang, Y.; Fan, L.; Luo, G.; Lin, Y.; Xie, Y.; Wei, Y. Counter electrodes in dye-sensitized solar cells. *Chem. Soc. Rev.* 2017, 46, 5975-6023.

[35] Thomas, S.; Deepak, T. G.; Anjusree, G. S.; Arun, T. A.; Nair, S. V.; Nair, A. S. A review on counter electrode materials in dye-sensitized solar cells. *J. Mater. Chem. A* 2014, 2, 4474-4490.

[36] Li, Q.; Wu, J.; Tang, Q.; Lan, Z.; Li, P.; Lin, J.; Fan, L. Application of microporous polyaniline counter electrode for dye-sensitized solar cells. *Electrochem. Commun.* 2008, 10, 1299-1302.

[37] Zhang, D. W.; Li, X. D.; Li, H. B.; Chen, S.; Sun, Z.; Yin, X. J.; Huang, S. M. Graphene-based counter electrode for dye-sensitized solar cells. *Carbon* 2011, 49, 5382-5388.

[38] Khan, M. Z. H.; Al-Mamun, M. R.; Halder, P. K.; Aziz, M. A. Performance improvement of modified dye-sensitized solar cells. *Renew. Sustain. Energy Rev.* 2017, 71, 602-617.

[39] Cavallo, C.; Di Pascasio, F.; Latini, A.; Bonomo, M.; Dini, D. Nanostructured semiconductor materials for dye-sensitized solar cells. *J. Nanomater.* 2017, 2017, 1-31.

[40] Wang, P.; Zakeeruddin, S. M.; Humphry-Baker, R.; Moser, J. E.; Grätzel, M. Molecular-scale interface engineering of TiO_2 nanocrystals: improve the efficiency and stability of dye-sensitized solar cells. *Adv. Mater.* 2003, 15, 2101-2104.

[41] Baxter, J. B.; Aydil, E. S. Dye-sensitized solar cells based on semiconductor morphologies with ZnO nanowires. *Sol. Energy Mater. Sol. Cells* 2006, 90, 607-622.

[42] Sabet, M.; Salavati-Niasari, M. Deposition of cadmium selenide semiconductor nanostructures on TiO2 surface via different chemical methods and investigation of their effects on dye sensitized solar cell efficiency. *Mater. Sci. Semicond. Process.* 2014, 27, 619-633.

[43] Sabet, M.; Salavati-Niasari, M. Deposition of lead sulfide nanostructure films on TiO_2 surface via different chemical methods due to improving dye-sensitized solar cells efficiency. *Electrochim. Acta* 2015, 169, 168-179.

[44] Zainudin, S. N. F.; Abdullah, H.; Markom, M. Electrochemical studies of tin oxide based-dye-sensitized solar cells (DSSC): a review. *J. Mater. Sci.: Mater. Electron.* 2019, 30, 5342-5356.

[45] Le Viet, A.; Jose, R.; Reddy, M. V.; Chowdari, B. V. R.; Ramakrishna, S. Nb_2O_5 photoelectrodes for dye-sensitized solar cells: choice of the polymorph. *J. Phys. Chem. C* 2010, 114, 21795-21800.

[46] Wu, J.; Lan, Z.; Hao, S.; Li, P.; Lin, J.; Huang, M.; Fang, L.; Huang, Y. Progress on the electrolytes for dye-sensitized solar cells. *Pure Appl. Chem.* 2008, 80, 2241-2258.

[47] Hoshikawa, T.; Ikebe, T.; Kikuchi, R.; Eguchi, K. Effects of electrolyte in dye-sensitized solar cells and evaluation by impedance spectroscopy. *Electrochim. Acta*, 2006, 51, 5286-5294.

[48] Yang, H.; Yu, C.; Song, Q.; Xia, Y.; Li, F.; Chen, Z.; Li, X.; Yi, T.; Huang, C. High-temperature and long-term stable solid-state electrolyte for dye-sensitized solar cells by self-assembly. *Chem. Mater.* 2006, 18, 5173-5177.

[49] Li, F.; Cheng, F.; Shi, J.; Cai, F.; Liang, M.; Chen, J. Novel quasi-solid electrolyte for dye-sensitized solar cells. *J. Power Sources* 2007, 165, 911-915.

[50] Cong, J.; Yang, X.; Kloo, L.; Sun, L. Iodine/iodide-free redox shuttles for liquid electrolyte-based dye-sensitized solar cells. *Energy Environ. Sci.* 2012, 5, 9180-9194.

[51] Kusama, H.; Arakawa, H. Influence of pyrazole derivatives in I^-/I_3^- redox electrolyte solution on Ru (II)-dye-sensitized TiO_2 solar cell performance. *Sol. Energy Mater. Sol. Cells* 2005, 85, 333-344.

[52] Daeneke, T.; Uemura, Y.; Duffy, N. W.; Mozer, A. J.; Koumura, N.; Bach, U.; Spiccia, L. Aqueous dye-sensitized solar cell electrolytes based on the ferricyanide-ferrocyanide redox couple. *Adv. Mater.* 2012, 24, 1222-1225.

[53] Li, B.; Wang, L.; Kang, B.; Wang, P.; Qiu, Y. Review of recent progress in solid-state dye-sensitized solar cells. *Sol. Energy Mater. Sol. Cells* 2006, 90, 549-573.

[54] Su'ait, M. S.; Rahman, M. Y. A.; Ahmad, A. Review on polymer electrolyte in dye-sensitized solar cells (DSSCs). *Sol. Energy* 2015, 115, 452-470.

[55] Mao, M.; Zhang, X. L.; Fang, X. Q.; Wu, G. H.; Dai, S. Y.; Song, Q. H.; Zhang, X. X. Highly efficient light-harvesting boradiazaindacene sensitizers for dye-sensitized solar cells featuring phenothiazine donor antenna. *J. Power Sources* 2014, 268, 965-976.

[56] Arteaga, D.; Cotta, R.; Ortiz, A.; Insuasty, B.; Martin, N.; Echegoyen, L. Zn (II)-porphyrin dyes with several electron acceptor groups linked by vinyl-fluorene or vinyl-thiophene spacers for dye-sensitized solar cells. *Dyes Pigm.* 2015, 112, 127-137.

[57] Sarikavak, K.; Kurtay, G.; Sevin, F.; Güllü, M. Molecular engineering of thienothiophene or dithienopyrrole-based π-spacers for dye-sensitized solar cells (DSSCs) with D-π-A architecture: A DFT/TD-DFT study. *Comput. Theor. Chem.* 2021, 1201, 113275.

[58] Chen, S. G.; Jia, H. L.; Ju, X. H.; Zheng, H. G. The impact of adjusting auxiliary donors on the performance of dye-sensitized solar cells based on phenothiazine D-D-π-A sensitizers. *Dyes Pigm.* 2017, 146, 127-135.

[59] Huang, R. Y.; Tsai, W. H.; Wen, J. J.; Chang, Y. J.; Chow, T. J. Spiro [fluorene-9, 9′-phenanthren]-10′-one as auxiliary acceptor of D-A-π-A dyes for dye-sensitized solar cells under one sun and indoor light. *J. Power Sources* 2020, 458, 228063-228072.

[60] Elmorsy, M. R.; Su, R.; Fadda, A. A.; Etman, H. A.; Tawfik, E. H.; El-Shafei, A. Co-sensitization of Ru (II) complex with terthiophene-based D-π-π-A metal-free organic dyes for highly efficient dye-sensitized solar cells: influence of anchoring group on molecular geometry and photovoltaic performance. *New J. Chem.* 2018, 14, 11430-11437.

[61] Naik, P.; Elmorsy, M. R.; Su, R.; Babu, D. D.; El-Shafei, A.; Adhikari, A. V. New carbazole based metal-free organic dyes with D-π-A-π-A architecture for DSSCs: synthesis, theoretical and cell performance studies. *Sol. Energy* 2017, 153, 600-610.

[62] Gao, F.; Cheng, Y.; Yu, Q.; Liu, S.; Shi, D.; Li, Y.; Wang, P. Conjugation of selenophene with bipyridine for a high molar extinction coefficient sensitizer in dye-sensitized solar cells. *Inorg. Chem.* 2009, 48, 2664-2669.

[63] Jiang, H.; Oniwa, K.; Islam, A.; Zhao, J.; Han, L.; Sun, Y. J.; Bao, M.; Asao, N.; Yamamoto, Y.; Jin, T. Efficient thieno [3, 2-a] carbazole-based organic dyes for dye-sensitized solar cells. *Tetrahedron*, 2015, 71, 6534-6540.

[64] Hua, Y.; Chang, S.; Huang, D.; Zhou, X.; Zhu, X.; Zhao, J.; Chen, T.; Wong, W. Y.; Wong, W. K. Significant improvement of dye-sensitized solar cell performance using simple phenothiazine-based dyes. *Chem. Mater.* 2013, 25, 2146-2153.

[65] Wang, S.; Guo, J.; He, L.; Wang, H.; Zhao, J.; Lu, C. Influence of thiophene and benzene unit in triphenylamine dyes on the performance of dye-sensitized solar cells. *Synth. Met.* 2013, 168, 1-8.

[66] Ren, X.; Jiang, S.; Cha, M.; Zhou, G.; Wang, Z. S. Thiophene-bridged double D-π-A dye for efficient dye-sensitized solar cell. *Chem. Mater.* 2012, 24, 3493-3499.

[67] Lin, J. T.; Chen, P. C.; Yen, Y. S.; Hsu, Y. C.; Chou, H. H.; Yeh, M. C. P. Organic dyes containing furan moiety for high-performance dye-sensitized solar cells. *Org. Lett.* 2009, 11, 97-100.

[68] Modelli, A.; Burrow, P. D. Electron attachment to dye-sensitized solar cell components: cyanoacetic acid. *J. Phys. Chem. A* 2011, 115, 1100-1107.

[69] Kula, S.; Szlapa-Kula, A.; Fabiańczyk, A.; Gnida, P.; Libera, M.; Bujak, K.; Siwy, M.; Schab-Balcerzak, E. Effect of thienyl units in cyanoacrylic acid derivatives toward dye-sensitized solar cells. *J. Photochem. Photobiol. B, Biol.* 2019, 197, 111555-111576.

[70] Gupta, A.; Armel, V.; Xiang, W.; Bilic, A.; Evans, R. A. New organic sensitizers using 4-(cyanomethyl) benzoic acid as an acceptor group for dye-sensitized solar cell applications. *Dyes Pigm.* 2015, 113, 280-288.

[71] Jin, L.; Shi, S.; Zhao, C.; Yu, X.; Lu, J.; Wang, Q.; Wei, Y. Y-shaped organic dyes with D_2-π-A configuration as efficient co-sensitizers for ruthenium-based dye sensitized solar cells. *J. Power Sources* 2021, 481, 228952-228959.

[72] Arkan, F.; Izadyar, M. The investigation of the central metal effects on the porphyrin-based DSSCs performance; molecular approach. *Mater. Chem. Phys.* 2017, 196, 142-152.

[73] Ilgün, C.; Sevim, A. M.; Çakar, S.; Özacar, M.; Gül, A. Novel Co and Zn-Phthalocyanine dyes with octa-carboxylic acid substituents for DSSCs. *Sol. Energy* 2021, 218, 169-179.

[74] Singh, S.; Maurya, I. C.; Sharma, S.; Kushwaha, S. P. S.; Srivastava, P.; Bahadur, L. Application of new natural dyes extracted from Nasturtium flowers (Tropaeolum majus) as photosensitizer in dye-sensitized solar cells. *Optik*, 2021, 243, 167331.

[75] Zhang, H.; Chen, Z. E.; Hu, J.; Hong, Y. Novel metal-free organic dyes containing linear planar 11, 12-dihydroindolo [2, 3-a] carbazole donor for dye-sensitized solar cells: effects of π spacer and alkyl chain. *Dyes Pigm.* 2019, 164, 213-221.

[76] Mukaddem, K. T.; Chater, P. A.; Devereux, L. R.; Al Bahri, O. K.; Jain, A.; Cole, J. M. Dye-anchoring modes at the dye··· TiO_2 interface of N3-and N749-sensitized solar cells revealed by glancing-angle pair distribution function analysis. *J. Phys. Chem. C* 2020, *124*, 11935-11945.

[77] Theil, F.; Zedler, L.; März, A.; Xie, W.; Csaki, A.; Fritzsche, W.; Cialla, D.; Schmitt, M.; Popp, J.; Dietzek, B. Evidence for SERRS enhancement in the spectra of ruthenium dye-metal nanoparticle conjugates. *J. Phys. Chem. C* 2013, 117, 1121-1129.

[78] Campbell, W. M.; Burrell, A. K.; Officer, D. L.; Jolley, K. W. Porphyrins as light harvesters in the dye-sensitised TiO_2 solar cell. *Coord. Chem. Rev.* 2004, 248, 1363-1379.

[79] Arkan, F.; Izadyar, M.; Nakhaeipour, A. A quantum chemistry study on the performance of porphyrin-based solar cell sensitisers; Zinc and anchor group position effects. *Mol. Phys.* 2015, 113, 3815-3825.

[80] Lu, J.; Li, H.; Liu, S.; Chang, Y. C.; Wu, H. P.; Cheng, Y.; Diau, E. W. G.; Wang, M. Novel porphyrin-preparation, characterization, and applications in solar energy conversion. *Phys. Chem. Chem. Phys.* 2016, 18, 6885-6892.

[81] Zou, J.; Yan, Q.; Li, C.; Lu, Y.; Tong, Z.; Xie, Y. Light-absorbing pyridine derivative as a new electrolyte additive for developing efficient porphyrin dye-sensitized solar cells. *ACS Appl. Mater. Interfaces* 2020, 12, 57017-57024.

[82] Campbell, W. M.; Jolley, K. W.; Wagner, P.; Wagner, K.; Walsh, P. J.; Gordon, K. C.; Schmidt-Mende, L.; Nazeeruddin, M. K.; Wang, Q.; Grätzel, M.; Officer, D. L. Highly efficient porphyrin sensitizers for dye-sensitized solar cells. *J. Phys. Chem. C* 2007, 111, 11760-11762.

[83] Sheng, Y.; Li, M.; Flores-Leonar, M. M.; Lu, W.; Yang, J.; Hu, Y. Rational design of SM315-based porphyrin sensitizers for highly efficient dye-sensitized solar cells: A theoretical study. *J. Mol. Struct.* 2020, 1205, 127567-127598.

[84] Li, S.; Zhang, Y.; Mei, S.; Kong, X.; Yang, M.; Hu, Z.; Wu, W.; He, J.; Tan, H. A molecular engineering srategy of phenylamine-based znc-porphyrin dyes for dye-sensitized solar cells: Synthesis, characteristics, and structure-performance relationships. *ACS Appl. Energy Mater.* 2021, 4, 9267-9275.

[85] Kimura, M.; Nomoto, H.; Masaki, N.; Mori, S. Dye molecules for simple co-sensitization process: Fabrication of mixed-dye-sensitized solar cells. *Angew. Chem. Int. Ed.* 2012, 51, 4371-4374.

[86] Urbani, M.; Ragoussi, M. E.; Nazeeruddin, M. K.; Torres, T. Phthalocyanines for dye-sensitized solar cells. *Coord. Chem. Rev.* 2019, 381, 1-64.

[87] Mori, S.; Nagata, M.; Nakahata, Y.; Yasuta, K.; Goto, R.; Kimura, M.; Taya, M. Enhancement of incident photon-to-current conversion efficiency for phthalocyanine-sensitized solar cells by 3D molecular structuralization. *J. Am. Chem. Soc.* 2010, 132, 4054-4055.

[88] Ince, M.; Yum, J. H.; Kim, Y.; Mathew, S.; Grätzel, M.; Torres, T.; Nazeeruddin, M. K. Molecular engineering of phthalocyanine sensitizers for dye-sensitized solar cells. *J. Phys. Chem. C* 2014, *118*, 17166-17170.

[89] Matsuzaki, H.; Murakami, T. N.; Masaki, N.; Furube, A.; Kimura, M.; Mori, S. Dye aggregation effect on interfacial electron-transfer dynamics in zinc phthalocyanine-sensitized solar cells. *J. Phys. Chem. C* 2014, *118*, 17205-17212.

[90] Ikeuchi, T.; Kudo, R.; Kitazawa, Y.; Mori, S.; Kimura, M. 3D structural optimization of Zinc phthalocyanine-based sensitizers for enhancement of open-circuit voltage of dye-sensitized solar cells. *Energies*, 2020, 13, 2288-2295.

[91] Gervaldo, M.; Fungo, F.; Durantini, E. N.; Silber, J. J.; Sereno, L.; Otero, L. Carboxyphenyl metalloporphyrins as photosensitizers of semiconductor film electrodes. A study of the effect of different central metals. *J. Phys. Chem. B* 2005, 109, 20953-20962.

[92] Ludin, N. A.; Mahmoud, A. A. A.; Mohamad, A. B.; Kadhum, A. A. H.; Sopian, K.; Karim, N. S. A. Review on the development of natural dye photosensitizer for dye-sensitized solar cells. *Renew. Sustain. Energy Rev.* 2014, 31, 386-396.

[93] Richhariya, G.; Kumar, A.; Tekasakul, P.; Gupta, B. Natural dyes for dye sensitized solar cell: A review. *Renew. Sustain. Energy Rev.* 2017, 69, 705-718.

[94] Omar, A.; Ali, M. S.; Abd Rahim, N. Electron transport properties analysis of titanium dioxide dye-sensitized solar cells (TiO_2-DSSCs) based natural dyes using electrochemical impedance spectroscopy concept: A review. *Sol. Energy* 2020, 207, 1088-1121.

[95] Chaiamornnugool, P.; Tontapha, S.; Phatchana, R.; Ratchapolthavisin, N.; Kanokmedhakul, S.; Sang-aroon, W.; Amornkitbamrung, V. Performance and stability of low-cost dye-sensitized solar cell based crude and pre-concentrated anthocyanins: Combined experimental and DFT/TDDFT study. *J. Mol. Struct.* 2017, 1127, 145-155.

[96] Zhou, H.; Wu, L.; Gao, Y.; Ma, T. Dye-sensitized solar cells using 20 natural dyes as sensitizers. *J. Photochem. Photobiol. A* 2011, 219, 188-194.

[97] Taya, S. A.; El-Agez, T. M.; El-Ghamri, H. S.; Abdel-Latif, M. S. Dye-sensitized solar cells using fresh and dried natural dyes. *Int. J. Mater. Sci. Appl.* 2013, 2, 37-42.

[98] Hao, S.; Wu, J.; Huang, Y.; Lin, J. Natural dyes as photosensitizers for dye-sensitized solar cell. *Sol. Energy* 2006, 80, 209-214.

[99] El-Agez, T. M.; El Tayyan, A. A.; Al-Kahlout, A.; Taya, S. A.; Abdel-Latif, M. S. Dye-sensitized solar cells based on ZnO films and natural dyes. *Int. J. Mater. Chem.* 2012, 2, 105-110.

[100] Gómez-Ortíz, N. M.; Vázquez-Maldonado, I. A.; Pérez-Espadas, A. R.; Mena-Rejón, G. J.; Azamar-Barrios, J. A.; Oskam, G. Dye-sensitized solar cells with natural dyes extracted from achiote seeds. *Sol. Energy Mater. Sol. Cells* 2010, 94, 40-44.

[101] Hosseinnezhad, M.; Gharanjig, K.; Moradian, S.; Saeb, M. R. In quest of power conversion efficiency in nature-inspired dye-sensitized solar cells: Individual, co-sensitized or tandem configuration? *Energy*, 2017, 134, 864-870.

[102] Narayan, M.; Raturi, A. Investigation of some common Fijian flower dyes as photosensi-tizers for dye sensitized solar cellsabstract. *Appl. Sol. Energy* 2011, 47, 112-117.

[103] Hussain, S. A. Development of dye sensitized solar cells using Botuje green leaves (Jathopha Curcas Linn). *Sci. J. Phys.* 2013, 2013, 1-4.

[104] Hernandez-Martinez, A. R.; Estevez, M.; Vargas, S.; Quintanilla, F.; Rodríguez, R. Natural pigment-based dye-sensitized solar cells. *J. Appl. Res. Technol.* 2012, 10, 38-47.

[105] Eckhardt, U.; Grimm, B.; Hörtensteiner, S. Recent advances in chlorophyll biosynthesis and breakdown in higher plants. *Plant Mol. Biol.* 2004, 56, 1-14.

[106] Wang, X. F.; Xiang, J.; Wang, P.; Koyama, Y.; Yanagida, S.; Wada, Y.; Hamada, K.; Sasaki, S. I.; Tmiaki, H. Dye-sensitized solar cells using a chlorophyll a derivative as the sensitizer and carotenoids having different conjugation lengths as redox spacers. *Chem. Phys. Lett.* 2005, 408, 409-414.

[107] Sabagh, S.; Izadyar, M.; Arkan, F. Insight into incident photon to current conversion efficiency in chlorophylls. *Int. J. Quantum Chem.* 2020, 121, 26483-26495.

[108] Frank, H. A.; Cogdell, R. J. Carotenoids in photosynthesis. *Photochem. Photobiol.* 1996, 63, 257-264.

[109] Wang, X. F.; Koyama, Y.; Nagae, H.; Yamano, Y.; Ito, M.; Wada, Y. Photocurrents of solar cells sensitized by aggregate-forming polyenes: Enhancement due to suppression of singlet-triplet annihilation by lowering of dye concentration or light intensity. *Chem. Phys. Lett.* 2006, 420, 309-315.

[110] Halidun, W. O. N. S.; Prima, E. C.; Yuliarto, B. *MATEC Web of Conferences;* Fabrication dye sensitized solar cells (DSSCs) using β-carotene pigment based natural dye, *EDP Sciences*, 2018, Vol. 159, pp 02052.

[111] Pratheesh, V. B.; Benny, N.; Sujatha, C. H. Isolation, stabilization and characterization of xanthophyll from marigold flower-Tagetes erecta-L. *Mod. Appl. Sci.* 2009, 3, 19-28.

[112] Yamazaki, E.; Murayama, M.; Nishikawa, N.; Hashimoto, N.; Shoyama, M.; Kurita, O. Utilization of natural carotenoids as photosensitizers for dye-sensitized solar cells. *Sol. Energy* 2007, 81, 512-516.

[113] Galeotti, F.; Barile, E.; Curir, P.; Dolci, M.; Lanzotti, V. Flavonoids from carnation (Dianthus caryophyllus) and their antifungal activity. *Phytochem. Lett.* 2008, 1, 44-48.

[114] Ververidis, F.; Trantas, E.; Douglas, C.; Vollmer, G.; Kretzschmar, G.; Panopoulos, N. Biotechnology of flavonoids and other phenylpropanoid-derived natural products. Part I: Chemical diversity, impacts on plant biology and human health. *Biotechnol. J.* 2007, 2, 1214-1234.

[115] Ververidis, F.; Trantas, E.; Douglas, C.; Vollmer, G.; Kretzschmar, G.; Panopoulos, N. Biotechnology of flavonoids and other phenylpropanoid-derived natural products. Part II: Reconstruction of multienzyme pathways in plants and microbes. *Biotechnol. J.* 2007, 2, 1235-1249.

[116] Sabagh, S.; Izadyar, M.; Arkan, F. Photovoltaic properties of the flavonoid-based photosensitizers: Molecular-scale perspective on the natural dye solar cells. *Int. J. Quantum Chem.* 2020, 120, 26171-26181.

[117] Karageorgou, P.; Manetas, Y. The importance of being red when young: anthocyanins and the protection of young leaves of Quercus coccifera from insect herbivory and excess light. *Tree Physiol.* 2006, 26, 613-621.

[118] Isaad, J.; El Achari, A. Colorimetric sensing of cyanide anions in aqueous media based on functional surface modification of natural cellulose materials. *Tetrahedron*, 2011, 67, 4939-4947.

[119] Patel, K.; Jain, A.; Patel, D. K. Medicinal significance, pharmacological activities, and analytical aspects of anthocyanidins 'delphinidin': A concise report. *J. Acute Dis.* 2013, 2, 169-178.

[120] Gokilamani, N.; Muthukumarasamy, N.; Thambidurai, M.; Ranjitha, A.; Velauthapillai, D. Grape pigment (malvidin-3-fructoside) as natural sensitizer for dye-sensitized solar cells. *Mater. Renew. Sustain. Energy,* 2014, 3, 1-7.

[121] Mohan, R.; Geetha, N.; Jennifer, D. H.; Sivakumar, V. Studies on natural dye (Pelargonidin) extraction from onion peel and application in dyeing of leather. *Int. J. Eng. Sci.* 2020, 7, 14-22.

[122] Rajan, V. K.; Hasna, C. K.; Muraleedharan, K. The natural food colorant Peonidin from cranberries as a potential radical scavenger-a DFT based mechanistic analysis. *Food Chem.* 2018, 262, 184-190.

[123] Maahury, M. F.; Martoprawiro, M. A.; Sutapa, I. W. Computational calculation potency of petunidin and peonidin as photosensitizer in dye-sensitized solar cell, *Int. J. Sci. Res.* 2019, 9, 812-816.

[124] Fossen, T.; Cabrita, L.; Andersen, O. M. Colour and stability of pure anthocyanins influenced by pH including the alkaline region. *Food Chem.* 1998, 63, 435-440.

[125] Wu, H.; Huang, Z.; Hua, T.; Liao, C.; Meier, H.; Tang, H.; Wang, L.; Cao, D. Metal-free organic dyes with di(1-benzothieno)[3,2-b:2′,3′-d] pyrrole as a donor for efficient dye-sensitized solar cells: Effect of mono-and bi-anchors on photovoltaic performance. *Dyes Pigm.* 2019, 165, 103-111.

[126] Lin, F. S.; Priyanka, P.; Fan, M. S.; Vegiraju, S.; Ni, J. S.; Wu, Y. C.; Li, Y. H.; Lee, G. H.; Ezhumalai, Y.; Jeng, R. J.; Chen, M. C. Metal-free efficient dye-sensitized solar cells based on thioalkylated bithiophenyl organic dyes. *J. Mater. Chem. C* 2020, 8, 15322-15330.

[127] Devadiga, D.; Selvakumar, M.; Shetty, P.; Sridhar Santosh, M.; Chandrabose, R. S.; Karazhanov, S. Recent developments in metal-free organic sensitizers derived from carbazole, triphenylamine, and phenothiazine for dye-sensitized solar cells. *Int. J. Energy Res.* 2021, 45, 6584-6643.

[128] Cherifi, K.; Cheknane, A.; Hilal, H. S.; Benghia, A.; Rahmoun, K.; Benyoucef, B. Investigation of triphenylamine-based sensitizer characteristics and adsorption behavior onto $ZnTiO_3$ perovskite (1 0 1) surfaces for dye-sensitized solar cells using first-principle calculation. *Chem. Phys.* 2020, 530, 110595.

[129] Jiang, S.; Chen, Y.; Li, Y.; Han, L. Novel D-D-π-A indoline-linked coumarin sensitizers for dye-sensitized solar cells. *J. Photochem. Photobiol. A* 2019, 384, 112031-112038.

[130] Arkan, F.; Izadyar, M. Optoelectronic properties and energy conversion of organic dye-based solar cells: Molecular approach. *Optik* 2020, 203, 163972-163981.

[131] Han, L.; Chen, Y.; Cui, Y.; Jiang, S. Phenothiazine dyes containing a 4-phenyl-2-(thiophen-2-yl) thiazole bridge for dye-sensitized solar cells. *Tetrahedron* 2020, 76, 131102-131109.

[132] Zhang, W.; Yang, X.; An, J.; Zhang, L.; Cai, B.; Yang, K. Effect of side substituents incorporated into π-bridges of quinoxaline-based sensitizers for dye-sensitized solar cells. *Energy Technol.* 2020, 8, 2000032-2000049.

[133] Patil, J. V.; Mali, S. S.; Shaikh, J. S.; Patil, A. P.; Patil, P. S.; Hong, C. K. Influence of reduced graphene oxide-TiO_2 composite nanofibers in organic indoline DN350 based dye sensitized solar cells. *Synth. Met.* 2019, 256, 116146-116154.

[134] Huang, Z. S.; Zang, X. F.; Hua, T.; Wang, L.; Meier, H.; Cao, D. 2,3-Dipentyldithieno[3,2-f:2′,3′-h]quinoxaline-Based organic dyes for efficient dye-sensitized solar cells: effect of π-bridges and electron donors on solar cell performance. *ACS Appl. Mater. Interfaces* 2015, 7, 20418-20429.

[135] Qian, X.; Zhu, Y. Z.; Chang, W. Y.; Song, J.; Pan, B.; Lu, L.; Gao, H. H.; Zheng, J. Y. Benzo[a] carbazole-based donor-π-acceptor type organic dyes for highly efficient dye-sensitized solar cells. *ACS Appl. Mater. Interfaces* 2015, 7, 9015-9022.

[136] Xu, M.; Hu, X.; Zhang, Y.; Bao, X.; Pang, A.; Fang, J. K. Novel organic dyes featuring spiro [dibenzo[3,4:6,7]cyclohepta[1,2-b]quinoxaline-10,9′-fluorene] (SDBQX) as a rigid moiety for dye-sensitized solar cells. *ACS Appl. Energy Mater.* 2018, 1, 2200-2207.

[137] Yen, Y. S.; Ni, J. S.; Hung, W. I.; Hsu, C. Y.; Chou, H. H.; Lin, J. T. S. Naphtho [2,3-c][1,2,5] thiadiazole and 2H-Naphtho[2,3-d][1,2,3] triazole-containing D-A-π-A conjugated organic dyes for dye-sensitized solar cells. *ACS Appl. Mater. Interface* 2016, 8, 6117-6126.

[138] Harikrishnan, M.; Sadhasivam, V.; Siva, A.; Anandan, S.; Subbiah, V.; Murugesan, S. Energy level tuning of novel star-shaped D-π-D-A-based metal-free organic dyes for solar cell application. *J. Phys. Chem. C* 2019, 123, 21959-21968.

[139] Mahmood, A. Triphenylamine based dyes for dye sensitized solar cells: A review. *Sol. Energy* 2016, 123, 127-144.

[140] Wu, Z. S.; Guo, W. J.; Zhang, J.; Liu, Y. D.; Song, X. C.; Jiang, Y.; Weng, Q.; An, Z. W. Novel 4, 4′-bis (alkylphenyl/alkyloxyphenyl)-2, 2′-bithiophene bridged

cyclic thiourea functionalized triphenylamine sensitizers for efficient dye-sensitized solar cells. *Sol. Energy* 2019, 186, 1-8.

[141] Yan, Y. Q.; Zhu, Y. Z.; Dai, P. P.; Han, J.; Yan, M.; Zheng, J. Y. Effect of triphenylethylene analogues on the performance of carbazole-based dye-sensitized solar cells. *Sol. Energy* 2020, 207, 428-435.

[142] Kitamura, T.; Ikeda, M.; Shigaki, K.; Inoue, T.; Anderson, N. A., Ai, X.; Lian, T.; Yanagida, S. Phenyl-conjugated oligoene sensitizers for TiO_2 solar cells. *Chem. Mater.* 2004, 16, 1806-1812.

[143] Ferdowsi, P.; Saygili, Y.; Jazaeri, F.; Edvinsson, T.; Mokhtari, J.; Zakeeruddin, S. M.; Liu, Y.; Grätzel, M.; Hagfeldt, A. Molecular engineering of simple metal-free organic dyes derived from triphenylamine for dye-sensitized solar cell applications. *ChemSusChem* 2020, 13, 212-220.

[144] Slimi, A.; Hachi, M.; Fitri, A.; Benjelloun, A. T.; Elkhattabi, S.; Benzakour, M.; Mcharfi, M.; Khenfouch, M.; Zorkani, I.; Bouachrine, M. Effects of electron acceptor groups on triphenylamine-based dyes for dye-sensitized solar cells: Theoretical investigation. *J. Photochem. Photobiol. A* 2020, 398, 112572.

[145] Mohankumar, V.; Pounraj, P.; Pandian, M. S.; Ramasamy, P. Tuning the lifetime from molecular engineering of carbazole donor based metal-free organic dyes for dye sensitized solar cells-A computational approach. *J. Mol. Struct.* 2019, 1195, 494-505.

[146] Nesheli, F. G.; Tajbakhsh, M.; Hosseinzadeh, B.; Hosseinzadeh, R. Design, synthesis and photophysical analysis of new unsymmetrical carbazole-based dyes for dye-sensitized solar cells. *J. Photochem. Photobiol. A* 2020, 397, 112521-112567.

[147] Ennehary, S.; Toufik, H.; Lazrak, M.; Bouzzine, S. M.; Lamchouri, F. Computational study of the effect of π-spacers on the optoelectronic properties of carbazole-based organic dyes. *J. Mol. Model.* 2021, 27, 1-10.

[148] Naik, P.; Su, R.; Elmorsy, M. R.; El-Shafei, A.; Adhikari, A.V. Investigation of new carbazole based metal-free dyes as active photo-sensitizers/co-sensitizers for DSSCs. *Dyes Pigm.* 2018, 149, 177-187.

[149] Ouared, I.; Rekis, M.; Trari, M. Phenothiazine based organic dyes for dye sensitized solar cells: A theoretical study on the role of π-spacer. *Dyes Pigm.* 2021, 190, 109330.

[150] Sambathkumar, S.; Zych, D.; Kavitha, E.; Ramesh, P.; Jagatheesan, R. Theoretical investigations on the electronic absorption properties of phenothiazine based organic materials for dye-sensitized solar cells. *Mater. Today: Proc* 2021, Doi: 10.1016/j.matpr.2021.03.715.

[151] Periyasamy, K.; Sakthivel, P.; Vennila, P.; Anbarasan, P. M.; Venkatesh, G.; Mary, Y. S. Novel D-π-A phenothiazine and dibenzofuran organic dyes with simple structures for efficient dye-sensitized solar cells. *J. Photochem. Photobiol. A* 2021, 413, 113269.

[152] Tian, H.; Yang, X.; Chen, R.; Pan, Y.; Li, L.; Hagfeldt, A.; Sun, L. Phenothiazine derivatives for efficient organic dye-sensitized solar cells. *Chem. Commun.* 2007, 36, 3741-3743.

[153] Xu, Z.; Li, Y.; Zhang, W.; Yuan, S.; Hao, L.; Xu, T.; Lu, X. DFT/TD-DFT study of novel T shaped phenothiazine-based organic dyes for dye-sensitized solar cells applications. *Spectrochim. Acta A Mol. Biomol. Spectrosc.* 2019, 212, 272-280.

[154] Slodek, A.; Zych, D.; Szafraniec-Gorol, G.; Gnida, P.; Vasylieva, M.; Schab-Balcerzak, E. Investigations of new phenothiazine-based compounds for dye-sensitized solar cells with theoretical. *Materials*, 2020, 13, 2292.

[155] El Assyry, A.; Lamsayah, M.; Warad, I.; Touzani, R.; Bentiss, F.; Zarrouk, A. Theoretical investigation using DFT of quinoxaline derivatives for electronic and photovoltaic effects. *Heliyon*, 2020, 6, 03620-03629.

[156] Wazzan, N.; Irfan, A. Promising architectures modifying the D-π-A architecture of 2, 3-dipentyldithieno [3, 2-f: 2′, 3′-h] quinoxaline-based dye as efficient sensitizers in dye-sensitized solar cells: a DFT study. *Mater. Sci. Semicond. Process.* 2020, 120, 105260-105272.

[157] Jiang, H.; Wu, Y.; Islam, A.; Wu, M.; Zhang, W.; Shen, C.; Zhang, H.; Li, E.; Tian, H.; Zhu, W. H. Molecular engineering of quinoxaline-based D-A-π-A organic sensitizers: taking the merits of a large and rigid auxiliary acceptor. *ACS Appl. Mater. Interface* 2018, 10, 13635-13644.

[158] Barati-darband, F.; Izadyar, M.; Arkan, F. Anchoring group and π-spacer effects on the dynamics and kinetics of the photovoltaic processes in the quinoxaline-based organic dye-sensitized solar cells. *J. Phys. Chem. C* 2018, 122, 23968-23977.

[159] Lewars, E. *Computational chemistry*; *Introduction to the theory and applications of molecular and quantum mechanics*, Springer Netherlands, 2011, Vol. 43, pp 44-45.

[160] McArdle, S.; Endo, S.; Aspuru-Guzik, A.; Benjamin, S. C.; Yuan, X. Quantum computational chemistry. *Rev. Mod. Phys.* 2020, 92, 015003-15053.

[161] Frisch, M. J.; Trucks, G. W.; Schlegel, H. B.; Scuseria, G. E.; Robb, M. A.; Cheeseman, J. R.; Scalmani, G.; Barone, V.; Petersson, G. A.; Nakatsuji, H.; Li, X.; Hratchian, H. P.; Izmaylov, A. F.; Bloino, J.; Zheng, G.; Sonnenberg, J. L.; Hada, M.; Ehara, M.; Toyota, K.; Fukuda, R.; Hasegawa, J.; Ishida, M.; Nakajima, T.; Honda, Y.; Kitao, O.; Nakai, H.; Vreven, T.; Montgomery, J. A.; Peralta, J. E.; Ogliaro, F.; Bearpark, M.; Heyd, J. J.; Brothers, E.; Kudin, K. N.; Staroverov, V. N.; Kobayashi, R.; Normand, J.; Raghavachari, K.; Rendell, A.; Burant, J. C.; Iyengar, S. S.; Tomasi, J.; Cossi, M.; Rega, N.; Millam, J. M.; Klene, M.; Knox, J. E.; Cross, J. B.; Bakken, V.; Adamo, C.; Jaramillo, J.; Gomperts, R.; Stratmann, R. E.; Yazyev, O.; Austin, A. J.; Cammi, R.; Pomelli, C.; Ochterski, J. W.; Martin, R. L.; Morokuma, K.; Zakrzewski, V. G.; Voth, G. A.; Salvador, P.; Dannenberg, J. J.; Dapprich, S.; Daniels, A. D.; Farkas, O.; Foresman, J. B.; Ortiz, J. V.; Cioslowski, J.; Fox, D. J. Gaussian 09, *Revision A.01, Inc. Wallingford CT*, 2009, 121, 150-166.

[162] Stangl, R.; Leendertz, C.; Haschke, J. Numerical simulation of solar cells and solar cell characterization methods: the open-source on demand program AFORS-HET. *Sol. Energy* 2010, 14, 319-352.

[163] Chakraborty, K.; Choudhury, M. G.; Paul, S. Numerical study of Cs_2TiX_6 (X= Br-, I-, F- and Cl-) based perovskite solar cell using SCAPS-1D device simulation. *Sol. Energy* 2019, 194, 886-892.

[164] Clugston, D. A., Basore, P. A. PC1D version 5: 32-bit solar cell modeling on personal computers. *IEEE* 1997, 207-210 Doi: 10.1109/PVSC.1997.654065.

[165] Soler, J. M.; Artacho, E.; Gale, J. D.; García, A.; Junquera, J.; Ordejón, P.; Sánchez-Portal, D. The SIESTA method for ab initio order-N materials simulation. *J. Phys. Condens. Matter Phys.* 2002, 14, 2745.

[166] Lu, T.; Chen, F. Multiwfn: a multifunctional wavefunction analyzer. *J. Comput. Chem.* 2012, 33, 580-592.

[167] Bhuvaneswari, G.; Annamalai, R. Development of a solar cell model in MATLAB for PV based generation system. *IEEE* 2011, 1-5 Doi: 10.1109/INDCON.2011.6139509.

[168] Giannozzi, P.; Baroni, S.; Bonini, N.; Calandra, M.; Car, R.; Cavazzoni, C.; Ceresoli, D.; Chiarotti, G. L.; Cococcioni, M.; Dabo, I.; Dal Corso, A. QUANTUM ESPRESSO: a modular and open-source software project for quantum simulations of materials. *J. Phys. Condens. Matter Phys.* 2009, 21, 395502-395520.

[169] Mishra, S.; Bhargava, K.; Deb, D. Numerical simulation of potential induced degradation (PID) in different thin-film solar cells using SCAPS-1D. *Sol. Energy* 2019, 188, 353-360.

[170] Te Velde, G. T.; Bickelhaupt, F. M.; Baerends, E. J.; Fonseca Guerra, C.; van Gisbergen, S. J.; Snijders, J. G.; Ziegler, T. Chemistry with ADF. *J. Comput. Chem.* 2001, 22, 931-967.

[171] Rackers, J. A.; Wang, Z.; Lu, C.; Laury, M. L.; Lagardère, L.; Schnieders, M. J.; Piquemal, J. P.; Ren, P.; Ponder, J. W. Tinker 8: software tools for molecular design. *J. Chem. Theory Comput.* 2018, 14, 5273-5289.

[172] Alden, K.; Read, M.; Timmis, J.; Andrews, P. S.; Veiga-Fernandes, H.; Coles, M. Spartan: a comprehensive tool for understanding uncertainty in simulations of biological systems. *PLoS Comput. Biol.* 2013, 9, 1002916-1002924.

[173] Schmidt, M. W.; Baldridge, K. K.; Boatz, J. A.; Elbert, S. T.; Gordon, M. S.; Jensen, J. H.; Koseki, S.; Matsunaga, N.; Nguyen, K. A.; Su, S.; Windus, T. L. General atomic and molecular electronic structure system. *J. Comput. Chem.* 1993, 14, 1347-1363.

[174] Godbout, N.; Salahub, D. R.; Andzelm, J.; Wimmer, E. Optimization of Gaussian-type basis sets for local spin density functional calculations. Part I. Boron through neon, optimization technique and validation. *Can. J. Chem.* 1992, 70, 560-571.

[175] Stratmann, R. E.; Scuseria, G. E.; Frisch, M. J. An efficient implementation of time-dependent density-functional theory for the calculation of excitation energies of large molecules. *J. Chem. Phys.* 1998, 109, 8218-8224.

[176] Löwdin, P. O. Quantum theory of many-particle systems. I. Physical interpretations by means of density matrices, natural spin-orbitals, and convergence problems in the method of configurational interaction. *Phys. Rev.* 1955, 97, 1474-1488.

[177] Foster, A. J.; Weinhold, F. Natural hybrid orbitals. *J. Am. Chem. Soc.* 1980, 102, 7211-7218.

[178] Xu, Y.; Schoonen, M. A. The absolute energy positions of conduction and valence bands of selected semiconducting minerals. *Am. Mineral.* 2000, 85, 543-556.

[179] Bagheri Novir, S.; Hashemianzadeh, S. M. Computational investigation of low band gap dyes based on 2-styryl-5-phenylazo-pyrrole for dye-sensitized solar cells. *Curr. Appl. Phys.* 2014, 14, 1401-1410.

[180] Arkan, F.; Izadyar, M.; Nakhaeipour, A. Improvement in charge transfer dynamic of the porphyrin-based solar cells in water: A theoretical study. *J. Renew. Sustain. Energy* 2017, 9, 023502.

[181] Sharma, V.; Jha, P. K. Enhancement in power conversion efficiency of edge-functionalized graphene quantum dot through adatoms for solar cell applications. *Sol. Energy Mater Sol. Cells* 2019, 200, 109908-109915.

[182] Mennucci, B.; Tomasi, J.; Cammi, R.; Cheeseman, J. R.; Frisch, M. J.; Devlin, F. J.; Gabriel, S.; Stephens, P. J. Polarizable continuum model (PCM) calculations of solvent effects on optical rotations of chiral molecules. *J. Phys. Chem. A* 2002, 106, 6102-6113.

[183] Roohi, H.; Noruozi, A. R.; Salemi, S.; Sharaki, J. Theoretical study of solvent effects on the conformational preference in CH_2FWH (W=O,S) using PCM and IPCM methods. *J. Mol. Liq.* 2008, 143, 119-124.

[184] Grensemann, H.; Gmehling, J. Performance of a conductor-like screening model for real solvents model in comparison to classical group contribution methods. *Ind. Eng. Chem. Res.* 2005, 44, 1610-1624.

[185] Guerard, J. J.; Arey, J. S. Critical evaluation of implicit solvent models for predicting aqueous oxidation potentials of neutral organic compounds. *J. Chem. Theory Comput.* 2013, 9, 5046-5058.

[186] Ferrara, P.; Apostolakis, J.; Caflisch, A. Evaluation of a fast implicit solvent model for molecular dynamics simulations. *Proteins* 2002, 46, 24-33.

[187] Wang, L. P.; Van Voorhis, T. A polarizable QM/MM explicit solvent model for computational electrochemistry in water. *J. Chem. Theory Comput.* 2012, 8, 610-617.

[188] Marques, M. A.; Oliveira, M. J.; Burnus, T. Libxc: A library of exchange and correlation functionals for density functional theory. *Comput. Phys. Commun.* 2012, 183, 2272-2281.

[189] Perdew, J. P.; Wang, Y. Accurate and simple analytic representation of the electron-gas correlation energy. *Phys. Rev. B* 1992, 45, 13244-13249.

[190] Vosko, S. H.; Wilk, L.; Nusair, M. Accurate spin-dependent electron liquid correlation energies for local spin density calculations: a critical analysis. *Can. J. Phys.* 1980, 58, 1200-1211.

[191] Hohenberg, P.; Kohn, W. Inhomogeneous electron gas. *Phys. Rev.* 1964, 136, 864-871.

[192] Perdew, J. P.; Burke, K.; Ernzerhof, M. Generalized gradient approximation made simple. *Phys. Rev. Lett.* 1996, 77, 3865-3868.

[193] Alexiadis, A.; Kassinos, S. On the use of the BLYP functional for the DFT calculation of graphite-hydrogen systems. *J. Nucl. Mater.* 2010, 396, 307-308.

[194] Xu, X.; Goddard, W. A. The X3LYP extended density functional for accurate descriptions of nonbond interactions, spin states, and thermochemical properties. *Proc. Natl. Acad. Sci.* 2004, 101, 2673-2677.

[195] Mortensen, J. J.; Kaasbjerg, K.; Frederiksen, S. L.; Nørskov, J. K.; Sethna, J. P.; Jacobsen, K. W. Bayesian error estimation in density-functional theory. *Phys. Rev. Lett.* 2005, 95, 216401-216404.

[196] Casida, M. E.; Jamorski, C.; Casida, K. C.; Salahub, D. R. Molecular excitation energies to high-lying bound states from time-dependent density-functional response theory: Characterization and correction of the time-dependent local density approximation ionization threshold. *J. Chem. Phys.* 1998, 108, 4439-4449.

[197] Patey, M. D.; Dessent, C. E. A PW91 density functional study of conformational choice in 2-phenylethanol, n-butylbenzene, and their cations: Problems for density functional theory?. *J. Phys. Chem. A* 2002, 106, 4623-4631.

[198] Becke, A. D. Density-functional exchange-energy approximation with correct asymptotic behavior. *Phys. Rev. A* 1988, 106, 3098-3100.

[199] Tao, J.; Perdew, J. P.; Staroverov, V. N.; Scuseria, G. E. Climbing the density functional ladder: Nonempirical meta-generalized gradient approximation designed for molecules and solids. *Phys. Rev. Lett.* 2003, 91, 146401-146404.

[200] Zhao, Y.; Truhlar, D. G. A new local density functional for main-group thermochemistry, transition metal bonding, thermochemical kinetics, and noncovalent interactions. *J. Chem. Phys.* 2006, 125, 194101-194119.

[201] Sun, J.; Xiao, B.; Ruzsinszky, A. Communication: Effect of the orbital-overlap dependence in the meta generalized gradient approximation. *J. Chem. Phys.* 2012, 137, 051101-051105.

[202] Sun, J.; Haunschild, R.; Xiao, B.; Bulik, I. W.; Scuseria, G. E.; Perdew, J. P. Semilocal and hybrid meta-generalized gradient approximations based on the understanding of the kinetic-energy-density dependence. *J. Chem. Phys.* 2013, 138, 044113-044121.

[203] Becke, A. D. Density-functional thermochemistry. I. The effect of the exchange-only gradient correction. *J. Chem. Phys.* 1992, 96, 2155-2160.

[204] Adamo, C.; Barone, V. Toward reliable density functional methods without adjustable parameters: The PBE0 model. *J. Chem. Phys.* 1999, 110, 6158-6170.

[205] Chai, J. D.; Head-Gordon, M. Long-range corrected hybrid density functionals with damped atom-atom dispersion corrections. *Phys. Chem. Chem. Phys.* 2008, 10, 6615-6620.

[206] Jensen, K. P. Bioinorganic chemistry modeled with the TPSSh density functional. *Inorg. Chem.* 2008, 47, 10357-10365.

[207] Zhao, Y.; Truhlar, D.G. The M06 suite of density functionals for main group thermochemistry, thermochemical kinetics, noncovalent interactions, excited states, and transition elements: two new functionals and systematic testing of four M06-class functionals and 12 other functionals. *Theor. Chem. Acc.* 2008, 120, 215-241.

[208] Boese, A. D.; Martin, J. M. Development of density functionals for thermochemical kinetics. *J. Chem. Phys.* 2004, 121, 3405-3416.

[209] Yanai, T.; Tew, D. P.; Handy, N. C. A new hybrid exchange-correlation functional using the Coulomb-attenuating method (CAM-B3LYP). *Chem. Phys. Lett.* 2004, 393, 51-57.

[210] Vydrov, O. A.; Scuseria, G. E. Assessment of a long-range corrected hybrid functional. *J. Chem. Phys.* 2006, 125, 234109.

[211] Khajehzadeh, M.; Moghadam, M. Molecular structure, FT IR, NMR, UV, NBO and HOMO-LUMO of 1-(3-(dimethylamino)propyl)-1-(4-fluorophenyl)-1,3-dihydroisobenzofuran-5-carbonitrile by DFT/B3LYP and PBEPBE methods with LanL2DZ and 6-311++G(d 2p) basis sets. *Spectrochim. Acta A Mol. Biomol. Spectrosc.* 2017, 180, 51-66.

[212] AL-Temimei, F. A.; Alkhayatt, A. H. O. A DFT/TD-DFT investigation on the efficiency of new dyes based on ethyl red dye as a dye-sensitized solar cell light-absorbing material. *Optik*, 2020, 208, 163920-163935.

[213] Arkan, F.; Izadyar, M.; Nakhaeipour, A. The role of the electronic structure and solvent in the dye-sensitized solar cells based on Zn-porphyrins: Theoretical study. *Energy* 2016, 114, 559-567.

[214] Urbani, M.; Grätzel, M.; Nazeeruddin, M. K.; Torres, T. Meso-substituted porphyrins for dye-sensitized solar cells. *Chem. Rev.* 2014, 114, 12330-12396.

[215] Wazzan, N.; Irfan, A. Theoretical study of triphenylamine-based organic dyes with mono-, di-, and tri-anchoring groups for dye-sensitized solar cells. *Org. Electron.* 2018, 63, 328-342.

[216] Wang, H.; Zhao, D.; Song, P.; Ma, F.; Li, Y. Effect of charge transport channel and interaction of IDT type dyes on photoelectric characteristics. *J. Mol. Liq.* 2020, 303, 112594-112606.

[217] Heng, P.; Mao, L.; Guo, X.; Wang, L.; Zhang, J. Accurate estimation of the photoelectric conversion efficiency of a series of anthracene-based organic dyes for dye-sensitized solar cells. *J. Mater. Chem. C* 2020, 8, 2388-2399.

[218] Zhang, W.; Heng, P.; Su, H.; Ren, T.; Wang, L.; Zhang, J. Rational design of high-efficiency organic dyes in dye-sensitized solar cells by multiscale simulations. *J. Phys. Chem. C* 2018, 122, 25219-25228.

[219] Pakravesh, F.; Izadyar, M.; Arkan, F. Effect of electron donor and acceptor on the photovoltaic properties of organic dyes for efficient dye-sensitized solar cells. *Physica B Condens. Matter.* 2021, 609, 412815-412822.

[220] Kuhn, H. A quantum-mechanical theory of light absorption of organic dyes and similar compounds. *J. Chem. Phys.* 1949, 17, 1198-1212.

[221] Al-Alwani, M. A.; Mohamad, A. B.; Ludin, N. A.; Kadhum, A. A. H.; Sopian, K. Dye-sensitised solar cells: Development, structure, operation principles, electron kinetics, characterisation, synthesis materials and natural photosensitisers. *Renew. Sustain. Energy Rev.* 2016, 65, 183-213.

[222] Kumara, N. T. R. N.; Lim, A.; Lim, C. M.; Petra, M. I.; Ekanayake, P. Recent progress and utilization of natural pigments in dye sensitized solar cells: A review. *Renew. Sustain. Energy Rev.* 2017, 78, 301-317.

[223] Arkan, F.; Izadyar, M. Molecular engineering of the organometallic perovskites/HTMs in the PSCs: Photovoltaic behavior and energy conversion. *Sol. Energy Mater. Sol. Cells* 2018, 180, 46-58.

[224] Arkan, F.; Izadyar, M. The role of solvent and structure in the kinetics of the excitons in porphyrin-based hybrid solar cells. *Sol. Energy* 2017, 146, 368-378.

[225] Arkan, F.; Izadyar, M. Structural modification as the pioneer strategy in competition of the porphyrin dye and perovskite solar cells: From dynamics to kinetics of the photovoltaic processes. *Appl. Phys. Lett.* 2019, 115, 183903-183907.

[226] Peymani, S.; Izadyar, M.; Arkan, F. Photovoltaic and spectroscopic properties of bacteriochlorin-based photosensitizer: molecular approach. *Res. Chem. Intermed.* 2021, 47, 1071-1085.

[227] Arkan, F.; Izadyar, M. Advances in molecular engineering of organic-inorganic/inorganic halide perovskites: Photochemical properties behind the energy conversion ability. *Sol. Energy* 2019, 194, 51-60.

[228] Prakasam, M.; Anbarasan, P. M. Stilbene based organic dye as efficient sensitizer for NLO and dye-sensitized solar cells: A first principle study. *Mater. Today:* Proc. 2019, 9, 156-163.

[229] Khan, M. U.; Ibrahim, M.; Khalid, M.; Jamil, S.; Al-Saadi, A. A.; Janjua, M. R. S. A. Quantum chemical designing of indolo [3, 2, 1-jk] carbazole-based dyes for highly efficient nonlinear optical properties. *Chem. Phys. Lett.* 2019, 719, 59-66.

[230] Gardeh, M. G.; Hashemianzadeh, S. M. Study of charge injection in Ortho-, Meta- and Para-NKX-2553 in dye-sensitized solar cells: A theoretical approach. *Optik*, 2020, 203, 163987-164016.

[231] Li, Y.; Liu, J.; Liu, D.; Li, X.; Xu, Y. D-A-π-A based organic dyes for efficient DSSCs: A theoretical study on the role of π-spacer. *Comput. Mater. Sci.* 2019, 161, 163-176.

[232] Wu, H.; Ma, T.; Wu, C.; Yan, L.; Su, Z. Effect of polyoxometalate in organic-inorganic hybrids on charge transfer and absorption spectra towards sensitizers. *Dyes Pigm.* 2017, 142, 379-386.

[233] Li, Y.; Li, X.; Xu, Y. Theoretical screening of high-efficiency sensitizers with D-π-A framework for DSSCs by altering promising donor group. *Sol. Energy* 2020, 196, 146-156.

[234] Le Bahers, T.; Adamo, C.; Ciofini, I. A qualitative index of spatial extent in charge-transfer excitations. *J. Chem. Theory Comput.* 2011, 7, 2498-2506.

[235] He, L. J.; Chen, J.; Bai, F. Q.; Jia, R.; Wang, J.; Zhang, H. X. Fine-tuning π-spacer for high efficiency performance DSSC: a theoretical exploration with D-π-A based organic dye. *Dyes Pigm.* 2017, 141, 251-261.

[236] Ali, B. A.; Allam, N. K. Propping the optical and electronic properties of potential photo-sensitizers with different π-spacers: TD-DFT insights. *Spectrochim. Acta A Mol. Biomol. Spectrosc.* 2018, 188, 237-243.

[237] Saengsuwan, N.; Klinhom, N.; Nakwanich, B.; Sriyab, S.; Prompinit, P.; Suramitr, S.; Hannongbua, S. Photophysical properties of various substituted thiophene-based heterocyclic chalcone: Experimental and DFT studies. *Chiang Mai J. Sci.* 2019, 46, 1176-1190.

[238] Zhang, H.; Iqbal, Z.; Chen, Z. E.; Hong, Y. Effects of various heteroatom donor species on the photophysical, electrochemical and photovoltaic performance of dye-sensitized solar cells. *Electrochim. Acta*, 2018, 290, 303-311.

[239] Serrano, L. A.; Park, K. W.; Ahn, S.; Wiles, A. A.; Hong, J.; Cooke, G. Coplanar donor-π-acceptor dyes featuring a furylethynyl spacer for dye-sensitized solar cells. *Materials* 2019, 12, 839.

[240] Wu, Z. S.; Song, X. C.; Liu, Y. D.; Zhang, J.; Wang, H. S.; Chen, Z. J.; Liu, S.; Weng, Q.; An, Z. W.; Guo, W. J. New organic dyes with varied arylamine donors as effective co-sensitizers for ruthenium complex N719 in dye sensitized solar cells. *J. Power Sources* 2020, 451, 227776.

[241] Divya, V. V.; Suresh, C. H. Density functional theory study on the donating strength of donor systems in dye-sensitized solar cells. *New J. Chem.* 2020, 44, 7200-7209.

[242] Yen, Y. S.; Chen, Y. C.; Chou, H. H.; Huang, S. T.; Lin, J. T. Novel organic sensitizers containing 2,6-difunctionalized anthracene unit for dye sensitized solar cells. *Polymers* 2012, 4, 1443-1461.

[243] Xiao, Z.; Chen, B.; Cheng, X. Novel red light-absorbing organic dyes based on indolo[3,2-b] carbazole as the donor applied in co-sensitizer-free dye-sensitized solar cells. *Materials* 2021, 14, 1716.

[244] Kesavan, R.; Attia, F.; Su, R.; Anees, P.; El-Shafei, A.; Adhikari, A. V. Asymmetric dual anchoring sensitizers/cosensitizers for dye sensitized solar cell application: an insight into various fundamental processes inside the cell. *J. Phys. Chem. C* 2019, *123*, 24383-24395.

[245] Lu, T. F.; Li, W.; Chen, J.; Tang, J.; Bai, F. Q.; Zhang, H. X. Promising pyridinium ylide based anchors towards high-efficiency dyes for dye-sensitized solar cells applications: Insights from theoretical investigations. *Electrochim. Acta* 2018, 283, 1798-1805.

[246] Ganesan, P.; Yella, A.; Holcombe, T. W.; Gao, P.; Rajalingam, R.; Al-Muhtaseb, S. A.; Grätzel, M.; Nazeeruddin, M. K. Unravel the impact of anchoring groups on the photovoltaic performances of diketopyrrolopyrrole sensitizers for dye-sensitized solar cells. *ACS Sustain. Chem. Eng.* 2015, 3, 2389-2396.

[247] Quan, Y. Y.; Li, Q.; Wang, Z.; Ma, H.; Dong, J.; Huang, Z. S. Effect of electron donor and acceptor in dithienopyrrolobenzothiadiazole-based organic dyes for efficient quasi-solid-state dye-sensitized solar cells. *Dyes Pigm.* 2020, 173, 107999-10808.

[248] Mao, L.; Wu, Y.; Jiang, J.; Guo, X.; Heng, P.; Wang, L.; Zhang, J. Rational design of phenothiazine-based organic dyes for dye-sensitized solar cells: the influence of π-spacers and intermolecular aggregation on their photovoltaic performances. *J. Phys. Chem. C* 2020, 124, 9233-9242.

[249] Li, F.; Zhu, Y. Z.; Zhang, S. C.; Gao, H. H.; Pan, B.; Zheng, J. Y. Pyrrolo [3, 2, 1-kl] phenothiazine-based D-π-A type organic dyes for efficient dye-sensitized solar cells. *Dyes Pigm.* 2017, 139, 292-299.

[250] Ezhumalai, Y.; Lin, F. S.; Fan, M. S.; Prabakaran, K.; Ni, J. S.; Wu, Y. C.; Lee, G. H.; Chen, M. C.; Ho, K. C. Thioalkyl-functionalized bithiophene (SBT)-based organic sensitizers for high-performance dye-sensitized solar cells. *ACS Appl. Mater. Interfaces* 2020, 12, 15071-15079.

[251] Fan, W. J.; Zhao, D.; Liu, N.; Tan, D. Z.; Chen, Y. G. Molecular design of novel indacenodithiophene-based organic dyes for efficient dye-sensitized solar cells applications. *Int. J. Quantum Chem.* 2020, 120, 26147-26157.

[252] Pakravesh, F.; Izadyar, M.; Arkan, F. Molecular engineering of triphenylamine-based metal-free organic dyes for dye-sensitized solar cells. *Int. J. Quantum Chem.* 2021, 121, 26620-26639.

[253] Raju, T. B.; Vaghasiya, J. V.; Afroz, M. A.; Soni, S. S.; Iyer, P. K. Effect of mono- and di-anchoring dyes based on o, m-difluoro substituted phenylene spacer in liquid and solid state dye sensitized solar cells. *Dyes Pigm.* 2020, 174, 108021-108028.

[254] Zhang, X.; Bolag, A.; Yun, W.; Eerdun, C.; Du, Y.; Ning, J.; Alata, H. Dye sensitized solar cells based on diphenylpyranylidene sensitizers with mono and dual cyanoacrylic acid anchoring groups. *Chem. Lett.* 2019, 48, 204-207.

[255] Mohankumar, V.; Pounraj, P.; Pandian, M. S.; Ramasamy, P. Theoretical investigation on flavones and isoflavones-added triphenylamine-based sensitizers for DSSC application. *Braz. J. Phys.* 2019, 49, 103-112.

[256] Manoharan, S.; Asiri, A. M.; Anandan, S. Impact of anchoring groups for improving the binding nature of organic dyes toward high efficient dye sensitized solar cells. *Sol. Energy* 2016, 126, 22-31.

[257] Elsherbiny, D.; Yildirim, E.; El-Essawy, F.; Abdel-Megied, A.; El-Shafei, A. Structure-property relationships: Influence of number of anchoring groups in triphenylamine-carbazole motifs on light harvesting and photovoltaic performance for dye-sensitized solar cells. *Dyes Pigm.* 2017, 147, 491-504.

[258] Liu, H.; Liu, L.; Fu, Y.; Liu, E.; Xue, B. Theoretical design of D-π-A-A sensitizers with narrow band gap and broad spectral response based on boron dipyrromethene for dye-sensitized solar cells. *J. Chem. Inf. Model.* 2019, 59, 2248-2256.

[259] Yang, W.; Cao, D.; Zhang, H.; Yin, X., Liao, X.; Huang, J.; Wu, G.; Li, L.; Hong, Y. Dye-sensitized solar cells based on (D- π- A)$_3$L2 phenothiazine dyes containing auxiliary donors and flexible linkers with different length of carbon chain. *Electrochim. Acta* 2018, 283, 1732-1741.

[260] Zhu, W. Q.; Mao, Y. B.; Chang, Y. Z.; Fan, W. J.; Tan, D. Z.; Chen, Y. G. Computational study of anthracene-based organic dyes for dye-sensitized solar cells: Effects of auxiliary electron donors. *J. Electron. Mater.* 2020, 49, 6317-6324.

[261] Dai, P.; Yang, L.; Liang, M.; Dong, H.; Wang, P.; Zhang, C.; Sun, Z.; Xue, S. Influence of the terminal electron donor in D-D-π-A organic dye-sensitized solar cells: dithieno[3,2-b:2′,3′-d]pyrrole versus bis (amine). *ACS Appl. Mater. Interfaces* 2015, 7, 22436-22447.

[262] Han, F.; Wang, Y.; Wan, Z.; Jia, C.; Luo, J.; Yao, X. Enhanced photovoltaic performances of dye-sensitized solar cells sensitized with D-D-π-A phenothiazine-based dyes. *Synth. Met.* 2016, 221, 95-102.

[263] Mane, S. B.; Cheng, C. F.; Sutanto, A. A.; Datta, A.; Dutta, A.; Hung, C. H. D-A-π-A organic dyes for dye-sensitized solar cells: Effect of π-bridge length between two acceptors on photovoltaic properties. *Tetrahedron*, 2015, 71, 7977-7984.

[264] Wu, Y.; Zhu, W. H.; Zakeeruddin, S. M.; Grätzel, M. Insight into D-A-π-A structured sensitizers: a promising route to highly efficient and stable dye-sensitized solar cells. *ACS Appl. Mater. Interfaces* 2015, 7, 9307-9318.

[265] Xie, Y.; Zhou, H.; Zhang, S.; Ge, C.; Cheng, S. Influence of the auxiliary acceptor and π-bridge in triarylamine dyes on dye-sensitized solar cells. *Photochem. Photobiol. Sci.* 2019, 18, 2042-2051.

[266] Ji, H. X.; Huang, Z. S.; Wang, L.; Cao, D. Quinoxaline-based organic dyes for efficient dye-sensitized solar cells: Effect of different electron-withdrawing auxiliary acceptors on the solar cell performance. *Dyes Pigm.* 2018, 159, 8-17.

[267] Wu, H.; Liao, C.; Tang, H.; Wang, L.; Meier, H.; Cao, D. Metal-free organic dyes with di(1-benzothieno)[3,2-b:2′,3′-d] pyrrole as an auxiliary donor for efficient dye-sensitized solar cells: Effect of the molecular engineering on the photovoltaic performance. *Dyes Pigm.* 2019, 171, 107676-107687.

[268] Putri, S. K.; Kim, Y. H.; Kim, J. H.; Chang, D. W. Synthesis of low bandgap small molecules containing fluorinated benzothiadiazole and phenothiazine for photovoltaic applications. *Mol. Cryst. Liq. Cryst.* 2017, 653, 27-32.

[269] Tingare, Y. S.; Su, C.; Shen, M. T.; Tsai, S. H.; Ho, S. Y.; Li, W. R. New oxindole-bridged acceptors for organic sensitizers: Substitution and performance studies in dye-sensitized solar cells. *Molecules* 2020, 25, 2159-2171.

[270] Higashino, T.; Kurumisawa, Y.; Alemayehu, A. B.; Einrem, R. F.; Sahu, D.; Packwood, D.; Kato, K.; Yamakata, A.; Ghosh, A.; Imahori, H. Heavy metal effects on the photovoltaic properties of metallocorroles in dye-sensitized solar cells. *ACS Appl. Energy Mater.* 2020, 3, 12460-12467.

[271] Harrath, K.; Talib, S. H.; Boughdiri, S. Theoretical design of metal-phthalocyanine dye-sensitized solar cells with improved efficiency. *J. Mol. Model.* 2018, 24, 1-11.

[272] Peymani, S.; Izadyar, M.; Arkan, F. The effect of the central metal ion on photovoltaic properties of bacteriochlorin derivatives. *Mater. Today Commun.* 2021, 27, 102367-102380.

[273] Barati-Darband, F.; Izadyar, M.; Arkan, F. Solvent effects on intra-/intermolecular charge transfer in indoloquinoxaline-based dyes. *J. Phys. Chem. A* 2019, 123, 2831-2842.

[274] Chitumalla, R. K.; Jang, J. Density functional theory study on ruthenium dyes and dye@TiO_2 assemblies for dye sensitized solar cell applications. *Sol. Energy* 2018, 159, 283-290.

[275] Wubie, G. Z.; Lu, M. N.; Desta, M. A.; Weldekirstos, H. D.; Lee, M. M.; Wu, W. T.; Li, S. R.; Wei, T. C.; Sun, S. S. Structural engineering of organic D-A-π-A dyes incorporated with a dibutyl-fluorene moiety for high-performance dye-sensitized solar cells. *ACS Appl. Mater. Interfaces* 2021, 13, 23513-23522.

Chapter 10

Present and Future Perspectives of Photo-Responsive Molecularly Imprinted Polymers

T. Sajini[1] and Beena Mathew[2,*]
[1]Mahatma Gandhi University, Changanassery, Kerala, India
[2]School of Chemical Sciences, Mahatma Gandhi University, Kottayam, Kerala, India

Abstract

Molecular imprinting technology is becoming a prominent method in modern era where artificial receptor cavities of target molecule were fabricated on a polymer network. The resultant polymer possessing complementary sites of target molecule was designated as molecularly imprinted polymers (MIPs). Recently, stimuli-responsive molecularly imprinted polymers (SR-MIPs) gained much attention since they symbolize a novel creation of artificial, self-regulating receptors and exhibit prospective applications in different fields. Among the various SR-MIPs, photo-responsive materials possess unique characteristic features and hence are used in nonlinear optics, erasable memory storage and processing, and in electro-optical displays. The present chapter gives some basic aspects of photo-responsive molecularly imprinted polymers (PR-MIPs) and their recent developments in science and technology. Several significant applications of PR-MIPs are highlighted in this chapter.

* Corresponding Author's Email beenamathew@mgu.ac.in.

In: Photosensitizers and Their Applications
Editors: Davor Margetic and Renjith Thomas
ISBN: 978-1-68507-880-5

Keywords: molecular imprinting, photo-responsive materials, azobenzene monomers, applications

Introduction

Since the technique was first presented by Wulff and Sarhan in 1972, molecular imprinting has prevailed in many research groups, as evidenced by increasing number of studies in recent years (Wulff and Vietmeier, 1989). The production of MIPs is reproducible, fast, cheap, stable and economical, but there are also some challenges (Öpik et al., 2009). Choosing the right template, preparation initiator, solvent composition, functional monomers and crosslinkers must be carefully planned and the quality of the final MIPs tested (Alexander et al., 2006).

Molecular imprinting is an excellent method for fabricating artificial receptor on polymer matrix with a given recognition capability of template molecule. Highly selective recognition cavities were formed through the direct orientation of structural components of template molecule by self-composition mechanisms. As a technique, it is described as a method for the formation of artificial, tailored binding cavities of complementary sizes, shapes and functional groups of template molecules.

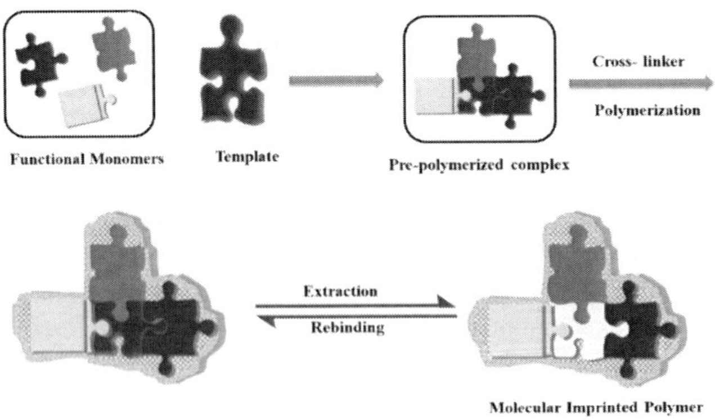

Scheme 1. Fabrication process of molecularly imprinted polymers.

Molecularly imprinted polymers (MIPs) are produced by co-polymerizing crosslinked monomers into a complex of preformed crosslinked monomers and functional monomers via a covalent, non-covalent or semi-covalent interactions (Mosbach, 1994). During polymerization, the curing agents are removed from the printed material and left in special cavities that complement their size, shape and chemical functionality. The binding cavities are tailor-made for *in situ* copolymerization, in which the crosslinkers and the functional monomer interact with the printed molecular template. A schematic representation of MIP fabrication is depicted in Scheme 1.

Molecular imprinting comprises the fabrication of polymers consisting of specific adsorption sites based on molecular and biomolecular models. The primary attraction of molecular imprinting and MIP are its simplicity in terms of the ingredients, materials and unit activities needed. MIP exhibits unique characteristic features such as structure predictability, application universality and specific target recognition. They have got widespread application in various fields of science and technologies such as purification, chemo-biosensing, separation science, drug delivery, artificial antibodies, catalysis, enantiomeric recognition and degradation (Lingxin et al., 2016). The significant properties like higher physical stability, straightforward preparation, low cost and remarkable robustness help them to act as better adsorbents in various applications.

Among the various MIPs, stimuli-responsive materials achieved widespread attention recently. The present chapter comprises some basic aspects of photo-responsive MIPs, mainly focusing on azo-moiety containing MIPs and their applications.

Stimuli-Responsive Molecularly Imprinted Polymers

Stimuli-responsive molecularly imprinted polymers (SR-MIPs) have currently obtained a massive consideration due to the fact they constitute an innovative group of smart and self-regulated synthetic receptors and feature proven tremendous capability in various applications. Incorporating stimuli-responsive molecular functions in the receptor cavities of stiff polymer materials by molecular imprinting offers a new way of making stimuli-responsive materials (SRM) (Xu, Lu, et al., 2013). The SRMs are able to undergo modification in response to environmental stimuli, such as pH and temperature, as well as ionic strength, electric power, or photon exposure (Xu, Li, et al., 2013). One of the most commonly used external stimuli for determining responsiveness of materials is photo-irradiation, which is easy to apply and control (Gong, Lam and Yu, 2006). Photo-responsive substances have attracted a lot of interest due to their applicability in various optical programs which includes nonlinear optics, erasable memory storage and processing, and electro-optical applications (Nicoletta et al., 2012).

Photo-Responsive Molecularly Imprinted Polymers (PR-MIPs)

By treating the material with photosensitive molecules such as cinnamic acid (CA), cinnamylidene or azo compounds, photo-responsive materials can be obtained. Due to their fast response when exposed to appropriate wavelengths of light, azobenzenes are among the most widely used photosensitive molecules (Barrett et al., 2007). Azobenzene derivatives are capable of changing their geometry upon the absorption of photon. As the name implies, azobenzene is composed of two aromatic rings linked by an azo bond (−N = N−) (Iqbal and Samiullah, 2013). Various kinds of azo derivatives can be synthetized by substituting different functionalities to the aromatic ring which will lead to alteration in the geometry and mechanism of electron donating and electron withdrawing. The photo-triggered conformational changes make azobenzenes as excellent materials for several photo-responsive applications such as in optics and photonics, information storage, image processing, biology, energy conservation and actuation.

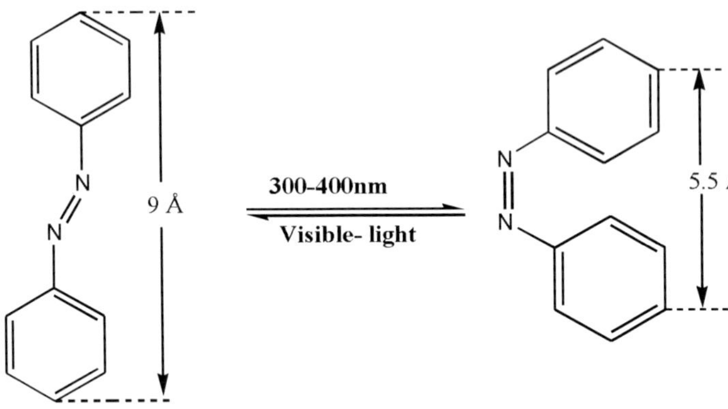

Figure 1. Molecular transformation of azobenzene induced by photo-irradiation.

Upon photon absorption, azobenzenes undergo reversible and clean conformational changes, as shown schematically in Figure 1. A photoisomerization process causes significant variations in the molecule's physical and chemical properties. The azobenzene can be isomerized in two ways: either rotated around its N = N bond or inverted around a nitrogen atom. There is about 12 kcal mol^{-1} thermodynamic stability difference between the *trans* and *cis* forms, while the cis form has an isomerization activation barrier, stabilizing it kinetically (Mina and Takumu, 2011). When exposed to UV light, the *trans* form will be transformed to the *cis* while becoming smaller in molecular size (Yu, Nakano and Ikeda, 2003). When the light source is turned off or visible light is returned, this *cis* form thermally reverts to the more stable *trans* form. Thus, the azobenzenes reverse their geometry upon irradiation with a decreasing space among the para carbon atoms from 9 Å to 5 Å and an increasing dipole moment from 0.5 D to 3.1 D. Photochemistry of these systems is extremely clean, resulting in their remarkable photo-controlled behavior and multiple photo-responsive features (Iqbal and Samiullah, 2013).

The significant properties exhibited by these isomers, help in different photo-controlled processes such as, adsorption, solubility and sol-gel transition of polymer, wettability, permeability, swelling, enzyme activity, membrane dimensions and potential, photo-mechanical cycle, etc. (Nicoletta et al., 2012).

The fabrication process of PR-MIPs encompasses the supply of photo-responsive functional monomers which replace the conventionally used monomers like methacrylic acid, acrylamide and vinyl pyridine. In Scheme 2, the reversible release and uptake of substrate by photo-regulated macromolecules is shown schematically.

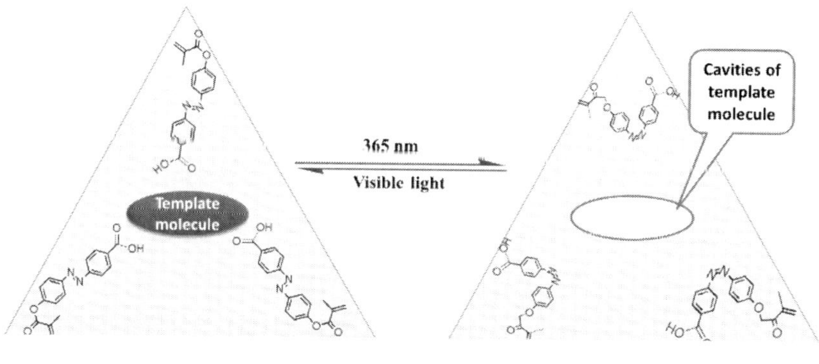

Scheme 2. Diagram illustrating a photoregulated substrate release and uptake mechanism in PR-MIP.

When irradiated by UV light, the interior sites of the PR-MIPs do not equal with the target materials in shape, size, or chemical composition, causing them to be released from matrix. As soon as the recognition site is irradiated with visible light, it returns to its original state, capable of selectively adsorbing the template again.

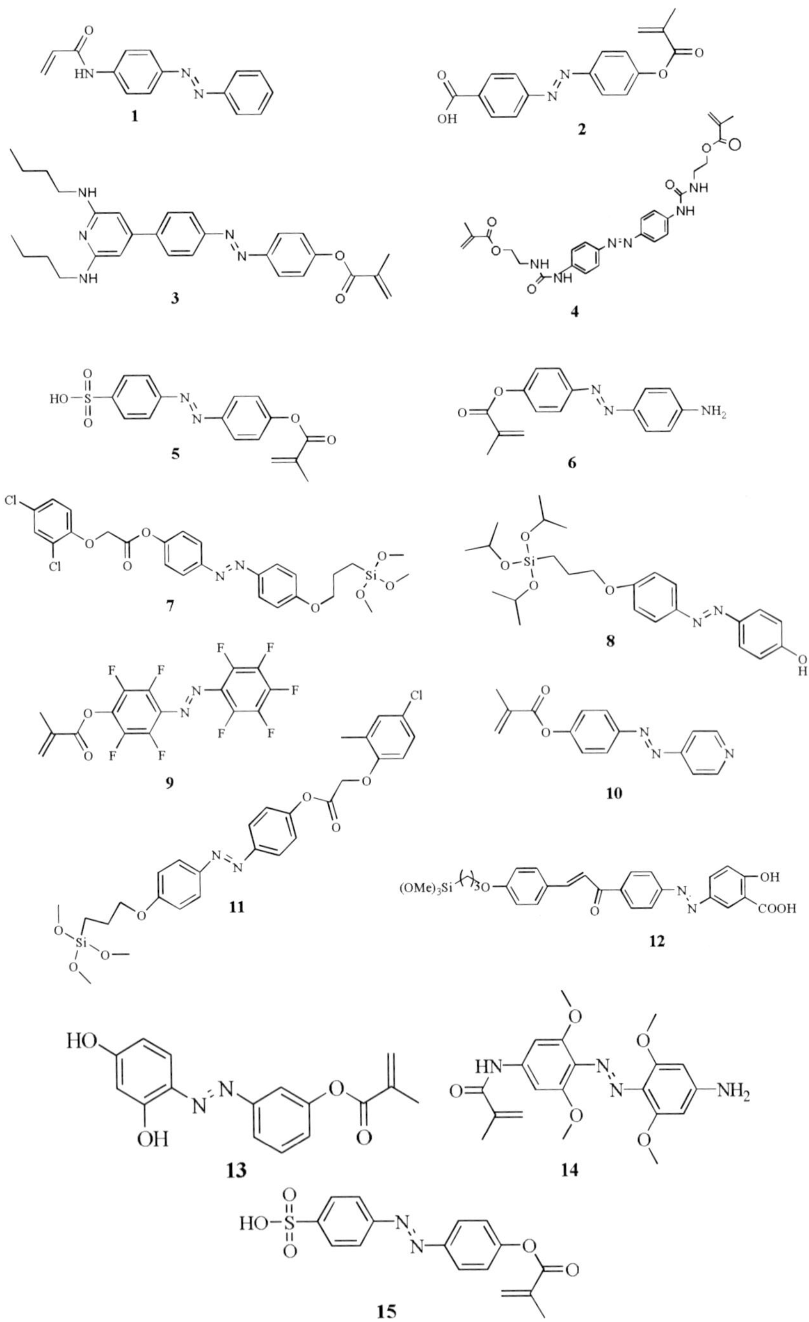

Figure 2. Azobenzene derivatives of functional monomers.

Table 1. List of azobenzene functional monomer

Sl No	Functional monomer	Template	Reference
1	*p*-Phenylazoacrylanilide (PhAAAn)	Dansylamide	(Minoura et al., 2003)
2	4-[(4-Methacryloyloxy)phenylazo]benzoic acid (MPABA)	Caffeine	(Gong, Lam and Yu, 2006)
3	4-{4-[2,6-Bis(*n*-butylamino)pyridine-4-yl]-phenylazo}-phenyl methacrylate	Porphyrin	(Takeuchi et al., 2007)
4	Di(ureidoethylenemethacrylate)azobenzene	Bis(TBA)-N-Z-L-Glutamate	(Gomy and Schmitzer, 2007)
5	4-[(4-Methacryloyloxy)phenylazo] benzenesulfonic acid (MAPASA)	Paracetamol & 1,3,5-Benzenetriol	(Gong, Wong and Lam, 2008)
6	4-Amino-4-methacrylatylazobenzene (AMAAB)	-	(Tang et al., 2010)
7	4-{[4-(3-(Trimethoxysilyl)propoxy)phenyl] diazenyl}phenyl 2-(2,4)acetate	2,4-Dichlorophenoxy acetic acid	(Jiang et al., 2011)
8	4-Hydroxyl-4-[(triisopropoxysilyl) propyloxy]azobenzene	Ibuprofen	(Tang, Gong, Lam, et al., 2011)
9	(4-Methacryloyloxy)nonafluoroazobenzene	2,3,4,5,7,8,9,10-Octafluorophenazine (PAF)	Tang, Gong, Hon, et al., 2011)
10	4-((4-Methacryloyloxy) phenylazo)pyridine	2,4-Dichloro phenoxyacetic acid	(Fang et al., 2012)
11	(4-Chloro-2-methylphenoxy) acetyloxy- 4'-[(trimethoxysilyl)propyloxy]azobenze	(4-Chloro-2-methylphenoxy)acetic acid	(Zhou, Zhong and Jiang, 2012)
12	2-Hydroxy-5-{4-[3-(4-trimethoxysilyl) propyloxy-phenyl)-acryloyl]-phenylazo} benzoic acid	Ibuprofen	(Li et al., 2013)
13	5-[(4,3-(Methacryloyloxy) phenyl) diazenyl] dihydroxy aniline (MAPDHA)	Azathioprine	(Alaei et al., 2018)
14	N-(4-((4-Amino-2,6-dimethoxyphenyl) diazenyl)-3,5-dimethoxyphenyl) methacrylamide (ADDDM)	Acyclovir	(Liu et al., 2018)
15	4-[(4-Methacryloyloxy)phenylazo] benzenesulfonic acid	Bisphenol A	(Gong, Wong, Lam, 2008)

Therefore, reversible photo-switchable chromophores can be impregnated into the binding sites of substrates to facilitate photoregulated release and uptake. Several variables contribute to the outcome of PR-MIP preparation, but most importantly the photo-responsive monomer is the significant step. These photo-responsive functional monomers consist of a photo-responsive moiety, a recognition part, and a polymerizable moiety (Arabi et al., 2021). In general, the most studied photo-responsive functional monomers are derivatives of azobenzene. An array of functional monomers containing azobenzene has been developed and is depicted in Table 1 and their structures are illustrated in Figure 2. Azobenzene bearing functional monomers have been synthesized by mixing azobenzene with common molecular imprinting monomers such as methacrylic acid, acrylamide, or 4-vinylpyridine.

To prepare PR-MIPs, a variety of techniques have been developed and used with the development of molecular imprinting. Among the most popular and general methods of producing MIP, bulk polymerization is known for its fast reaction rate and simplicity, as well as the purity of the product. There is no doubt that bulk polymerization strategies have also been used to fabricate PR-MIPs.

Conventional PR-MIPs

The photo-responsive polymers synthesised via conventional methodologies, where the effect of imprinting process is obtained by the entrapment of target molecule within the bulk part of the polymer network, are known as conventional PR-MIPs. Sol-gel synthesis, precipitation polymerization and bulk polymerization are some methods adopted for the fabrication of conventional PR-MIPs.

PR-MIPs

Bulk Polymerization

Among the various fabrication methods, bulk polymerization is the most widely used which undergone via a free radical polymerization mechanism. This technique comprises grinding and sieving of polymer particles to micrometer-sized polymers with wide applications.

The first fabrication of azo-benzene derivative of functional monomer was done by Minoura et al. They prepared the MIP membranes with *p*-phenylazoacrylanilide as the photoregulated functional monomer using bulk polymerization technique, which is used for the detection of dansylamide template molecule (Minoura et al., 2003). Light of different wavelengths can be used to change the affinity sites of selected recognition of the template on photosensitive membranes. Later Gong et al. produced azobenzene bearing functional monomer, 4-[(4-methacryloyloxy)phenylazo]benzoic acid (MPABA) which comprises a methacrylic polymerizable unit (Gong, Lam and Yu, 2006). After MPABA is incorporated into a rigid crosslinked polymer matrix, its *trans/cis* photoisomerization properties are preserved and used for the recognition of caffeine molecule. The photoregulated processes of substrate binding and release can generally be repeated, but the extent of binding and release is gradually decreasing. A complete study of adsorption selectivity between the *trans* and *cis*-forms was not conducted here. The low water compatibility of MPABA limits its use in applications in biomedical fields such as in drug delivery systems, since it can only dissolve in extremely polar solvents, like DMF and DMSO.

A photo-sensitive functional monomer having diaminopyridine and azobenzene moieties, 4-{4-[2,6-bis(*n*-butylamino)pyridine-4-yl]-phenylazo} phenyl methacrylate, for the detection of porphyrin derivatives as the template molecule was fabricated by Takeuchi et al. (Takeuchi et al., 2007). By forming multiple hydrogen bonds between the template molecule and the functional monomer, they were able to accumulate functional monomers in appropriate positions for polymerization to yield specific imprinted sites which are complementary to the template molecule. It appears that the binding affinity of the imprinted cavities was controlled by UV irradiation, implying that azobenzene groups were acting as photosensitizers inside the cavities and that *trans-cis* isomerization may have mediated the control.

Researchers at Jiang's group created a photoresponsive monomer containing a siloxane polymerisable group and azobenzene part, earning it the title of being an organic-inorganic polymerisable monomer (Jiang et al., 2011). The fabricated photo-responsive monomer was used to synthesize PR-MIPs which possess great affinity towards 2,4-dichlorophenoxyacetic acid (2,4-D) via hydrogen bonding interaction. Spectral analysis of the azobenzene chromophore can quantify the 2,4-D concentration through the rate of *trans*-to *cis* photoisomerization.

Tang et al. prepared a novel fluorine-incorporated photo-sensitive functional monomer, (4-methacryloyloxy) nonafluoroazobenzene (MANFAB) (Tang, Gong, Hon, et al., 2011). Fluorine-substituted azobenzene chromophores retain their photoisomerization properties regardless of how they are dispersed in the crosslinked matrix of a 3-dimensional polymer. This system was used for the molecular recognition of PAF (2,3,4,5,7,8,9,10-octafluorophenazine).

Schmitzer et al. developed an effective method for photosensitive recognition of methotrexate using a new azo-monomer, di (ureidoethylenemethaacrylate) azobenzene, that also acts as a cross-linker (Gomy and Schmitzer, 2007). They explained that this new crosslinking monomer combines the functionality of an interactive monomer with crosslinking capabilities. Under mild conditions, this new crosslinking agent can be readily copolymerized and has been successfully applied to the non-covalent synthesis of MIP that allows for more functionality to be introduced without substantially compromising performance.

As a consequence, after bulk polymerization monolithic polymers must be crumpled, ground, and filtered to a suitable size, resulting in particles of uneven shapes and sizes, and several effective binding cavities are damaged and turned into low adsorption sites. Consequently, their application has been significantly limited due to the rather restricted physical formats (e.g., bulk membranes, bulk monoliths).

Precipitation Polymerization

The method of precipitation polymerization can be used to produce microspheres of uniform size, since the chains that form in the dilute reaction system do not overlap or coalesce as they grow. The polymer precipitates out of the solution as microspherical particles. Precipitation polymerization is more difficult and highly controlled than bulk polymerization because the latter requires organic solvents and harsher reaction conditions (Lingxin et al., 2016).

A precipitation polymerizing method for fabricating PR-MIPs hydrogel materials applicable to biocompatible aqueous media has been developed by Gong et al. They fabricated an azobenzene- bearing functional monomer 4-[(4-methacryloyloxy) phenylazo] benzenesulfonic acid (MAPASA) which is soluble in aqueous media (Gong, Wong and Lam, 2008). In their study, the MAPASA-bearing polyacrylamide hydrogel showed good visual transparency in aqueous media, a good attraction for substrates, and the best response to light.

With the help of precipitation polymerization Fang et al. fabricated MIP microspheres using photo-responsive functional monomer, 4-{(4-methacryloyloxy)-phenylazo}pyridine. It exhibited high recognition capability towards the template, 2,4-D (Fang et al., 2012). The results demonstrated a strong molecular imprint towards the template, slow template rebinding kinetics, a high degree of selectivity, and a noticeable molecular imprinting effect.

As such, precipitation polymerization, one of the most effective mechanisms for synthesis of spherical particles, provides distinct preparatory step and allows outstanding control of size of particle, which are necessary to create excellent quality, homogeneous and spherical PR-MIPs.

Sol-Gel Method

Several researchers have already established the sol-gel method as an effective and convenient method for preparing MIPs. Most sol-gel materials are formed by hydrolyzing silanes with acid or base catalysts and then polycondensing the resulting silanols into polymers. Among the various materials, the amorphous sol-gel MIPs possess some significance. Under extremely severe conditions of removal, such as hydrolysis in strong acids and bases or combustion, the template can be removed more thoroughly from sol-gel materials. In addition, sol-gel materials are easier to control in terms of porosity, thickness, and surface area.

Using a siloxane polymerizable group and azobenzene moiety, Zhang et al. synthesized a novel photoreactive functional monomer. The synthesized functional monomer was successfully converted into PR-MIPs for the detection of (4-chloro-2-methylphenoxy) acetic acid (MCPA) (Zhou, Zhong and Jiang, 2012). The interaction of template molecules and functional monomers was investigated using computational molecular modeling. MIP design was rational as evidenced by the computational results.

Atom Transfer Radical Precipitation Polymerization

An efficient, facile, and universal method has been demonstrated by Fang et al. to prepare tightly dispersed, water-soluble, and dual-stimulus-responsive MIP microspheres (Fang et al., 2012). Here, the azobenzene bearing MIP layer is successively grafted into the pre-formed polymer microspheres with thermoresponsive poly(*N*-isopropylacrylamide) (PNIPAAm) brushes using surface-catalysed RAFT polymerization. For the fabrication of narrowly distributed photoresponsive core-shell MIP microspheres, an azo-containing MIP layer is attached to hollow core MIP beads with thin size distributions. A significant improvement of their surface hydrophilicity has been demonstrated by the introduction of PNIPAAm brushes onto core-shell microspheres, resulting in pure water-compatibility.

Surface Molecularly Imprinted Polymers

The conventional imprinting protocol is effective and simple, however it possesses major difficulties: (i) most of the MIPs are highly cross-linked bulky polymers with uneven shape, (ii) mechanical crushing and grinding methods are commonly used and the removal of target molecules at the internal cavities of bulky polymeric network is challenging which diminishes the capacity of rebinding analytes (Michael et al., 2000)(Mukti and Dave, 1998), (iii) slowing the kinetics of template adsorption (Carter and Rimmer, 2004)(Carter and Rimmer, 2002), (iv) the stiff polymeric network diminishes the conformational choice of template molecular recognition (Espinosa-García et al., 2007) (Aburto and Borgne, 2004), (v) the uncontrollable random polymerization constantly suffers from the heterogeneity of the imprinted cavities (Rampey et al., 2004)(Katz and Davis, 2000)(Bass and Katz, 2003), and (vi) MIP materials are usually insufficient for chemical detections or bioassays because the analytes bind poorly on the surface of the transducer, resulting in lack of signal output (Holthoff and Bright, 2007). These limitations and heterogeneous binding sites of non-covalent MIP can be endured in many analytical applications, some artificial approaches have been presented to enhance the homogeneous formation of binding cavities of MIP (Umpleby et al., 2001)(Kirsch et al., 2000)(Patel, Fouace and Steinke, 2004). In order to overcome these disadvantages, photo-responsive surface molecularly imprinted polymers (PR-SMIPs) were developed. In comparison with traditional PR-MIPs, PR-SMIPs possess enhanced accessibility of template molecules, strong fixation of binding sites, ease of template removal, regular and controlled size etc.

Using a simple surface polymerization synthesis method, Yang's group developed a simple recognition method for detecting traces of bisphenol A (BPA) in the environment. Here azobenzene-bearing functional monomer 4-[(4-methacryloyloxy)-phenylazo]benzenesulfonic acid having water solubility capability was used as the photo-responsive functional monomer (Yang et al., 2014). Microspheres prepared with PR-SMIPs exhibited good photoresponsive properties with high specific affinity for BPA. Additionally, they displayed fast binding kinetics and demonstrated high recognition capability. Microspheres of PR-SMIPs were able to bind and release BPA upon alternate irradiation at 365 and 440 nm.

A poly(ether sulfone) microfiber with photo-sensitive surfaces was synthesized by Wang et al. for the specific detection of 4-hydrobenzoic acid (4-HA) (Wang et al., 2012). Irradiation at 450 nm and 365 nm selectively bound 4-HA to the molecularly imprinted microfibers. Besides being simple, convenient, and effective, the method also works for polysulfone and poly(ether ketone) modification. Adding azo moieties to all three benzene-containing polymers allowed them to be significantly photoresponsive.

Li et al. synthesised an azobenzene bearing PR-MIPS supported on silica microspheres (Li et al., 2013). The obtained pre-polymerized azobenzene-bearing functional monomer comprises a carboxyl and hydroxyl moieties for effective template interaction. The homogeneous binding cavities formation in azo-bearing MIP-silica particles showed high mechanical stability and exhibited effective recognition towards ibuprofen molecule.

Shoufang et al. fabricated a dual responsive MIPs (DR-MIPs) which exhibited both photonic and magnetic character (Xu, Li, et al., 2013). They used MPABA as the photo-responsive functional monomer. The obtained DR-MIPs of Fe_3O_4@MIP displayed high binding for caffeine template molecule. For enriching caffeine from water samples and beverages, novel DR-MIPs was used. The concept of photonic and magnetic DR-MIPs allows for the photo-regulated diffusion of pollutants into and out of environmental samples, while its magnetic characteristics enable it to be easily separated. Controlled-release of drugs is also possible using the photonic and magnetic DR-MIPs.

An imprinted molecular polymer with dual responsiveness was synthesized by Chen's group for the detection of four sulfonamides in aqueous media (Chen, Dong and Zhao, 2016). They used a surface imprinting polymerization technique consuming superparamagnetic Fe_3O_4 nanomaterials functionalized with a silica layer as a support, and 4-[(4-methacryloyloxy) phenylazo]benzenesulfonic acid as the functional monomer. The MIP formed via water soluble functional monomer was used for the detection of sulfadiazine. These polymers exhibited selective binding towards sulfadiazine and structurally related compounds in aqueous media and acted as a good solid- phase extraction material with good recoveries.

Another DR-MIPs which exhibit photonic-magnetic characteristics were fabricated by Peng's group using a seed polymerization method, which successfully implemented to the rapid and selective removal of 17β-estradiol from real samples (Peng et al., 2016). The fabricated DR-MIPs exhibited a sandwich micro-spherical assembly containing Fe_3O_4 core, SiO_2 middle layer, and MIP shell with thickness of 25 nm. The entire work rationally combined photonic response, magnetic separation and surface imprinting, which provided the MIPs with significant advantages of high adsorption capacity and fast binding kinetics, appropriate separation and recycling efficiency, and modest rapid eco-benign adsorption/elution processes for target molecules.

Fan's group investigated a molecularly imprinted polymer based on 4-methacryloyloxy azobenzene as the functional monomer for the recognition and selective separation of the branched cyclodextrins, 6-O-α-d-maltosyl-β-cyclodextrin (G2-β-CD) (Fan et al., 2017). Using HPLC, it was found that after MIP, G2- β -CD could reach a purity of 90.8%. They discovered that the PR-MIPs behaves as an easy and effective technique for the selective and specific separation of the branched cyclodextrins.

A novel photo-responsive magnetic electrochemical sensor for the specific detection and removal of paracetamol from biological samples was developed by Wei's group (Wei et al., 2017). In their work, nanosized PR-MIPs were prepared on the surface of magnetic Fe_3O_4 nanoparticles using living radical polymerization of azobenzene. Combining photocontrolled selective extraction, specifically electrochemical response and interfacial stability from magnetic adsorption, the PR-MIP sensor displayed substantial advantages for instantaneous separation and improved recognition of trace paracetamol in biological samples.

Alaei's group developed a novel PR-MIPs with conjugated hyper-branched polymers based on functionalized magnetic nanoparticles for the specific recognition and separation of azathioprine from complex samples (Alaei et al., 2018). As a result of the synthesis of these polymers, azathioprine could be successfully extracted from biological and pharmaceutical samples with good recovery, high accuracy, and precision, confirming their tremendous potential for the assessment of trace amounts of a given analyte. They offered valuable insight and an in-depth analysis of the possibility of consuming hyperbranched polymers- magnetic nanoparticles together with stimuli-responsive imprinted ingredients in smart separation systems, which had enormous potential to foster sustainable environmental enhancement.

Baimani's group introduced an approach with polyester dendrimer through the imprinting method to increase the number of recognition cavities in the polymer network to develop its identification capability. They synthesised a PR-MIPs on modified magnetic nano-particles comprising polyester dendrimer exploiting the reactivity between allyl glycidyl ether and acrylic acid for the high-yielding assembly using surface polymerization (Baimani et al., 2019). The PR-MIPs were fabricated using methyl-prednisoloneacetate as the target molecule and a 5-[(4,3-(methacryloyloxy) phenyl) diazenyl] dihydroxy aniline as the azobenzene containing water soluble functional monomer.

Gong reported a novel photo-sensitive functional monomer, tetra-*ortho*-methyl substituted azobenzene, (4-[(4-methacryloyloxy)-2,6-dimethyl-phenylazo]-3,5-dimethyl benzenesulfonic acid (MADPADSA) for the detection of 4-ethylphenol (4-EP) using $NaYF_4$:Yb^{3+}, Er^{3+} as a template molecule (Gong et al., 2019). The $NaYF_4$:Yb^{3+}, Er^{3+} substrate could effectively enhance the *trans* to *cis* isomerization rate of the PR-MIPs on its surface, which was quicker than that of the analogous azobenzene functional monomer.

Combining stimulus-responsive polymers with surface molecular imprinting, Chen et al. developed photo-responsive surface molecularly imprinted polymers (PS-co-PMAA@PSMIPs) with uniform layers of thickness (Chen et al., 2020). PS-co-PMAA@PSMIPs were capable of selectively recognition of profenofos (PFF). Azobenzene's photoresponsive properties were well-preserved here.

Molecularly imprinted photo-responsive polymers based on azobenzene are limited in their applications by UV light phototoxicity. Liu's group developed a solution to this problem by fabricating a tetra-*ortho*-methoxy-substituted azobenzene, *N*-(4-((4-amino-2,6-dimethoxyphenyl)diazenyl)-3,5-dimethoxyphenyl) methacrylamide (ADDDM). Here, photoswitching was induced by all visible-light irradiation (440 nm for *trans* form to *cis* form and 630 nm for *cis* form to *trans* form) in *N,N*-dimethylformamide and tetrahydrofuran (1:9, v/v) (Liu et al., 2018). Using ADDDM as the monomer, they fabricated a visible-light-responsive surface molecularly imprinted polymer (VS-MIP) on silica microspheres for the detection of acyclovir (ACV). VS-MIP exhibited a higher drug loading capacity, better specificity, fast drug release rate and photoisomerization rate constant to ACV than its analogous visible-light-responsive surface molecularly non-imprinted polymer (VS-NIP).

Conclusion and Perspectives

A PR-MIP is capable of being remotely controlled because it can react to changes in light, or distant stimulation. Through adjusting the variables of light irradiation, namely wavelength, intensity, and duration, the physical and chemical variations occurring in PR-MIPs can be transformed into large-scale motion. It is possible to manufacture PR-MIPs by incorporating a photosensitive moiety into the polymer chain. Many different fields can benefit from the application of PR-MIPs particles. On the other hand, there are two reasons that prevent photo-responsive MIPs from being widely used; (a) highly cross-linked PR-MIPs slow down the photo-induced *trans-cis* isomerization of azobenzene containing functional monomers, which does not meet the criteria for efficient rapid analysis, so its use in environmental analysis is restricted.; and (b) visible light can penetrate human tissue only to a limited extent, so its use in drug delivery is limited. A method developed by

Gong et al., who described a simple process for detecting the presence of melamine in dairy products that offers the benefits of easy sample preparation, a fast detection step, good sensitivity and inexpensive equipment. PR-MIP demonstrated new applications in chemosensing, food safety, and environmental analysis as a result of this kind of research.

In order to reach mass production, PR-MIPs need more research. It is also a common problem of these smart materials that the photosensitive moieties can be photodegraded, denatured, or bleached by repeated photo-switching cycles. The PR-MIPs suitable for biotechnology and cell culture applications are extremely limited. Cells and tissues can scatter UV light violently and UV irradiation can prompt apoptotic processes in cell cultures. Some high-energy UV lamps damage the operator's DNA and RNA through UV irradiation. In addition, UV light can destroy oxygen molecules and produce ozone which can cause pulmonary damage even at small concentrations. There is still a long way to further develop and advance PR-MIP's versatile functionalities.

References

Aburto, J. and Borgne, S. Le (2004) 'Selective adsorption of dibenzothiophene sulfone by an imprinted and stimuli-responsive chitosan hydrogel', *Macromolecules*. 37(8), pp. 2938-2943. doi: 10.1021/ma049961e.

Alaei, H. S. et al. (2018) 'Photo-regulated ultraselective extraction of Azatioprine using a novel photoresponsive molecularly imprinted polymer conjugated hyperbranched polymers based magnetic nano-particles', *Polymer*. 148, pp. 191-201. doi: 10.1016/j.polymer.2018.06.013.

Alexander, C. et al. (2006) 'Molecular imprinting science and technology: a survey of the literature for the years 2004-2011', *J. Mol. Recogn.* 19(6), pp. 106-180. doi: 10.1002/jmr.2347.

Arabi, M. et al. (2021) 'Molecular Imprinting: Green Perspectives and Strategies.', *Adv. Mater.* 33(30). https://sci-hub.se/https://doi.org/10.1002/adma.202100543.

Baimani, N. et al. (2019) 'Ultrasensitive separation of methylprednisolone acetate using a photoresponsive molecularly imprinted polymer incorporated polyester dendrimer based on magnetic nanoparticles', *J. Sep. Sci.* 42(7), pp. 1468-1476. doi: 10.1002/jssc.201801093.

Barrett, C. J. et al. (2007) 'Photo-mechanical effects in azobenzene-containing soft materials', *Soft Matter*. 3, pp. 1249-1261. doi: 10.1039/b705619b.

Bass, J. D. and Katz, A. (2003) 'Thermolytic synthesis of imprinted amines in bulk silica', *Chem. Mater.* 15(14), pp. 57-2763. doi: 10.1021/cm021822t.

Carter, S. R. and Rimmer, S. (2002) 'Molecular recognition of caffeine by shell molecular imprinted core-shell polymer particles in aqueous media', *Adv. Mater.* 14(9), pp. 667-670.

Carter, S. R. and Rimmer, S. (2004) 'Surface molecularly imprinted polymer core-shell particles', *Adv. Funct. Mater.* 14(6), pp. 553-561. doi: 10.1002/adfm.200305069.

Chen, F., Dong, Y. and Zhao, Y. (2016) 'Synthesis and characterization of photo-responsive magnetic molecularly imprinted microspheres for the detection of sulfonamides in aqueous solution', *J. Sep. Sci.* 39(24), pp. 4866-4875. doi: 10.1002/jssc.201600983.

Chen, M. et al. (2020) 'Photoresponsive Surface Molecularly Imprinted Polymers for the Detection of Profenofos in Tomato and Mangosteen', *Front. Chem.* 8, p. 905. doi: 10.3389/FCHEM.2020.583036.

Espinosa-García, B. et al. (2007) 'Molecularly imprinted chitosan-genipin hydrogels with recognition capacity toward o-xylene', *Biomacromol.* 8(11), pp. 3355-3364. doi: 10.1021/bm700458a.

Fan, H. et al. (2017) 'Preparation of photoirradiation molecular imprinting polymer for selective separation of branched cyclodextrins', *Molecules.* 22(2), p. 288. doi: 10.3390/molecules22020288.

Fang, L. et al. (2012) 'Azobenzene-containing molecularly imprinted polymer microspheres with photo- and thermoresponsive template binding properties in pure aqueous media by atom transfer radical polymerization', *Langmuir.* 28, pp. 9767-9777. doi: 10.1021/la301314w.

Gomy, C. and Schmitzer, A. R. (2007) 'Synthesis and photoresponsive properties of a molecularly imprinted polymer', *Org. Lett.* 40, pp. 6-9.

Gong, C. et al. (2019) 'A photoresponsive molecularly imprinted polymer with rapid visible-light-induced photoswitching for 4-ethylphenol in red wine', *Mater. Sci. Eng. C*, 96, pp. 661-668. doi: 10.1016/j.msec.2018.11.089.

Gong, C. B., Lam, M. H.-W. and Yu, H. X. (2006) 'The fabrication of a photoresponsive molecularly imprinted polymer for the photoregulated uptake and release of caffeine', *Adv. Funct. Mater.* 16(13), pp. 1759-1767. doi: 10.1002/adfm.200500907.

Gong, C., Wong, K. L. and Lam, M. (2008) 'Photoresponsive molecularly imprinted hydrogels for the photoregulated release and uptake of pharmaceuticals in the aqueous media', *Chem. Mater.* 20(4), pp. 1353-1358. doi: 10.1021/cm7019526.

Holthoff, E. L. and Bright, F. V. (2007) 'Molecularly templated materials in chemical sensing', *Anal. Chim. Acta.* 594(2), pp. 147-161. doi: 10.1016/j.aca.2007.05.044.

Iqbal, D. and Samiullah, M. H. (2013) 'Photo-responsive shape-memory and shape-changing liquid-crystal polymer networks', *Materials.* 6, pp. 116-142. doi: 10.3390/ma6010116.

Jiang, G. S. et al. (2011) 'Synthesis of molecularly imprinted organic-inorganic hybrid azobenzene materials by sol-gel for radiation induced selective recognition of 2,4-dichlorophenoxyacetic acid', *Radiat. Phys. Chem.* 80(2), pp. 130-135. doi: 10.1016/j.radphyschem.2010.07.020.

Katz, A. and Davis, M. E. (2000) 'Molecular imprinting of bulk, microporous silica', *Nature.* 403(6767), pp. 286-289. doi: 10.1038/35002032.

Kirsch, N. et al. (2000) 'Enhancement of selectivity of imprinted polymers via post-imprinting modification of recognition sites', *Polymer.* 41(15), pp. 5583-5590. doi: 10.1016/s0032-3861(99)00782-x.

Li, C. et al. (2013) 'Silica particles coated with azobenzene-containing photoresponsive molecule-imprinted skin layer', *Colloid. Polym. Sci.* 291(9), pp. 2049-2059. doi: 10.1007/s00396-013-2944-7.

Lingxin, S. et al. (2016) 'Molecular imprinting: perspectives and applications', *Chem. Soc. Rev.* 45, pp. 2137-2211. doi: 10.1039/C6CS00061D.

Liu, L. et al. (2018) 'Visible-light-responsive surface molecularly imprinted polymer for acyclovir through chicken skin tissue', *Appl. Bio Mater.* 1(3), pp. 845-852. doi: 10.1021/acsabm.8b00275.

Michael, M. et al. (2000) 'Catalytic silica particles via template-directed molecular imprinting', *Langmuir.* 16(4), pp. 1759-1765. doi: 10.1021/LA990809T.

Mina, H. and Takumu, H. (2011) 'Correlation between the molecular structure and *trans* ↔ *cis* isomerization characteristics of azobenzenes', *Sci. China Chem.* 54(12), pp. 1955-1961. doi: 10.1007/s11426-011-4413-9.

Minoura, N. et al. (2003) 'Molecularly imprinted polymer membranes with photoregulated template binding', *Chem. Mater.* 15(25), pp. 4703-4704. doi: 10.1021/cm0305608.

Mosbach, K. (1994) 'Molecular imprinting', *Trends Biochem. Sci.* 19(1), pp. 9-14. doi: 10.1016/0968-0004(94)90166-x.

Mukti S, R. and Dave, B. C. (1998) 'Selective intake and release of proteins by organically-modified silica sol–gels', *J. Am. Chem. Soc.* 120(50), pp. 13270-13271. doi: 10.1021/ja982468w.

Nicoletta, F. P. et al. (2012) 'Light responsive polymer membranes: a review', *Membranes.* 2, pp. 134-197. doi: 10.3390/membranes2010134.

Öpik, A. et al. (2009) 'Molecularly imprinted polymers: a new approach to the preparation of functional materials', *Proc. Estonian Acad. Sci.* 58(1), pp. 3-11. doi: 10.3176/proc.2009.1.01.

Patel, A., Fouace, S. and Steinke, J. H. (2004) 'Novel stereoselective molecularly imprinted polymers via ring-opening metathesis polymerisation', *Anal. Chim. Acta.* 504(1), pp. 53-62. doi: 10.1016/j.aca.2003.08.069.

Peng, H. et al. (2016) 'Preparation of photonic-magnetic responsive molecularly imprinted microspheres and their application to fast and selective extraction of 17β-estradiol', *J. Chromatogr. A.* 1442, pp. 1-11. doi: 10.1016/j.chroma.2016.03.003.

Rampey, A. M. et al. (2004) 'Characterization of the imprint effect and the influence of imprinting conditions on affinity, capacity, and heterogeneity in molecularly imprinted polymers using the Freundlich isotherm-affinity distribution analysis', *Anal. Chem.* 76, pp. 1123-1133. doi: 10.1021/ac0345345.

Takeuchi, T. et al. (2007) 'Photoresponsive porphyrin-imprinted polymers prepared using a novel functional monomer having diaminopyridine and azobenzene moieties', *Org. Biomol. Chem.* 5(15), pp. 2368-2374. doi: 10.1039/b704830k.

Tang, Q. et al. (2010) 'Synthesis and characterization of photo- and pH-responsive nanoparticles containing amino-substituted azobenzene', *J. Mater. Chem.* 20(41), pp. 9133-9139. doi: 10.1039/c0jm01235a.

Tang, Q., Gong, C., Lam, M. H. W., et al. (2011) 'Photoregulated uptake and release of drug by an organic-inorganic hybrid sol-gel material', *J. Solgel Sci. Technol.* 59(3), pp. 495-504. doi: 10.1007/s10971-011-2518-4.

Tang, Q., Gong, C., Hon, M., et al. (2011) 'Preparation of a photoresponsive molecularly imprinted polymer containing fluorine-substituted azobenzene chromophores', *Sens. Actuators B Chem.* 156, pp. 100-107. doi: 10.1016/j.snb.2011.03.078.

Umpleby, R. J. et al. (2001) 'Recognition directed site-selective chemical modification of molecularly imprinted polymers', *Macromolecules*. 34(24), pp. 8446-8452. doi: 10.1021/ma010903s.

Wang, D. et al. (2012) 'Photoresponsive surface molecularly imprinted poly(ether sulfone) microfibers', *Langmuir*. 28(37), pp. 13284-13293. doi: 10.1021/la302687d.

Wei, Y. et al. (2017) 'Nanosized difunctional photo responsive magnetic imprinting polymer for electrochemically monitored light-driven paracetamol extraction', *Appl. Mater. Interfaces*, 9(50), pp. 44114-44123. doi: 10.1021/acsami.7b14772.

Wulff, G. and Vietmeier, J. (1989) 'Enzyme-analogue built polymers. 26. Enantioselective synthesis of amino-acids using polymers possessing chiral cavities obtained by an imprinting procedure with template molecules', *Macromol. Chem. Phys.* 190(7), pp. 1727-1735.

Xu, S., Li, J., et al. (2013) 'Photonic and magnetic dual responsive molecularly imprinted polymers: preparation, recognition characteristics and properties as a novel sorbent for caffeine in complicated samples', *Anal. Methods*. 5, pp. 124-133. doi: 10.1039/ c2ay25922b.

Xu, S., Lu, H., et al. (2013) 'Stimuli-responsive molecularly imprinted polymers: versatile functional materials', *J. Mater. Chem. C*, 1, pp. 4406-4422. doi: 10.1039/ c3tc30496e.

Yang, Y. et al. (2014) 'Ultrasensitive detection of bisphenol A in aqueous media using photoresponsive surface molecular imprinting polymer microspheres', *New J. Chem.* 38, pp. 1780-1788. doi: 10.1039/c3nj01598j.

Yu, Y., Nakano, M. and Ikeda, T. (2003) 'Directed bending of a polymer film by light-miniaturizing a simple photomechanical system could expand its range of applications', *Nature*. 425, pp. 145-145.

Zhou, X., Zhong, S. and Jiang, G. (2012) 'Computer simulation and synthesis of stimuli-responsive polymer by sol-gel for selective recognition of (4-chloro-2-methylphenoxy) acetic acid', *Polym. Int.* 61, pp. 1778-1785. doi: 10.1002/pi.4275.

Chapter 11

Pyrene-Based Photosensitizers and Their Applications

Urumpackal S. Kanchana, Elizabeth J. Diana and Thomas V. Mathew*

Department of Chemistry, St. Thomas College Palai,
Kerala, India

Abstract

Photosensitization reactions have been among the most popular research topics in recent decades. Photosensitizers (PS) are chemicals that produce Reactive Oxygen Species (ROS) when exposed to the right amount of light. Pyrene-based systems have their unique significance in this sector among diverse photosensitizers. Pyrene (a polycyclic aromatic hydrocarbon) and its derivatives own four benzene ring blended structures, which are highly electron rich and well-known for their distinctive photophysical capabilities. Long excited-state lifetime (50-90 ns), excimer formation capabilities, effective electron hole-pair dissociation, clear vibrational developments in the fluorescence spectra, ease of functionalization, high photoluminescence quantum yield in the solid-state (0.68) and fluorophore with excellent emission properties drew researchers to this compound for photosensitizer research. Pyrene's photosensitizability is due to the singlet excited state (S_1) extended lifetime around 400 ns, triplet state with less energy and vital O_2-induced enhanced intersystem crossing (EISC), all of which are supportive for generating ROS, particularly singlet oxygen ($^1O_2$) in a quantum yield of 0.71 through energy transfer. This class of pyrene-based systems has a

* Corresponding Author's Email: majothomas@gmail.com; majothomas@stcp.ac.in, Fax: + 91-4822-216313.

In: Photosensitizers and Their Applications
Editors: Davor Margetic and Renjith Thomas
ISBN: 978-1-68507-880-5

wide range of appositeness, such as a photosensitizing agent in Photodynamic Therapy (PDT), luminescence, photocatalysis, heterogeneous catalysis, electrochemical purposes, and biomedical application. Additionally, pyrene-based chemicals are employed as visible light photoinitiators of polymerization activation at low light intensities. Currently, multiple studies are being conducted to create pyrene-based compounds as photosensitizers, which have a wide range of applications.

Keywords: pyrene, metal-organic frameworks, photodynamic therapy

Introduction

Recently, photosensitizers (PSs) have been the subject of intense interest. They are the reactive oxygen species generating compounds by irradiation of suitable light [1]. Pyrene-based systems have great importance among various photosensitizers due to their photophysical and photochemical properties. Pyrene is a well-known polycyclic aromatic hydrocarbon with four benzene rings joined together. Pyrene has become one of the most widely studied organic molecules in photochemistry and photophysics. It also acts as an excellent chromophore with conjugated π-systems. Pyrene-based molecules have attracted researchers because of their characteristic photophysical properties such as long excited-state lifetime, effective electron hole-pair dissociation, abilities of excimer formation and long fluorescence lifetime [2, 3]. The optical properties of pyrene based molecules have made them applicable to chemical sensors [4], organic photovoltaics (OPVs), probes for micelle and cell membrane properties [5-9], organic light-emitting diodes (OLEDs) [10], ligands for luminescent coordination complexes, surface-coating materials for nanoparticles [11-15] and photosensitizers producing singlet oxygen ($^1O_2$) [16]. The photosenzitisability of pyrene is mainly due to the relatively long lifetime (about 400ns) of the singlet excited state (S_1), which permits an electron exchanging reaction with an oxygen molecule to generate singlet oxygen in a quantum yield of 0.71.

Metal-organic frameworks (MOFs) or porous coordination polymers (PCPs) are inorganic secondary units and organic linkers formed by coordination bonds [17, 18]. Due to the intrinsic porosity and tenability, they possess wide usage in various fields such as gas storage, chemical sensing,proton conduction, heterogeneous catalysis etc. Recently, MOFs are developed as photosensitizers to generate singlet oxygen, which find utility in various applications, mainly light-harvesting and photocatalysis [19-22]. For the utilization of these MOFs as photosensitizers, they should be blended with organic and metal-organic chromophores to relate them with photoactivity [23-25]. Pyrene and its derivatives are the most common photochromic compounds known for their specific photoelectronic and optical properties. These systems are with low energy triplet state and remarkable O_2 induced EISC (enhanced intersystem crossing) to produce singlet oxygen by energy transfer. Many pyrene-based systems have been developed for singlet oxygen generation [26-28]. Various electrophilic aromatic substitution reactions are possible at different positions in pyrene generating 1-substituted pyrenes, 1,6- and 1,8-disubstituted pyrenes,1,3,6,8-tetrasubstituted pyrenes, 1,3-disubstituted pyrenes, 4,5,9,10-tetrasubstituted pyrenes and 2,7-disubstituted pyrenes. Among these 1,3,6,8-tetrasubstituted pyrenes find much utilization in MOF synthesis. The structures of very few such systems are shown in Figure 1. Thereby, the development of such systems contributes more to the fields of various applications as in photocatalysis, photodynamic therapy (PDT) etc.

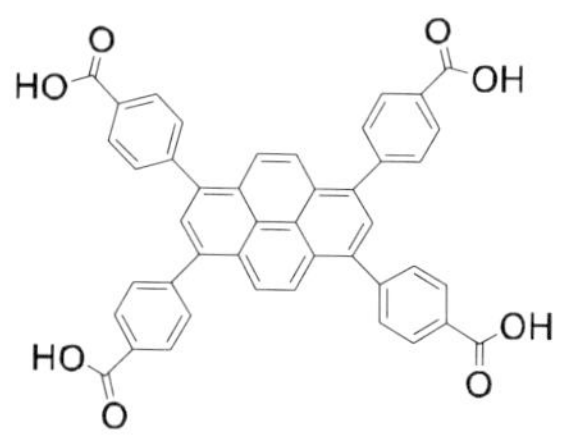

1,3,6,8,-Tetrakis(4-carboxylphenyl)pyrene (TBAPy)

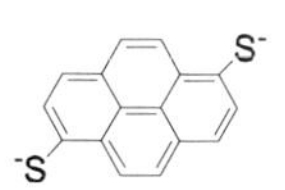

1,6-Pyrenedithiolate (PDT)

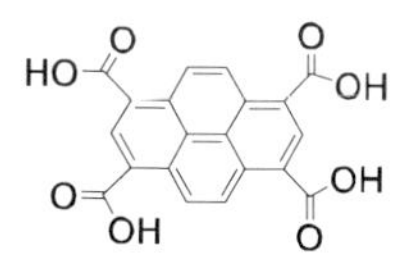

Pyrene-1,3,6,8-tetracarboxylic acid (PTCA)

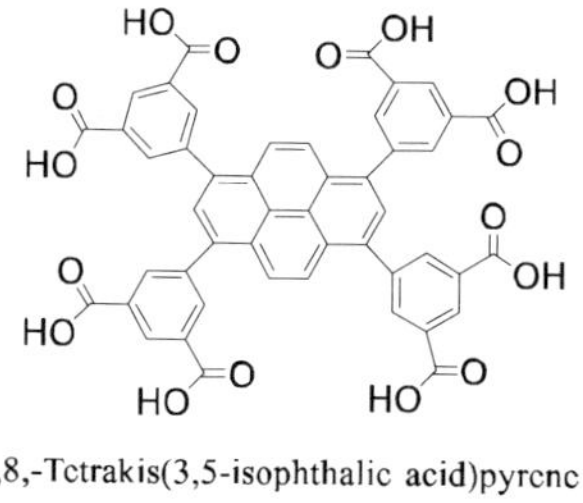

1,3,6,8,-Tetrakis(3,5-isophthalic acid)pyrene (TIAPy)

Figure 1. Structures of various pyrene-based ligands used in the synthesis of MOFs.

Photodynamic Therapy

Due to the tunable functionalities and properties, the pyrene-based systems also apply in the biomedical field [29, 30]. Pyrene-based photosensitizers are the significant chemicals used as PDT agents due to their fine photosensitizability. The three vital components involved in the PDT are photosensitizer (PS), light and substrate. By irradiating light, the photo-sensitizer is excited from ground state S0 to first singlet excited state S_1. Upon intersystem crossing (ISC), the first excited triplet state T_1 comes. From this first excited triplet state, PSs could directly react with substrates, mainly with oxygen, to generate reactive oxygen species (ROS), which cause cell destruction. Based on photodynamic therapy's photophysical and photochemical principles, these are classified into type I PDT and type II PDT (Scheme 1). In type I PDT, free radicals are produced with shortened lifetime by electron transfer or hydrogen abstraction between photosensitizers and substrates. These free radicals immediately react with water or oxygen, producing various products such as hydrogen peroxide, superoxide anion and hydroxyl radical. In type II PDT, cytotoxic singlet oxygen is generated due to energy transfer. In PDT, type I and type II processes occur in parallel, and their ratio depends on properties and concentrations of PSs and substrates.

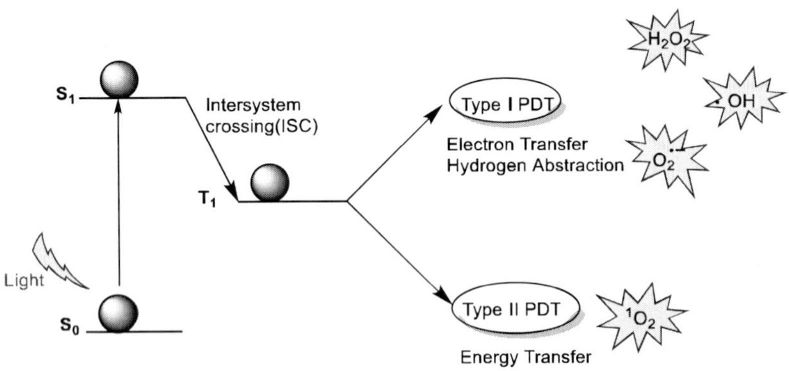

Scheme 1. Representation of photochemical basis of PDT.

Herein we discuss some of the recent pyrene-based systems as photosensitizers in PDT. In 2018, Yuasa *et al.* introduced glucosyl pyrene photosensitizers with moderate solubility in water. They synthesized two pyrene-based glycosyl photosensitizers such as 1-carboxypyren-2-yl C-glucoside (1) and 1-pyrenylmethyl O-glucoside (2) (Scheme 2) [31].

Among them, the second compound showed more application as PDT agent as it showed promising results regarding cell uptake on HeLa cells and decent PDT effect. Zhao and coworkers developed pyrene-based MOF nanoparticles for PDT by ligand substitution using NU-1000 and TCPP. The surface of the synthesized material is functionalised using PEG (polyethylene glycol) and folic acid for biocompatible and malignancy target effect [32]. The synthesized MOF nanoparticles exhibit low poisonous effect with excellent photodynamic effect in *invivo* and *invitro* studies.

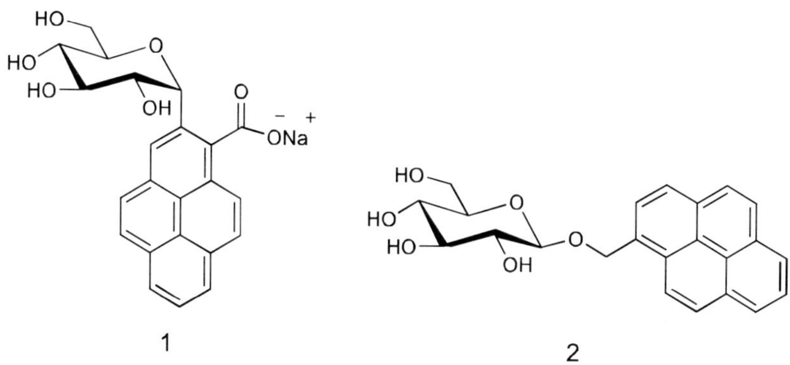

Scheme 2. Structures of the two glucosyl pyrene photosensitizers.

In 2019, Sun and the group synthesized six bis(terpyridine) iridium (III) complexes ($Ir(R_1$-tpy) $(R_2$-tpy$)^{3+}$) **Ir1**-**Ir6** (Scheme 3). Complexes **Ir4** and **Ir5** act as good PDT agents with long excited states of the lowest possible triplet excited state lifetime and generate the strongest singlet oxygen [33]. The pyrenyl group present in **Ir4** and **Ir5** plays a vital role in its PDT effects. Amongst these complexes, **Ir4** demonstrated uppermost $^1O_2$ quantum yield (0.81) in cell-free conditions, prevalent photocytotoxicity in opposition to SK-MEL-28 cells, an extremely capable generator of reactive oxygen species (ROS) *in vitro*, great phototherapeutic index and good photosensitizer at 400-700 nm. **Ir4** showed extremely intense *in vitro* PDT effect on MCF-7 breast cancer cells and significantly prevented the xenograft tumor growth upon 450-nm light activation with no harmfulness and minor complications.

In 2020, the same group developed bis-terpyridine ruthenium(II) [Ru(tpy-R_1)(tpy-R_2)]$^{2+}$photosensitizer complexes such as Ru1, Ru2 and Ru3 as varying R_1 and R_2 groups (Scheme 4) [34]. Ru3 contained pyren-1-yl as R_1 and 4-BODIPY-phenyl as R_2. Ru2 and Ru3 possessed the tetrahedral BODIPY and pyrenyl moieties which prolong the lifetime of triplet excited states.

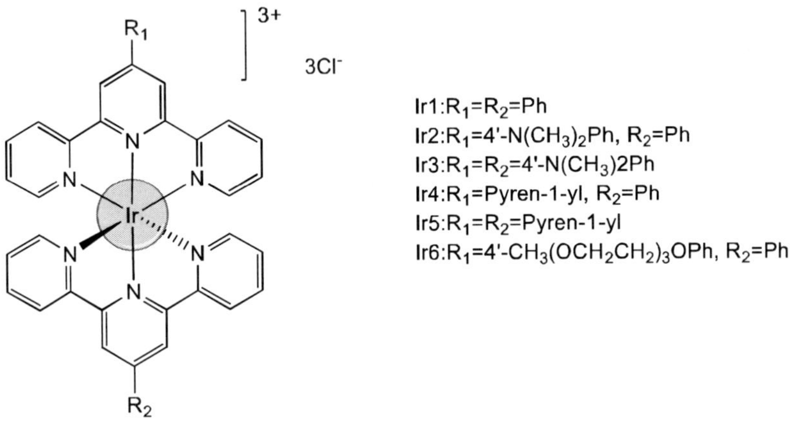

Scheme 3. Structure of the synthesized bis(terpyridine) iridium (III) complexes.

This leads to the singlet oxygen generation upon green light activation and subsequently triggered considerable phototoxicity towards A549 lung cancer cell line.

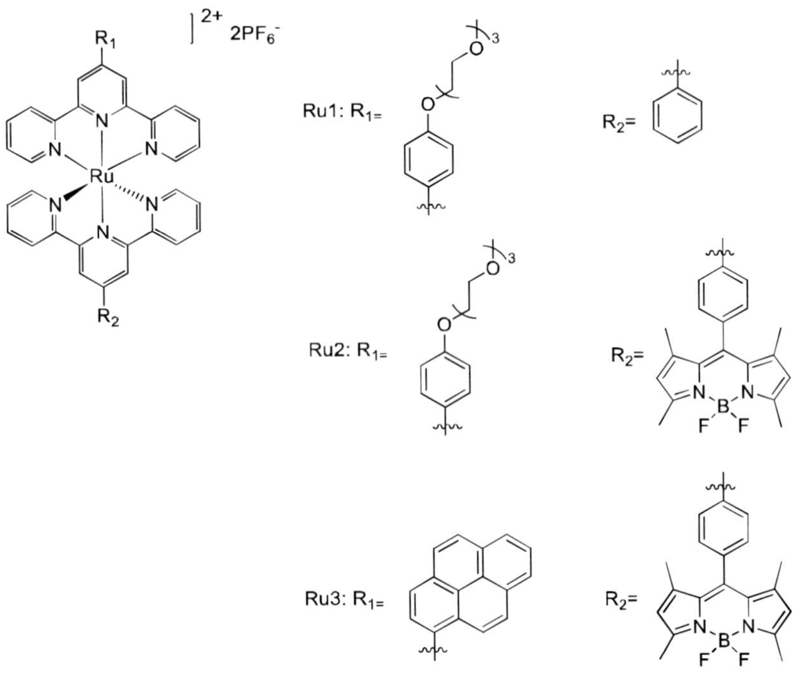

Scheme 4. Structures of Ru (II) complexes Ru1-Ru3.

In 2020, Franca and the team developed a new pyrene-based photosensitizer, 4-(3,6,8-tris(phenylethynyl)pyren-1-yl) butanoic acid (TPPy), which displays the antitumor photoactivity against murine melanoma cell line (B16F10) as a model for *in vitro* PDT [35]. Furthermore, the Galleria mellonella model was used to evaluate the toxicity *in vivo*. The studies revealed that TPPy is a promising photosensitizer designed for usage in PDT with possible antitumor photoactivity. In the same year, Sethuraman and his team synthesized pyrene derivatives with thiobarbituric acid, free from heavy atoms (Scheme 5). Under 455 nm LED irradiation, these PSs showed photodynamic activity in opposition to MCF-7 cancer cells [36]. Both pyrene derivatives showed highly intense intra-molecular charge transfer (ICT) bands in 460 nm and produced singlet oxygen state with adequate quantum yields in DMSO. Molecule (**3**) was observed to possess better ROS generation and PDT action against MCF-7 cells.

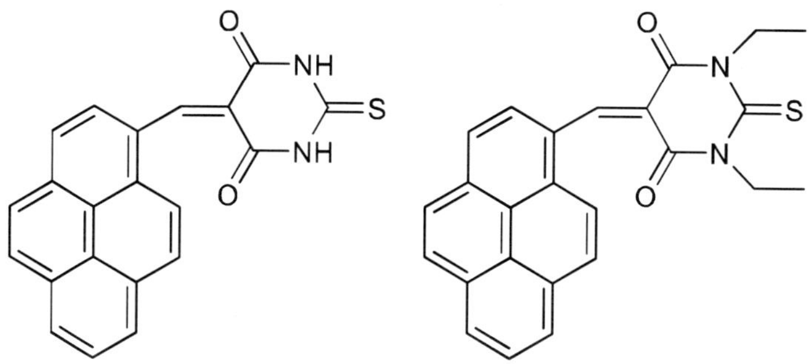

3 4

Scheme 5. Structures of thiobarbituric acid functionalized pyrene derivatives.

Photocatalysis

Pyrene-based photocatalysis has attracted wide attention. Currently, the degradation and elimination of chemical warfare agents (CWAs) is an important issue of global concern [37]. Sulfur mustard, bis(2-chloroethyl)sulfide is widely used CWA but it is a robust acute agent that produces harsh skin, eye, mucosal pain and irritation. Detoxification of sulfur mustard can be done by hydrolysis, dehydrohalogenation and oxidation. However, in practical use, first two methods are too slow. Selective oxidation of sulfur mustard is the best detoxification method and it can be achieved using a photosensitizer, which forms reactive singlet oxygen from ground state triplet oxygen. Here, the pyrene based systems can be utilized for sulphur mustard degradation. Farha and his team developed various organic-based photosensitizers for selective sulfur mustard oxidation [38]. In 2016, the same team developed NU-1000, a pyrene-based metal-organic framework, for oxidizing sulfur mustard deceptive 2-chloroethyl ethyl sulfide (CEES) using singlet oxygen to harmless product 2-chloroethyl ethyl sulfoxide (CEESO), without toxic sulfone generation [39]. This strategy utilizes the use of pyrene-based linkers as UV-photosensitizers.

Later in 2019, the same team demonstrated a highly effective photosensitizer, NU-400, a Zr_6 based MOF on a pyrene-2,7-dicarboxylic acid (Py-DCA) linker, for mustard gas degradation [40]. It generates singlet oxygen under UV light irradiation, allowing the oxidative degradation of sulfur mustard simulant CEES into a non-threatening sulfoxide product. It is more effective than NU-1000 photosensitizer because NU-400 can attain high singlet oxygen without oxygen saturation. For the photodegradation of pollutants, pyrene-based MOFs are used as in Azure B photodegradation [41]. In 2019, Zhou*et al.* introduced visible-light harvesting photosensitizers reactive oxygen species (ROS) generator from a pyrene-based MOF system, irradiated by visible light [42]. They developed two novel visible light responsive MOFs, PCN-822(M) (M = Zr, Hf), using the pyrene-based ligand, 4,4',4'',4'''-((2,7-di-*tert*-butylpyrene-4,5,9,10-tetrayl)tetrakis(ethyne-2,1-diyl))-tetrabenzoate (BPETB4) which is synthesized by modified Sonogashira coupling reaction. TG studies revealed both these MOFs possess outstanding thermo-chemical stability. PCN-822(Zr/Hf) exhibits broad adsorption of visible light from 225 - 650 nm. Among these, PCN-822(Hf) could act as a good photocatalyst, supporting oxidation of amines to imines. Zhang *et al.* reported the photocatalytic conversion of perfluoroalkyl iodides into olefins via (atom-transfer radical addition) ATRA [43]. Here the MOF is NU-1000 and the reaction proceeds by the photo induced energy-transfer pathway. For various other organic transformations, NU-1000 can also be used.

Choi *et al.* reported a MOF, $FeSC_{12}$-AIM, in an aqueous solution with nitrate and ascorbic acid which acts as an electron donor [44]. In this system, photochemical transformation of NO_3^- to NH_4^+ occurs due to the pyrene linker's photo-excitation to the iron thiolated cluster. The photocatalytic CO_2 reduction was also performed well using the photosensitizing action of pyrene-based MOFs [45]. Peters *et al.* discussed about a newly developed material produced by anchoring of NiSx in NU-1000 via AIM, i.e., NiS-AIM could catalytically produce H_2 [46]. Even for the catalytic reduction of water, nickel sulphide functionalized nodes are used. NU-1000 HER (hydrogen evolution reaction) photocatalysis can be improved further by the modification and incorporation of CdS nanoparticles along with reduced graphene oxide (RGO) [47]. Along with NU-1000-based MOFs, TBAPy-based MOFs also exhibit HER photocatalytic activity. TBAPy linkers combined with Cd, Bi, and Ti based metal clusters give rise to materials with high HER [48]. Recently, Nasrallah *et al.* produced two pyrene-based complexes i.e., pyrene modified photosensitizer (RuPy) and catalyst (ReCOPy), by a combination of π-π stacking and H- bond interactions to form a supramolecular assembly RuPy···ReCOPy. This novel system exhibits good photocatalytic activity in the conversion of CO_2 into CO. Generally, various photocatalytic organic reactions can be carried out using this pyrene -based MOFs [49].

Conclusion

This chapter describes pyrene-based photosensitizers and their recent application progress in photocatalysis and photodynamic therapy. Such pyrene-based systems are hopeful materials for biomedical functions because of their tunable porosity and flexible chemical functionality, along with unique optical and electronic properties of pyrene. Over the last few years, sustained studies have been done in this area and novel pyrene-based systems have emerged with extensive applications, though the synthesis of pyrene-based photosensitizers and the estimation of its coordination atmosphere are the main impediments. Hence, many researchers are enthusiastically focused on developing unique pyrene systems and their implementation for various purposes.

References

[1] DeRosa MC, Crutchley RJ. Photosensitized Singlet Oxygen and Its Applications. *Coord Chem Rev,* 2002, 233-234, 351-371. https://doi.org/ 10.1016/S0010-8545 (02)00034-6.

[2] Soc C, Kinik FP, Ortega-guerrero A, Ongari D, Ireland CP, Ortega-Guerrero A. Pyrene-Based Metal Organic Frameworks : From Synthesis to Applications. *Chem Soc Rev,* 2021, 50, 3143-3177. https://doi.org/10.1039/ d0cs00424c.

[3] Howarth AJ, Majewski MB, Wolf MO. Photophysical Properties and Applications of Coordination Complexes Incorporating Pyrene. *Coord Chem Rev,* 2014, 282-283, 139-149. https://doi.org/10.1016/ j.ccr.2014.03.024.

[4] Karuppannan S, Chambron J. Supramolecular Chemical Sensors Based on Pyrene Monomer - Excimer Dual Luminescence. *Chem Asian J,* 2011, 6, 964-984. https://doi.org/10.1002/asia.201000724.

[5] Somerharju P. Pyrene-Labeled Lipids as Tools in Membrane Biophysics and Cell Biology. *Chem Phys Lipids*, 2002, 116, 57-74. doi: 10.1016/s0009-3084(02)00020-8.

[6] Vanderkooi JM, Callis JB. Pyrene. A Probe of Lateral Diffusion in the Hydrophobic Region of Membranes. *Biochemistry,* 1974, 13, 4000-4006. https://pubs.acs.org/doi/10.1021/bi00716a028.

[7] Food, U. S. T -. 2039 (conformation D).

[8] Aguiar J, Carpena P, Ruiz CC. On the Determination of the Critical Micelle Concentration by the Pyrene 1 : 3 Ratio Method. *J Colloid Interf Sci,* 2003, 258, 116-122. https://doi.org/10.1016/S0021-9797(02)00082-6.

[9] Niko Y, Didier P, Mely Y, Konishi G, Klymchenko AS. Bright and Photostable Push-Pull Pyrene Dye Visualizes Lipid Order Variation between Plasma and Intracellular Membranes. *Sci Rep,* 2016, 6, 18870. https://doi.org/10.1038/srep18870.

[10] Figueira-Duarte TM, Müllen K. Pyrene-Based Materials for Organic Electronics. *Chem Rev,* 2011, 111, 7260-7314. https://doi.org/10.1021/ cr100428a.

[11] Montalti M, Zaccheroni N, Prodi L, Reilly NO, James SL. Enhanced Sensitized NIR Luminescence from Gold Nanoparticles via Energy Transfer from Surface-Bound Fluorophores. *J Am Chem Soc,* 2007, 129, 2418-2419.

[12] Thomas KG, Kamat PV, Dame N. Making Gold Nanoparticles Glow: Enhanced Emission from a Surface-Bound Fluoroprobe. *J Am Chem Soc,* 2000, 122, 2655-2656. https://doi.org/10.1021/ja9941835.

[13] Rosa MLa, Denisov SA, Jonusauskas G, Mcclenaghan ND, Credi A. Designed Long-Lived Emission from CdSe Quantum Dots through Reversible Electronic Energy Transfer with a Surface-Bound Chromophore. *Angew Chem Int Ed,* 2018, 57, 3104-3107. https://doi.org/ 10.1002/ange.201712403.

[14] Mathew TV, Kuriakose S. 4-(1-Pyrenyl)Butyric Acid-Functionalised Chitosan as a Matrix for AgNP: Photoresponsive and Thermal Properties. *J Polym Res,* 2013, 20, 291. https://doi.org/10.1007/s10965-013-0291-5.

[15] Mathew TV, Kuriakose S. Synthesis and Characterization of Sodium-Lithium Niobate Ceramic Structures and Their Composites with Biopolymers. *J Adv Ceram,* 2013, 2, 11-20. https://doi.org/10.1007/s40145-013-0035-3.

[16] Zoeller RA, Morand OH, Raetz CR. A possible role for plasmalogens in protecting animal cells against photosensitized killing. *J Biol Chem,* 1988, 263, 11590-11596. https://doi.org/10.1016/S0021-9258(18)38000-1.

[17] Li J, Sculley J, Zhou H. Metal À Organic Frameworks for Separations. *Chem Rev,* 2012, 112, 869-932. https://doi.org/10.1021/cr200190s.

[18] Li H, Eddaoudi M, O'Keeffe M, Yaghi OM. Design and Synthesis of an Exceptionally Stable and Highly Porous Metal-organic Framework. *Nature,* 1999, 402, 276-279. https://doi.org/10.1038/46248.

[19] Chen D, Xing H, Wang C, Su Z. Highly Efficient Visible-light-driven CO_2 Reduction to Formate by a New Anthracene-Based Zirconium MOF via Dual Catalytic Routes. *J Mater Chem. A Mater Energy Sustain,* 2016, 4, 2657-2662. https://doi.org/10.1039/C6TA00429F.

[20] Zhang T, Lin W. Metal–organic frameworks for artificial photosynthesis and photocatalysis. *Chem Soc Rev,* 2014, 43, 5982-5993. https://doi.org/ 10.1039/ c4cs00103f.

[21] Johnson JA, Luo J, Zhang X, Chen Y, Echeverría E, Torres FE, Zhang J. Porphyrin-Metalation-Mediated Tuning of Photoredox Catalytic Properties in Metal-Organic Frameworks. *ACS Catal,* 2015, 5, 5283-5291. https://doi.org/10.1021/acscatal.5b00941.

[22] Yuan S, Qin J, Xu H, Su J, Rossi D, Chen Y, Zhang L, Lollar C, Wang Q, Jiang H, Son D, Xu H, Huang Z, Zou X, Zhou H. $[Ti_8Zr_2O_{12}(COO)_{16}]$ Cluster: An Ideal Inorganic Building Unit for Photoactive Metal - Organic Frameworks. *ACS Cent Sci,* 2018, 4, 105-111. https://doi.org/10.1021/ acscentsci.7b00497.

[23] Liu Y, Howarth AJ, Hupp JT, Farha OK. Selective Photooxidation of a Mustard-Gas Simulant Catalyzed by a Porphyrinic Metal - Organic Framework. *Angew Chem Int Ed,* 2015, 127, 9129-9133. https://doi.org/ 10.1002/ange.201503741.

[24] Lu K, He C, Guo N, Chan C, Ni K, Weichselbaum RR, Lin W. Chlorin-Based Nanoscale Metal - Organic Framework Systemically Rejects Colorectal Cancers via Synergistic Photodynamic Therapy and Checkpoint Blockade Immunotherapy. *J Am Chem Soc*, 2016, 138, 12502-12510. https://doi.org/10.1021/jacs.6b06663.

[25] Park J, Jiang Q, Feng D, Zhou H. Metal - Organic Frameworks Hot Paper Controlled Generation of Singlet Oxygen in Living Cells with Tunable Ratios of the Photochromic Switch in Metal - Organic Frameworks. *Angew Chem Int Ed,* 2016, 128, 7304-7309. https://doi.org/10.1002/ ange.201602417.

[26] Buru CT, Majewski MB, Howarth AJ, Lavro RH, Kung C, Peters AW, Goswami S, Farha OK. Improving the Efficiency of Mustard Gas Simulant Detoxification by Tuning the Singlet Oxygen Quantum Yield in Metal - Organic Frameworks and Their Corresponding Thin Films. *ACS Appl Mater Interf,* 2018, 10, 23802-23806. https://doi.org/10.1021/acsami.8b05792.

[27] Atilgan A, Islamoglu T, Howarth AJ, Hupp JT, Farha OK. Detoxi Fi Cation of a Sulfur Mustard Simulant Using a BODIPY- Functionalized Zirconium-Based Metal - Organic Framework. *ACS Appl Mater Interfaces,* 2017, 9, 24555-24560. https://doi.org/10.1021/acsami.7b05494.

[28] Hu J, Era H, Elsegood MRJ, Yamato T. Synthesis and Photophysical Properties of Pyrene-Based Light-Emitting Monomers: Highly Pure-Blue-Fluorescent, Cruciform-Shaped Architectures. *Eur J Org Chem*, 2010, 1, 72-79. https://doi.org/ 10.1002/ejoc.200900806.

[29] Kamkaew A, Lim H, Lee B, Kiew V. BODIPY Dyes in Photodynamic Therapy. *Chem Soc Rev,* 2013, 42, 77-88. https://doi.org/10.1039/ c2cs35216h.

[30] Stacey OJ, Pope SJA. New avenues in the design and potential application of metal complexes for photodynamic therapy. *RSC Adv,* 2013, 3, 25550-25564. https://doi.org/10.1039/c3ra45219k.

[31] Kanamori T, Matsuyama A, Naito H, Tsuga Y, Ogura S, Okazaki S, Yuasa H. Water-Soluble Glucosyl Pyrene Photosensitizers : An Intramolecularly Synthesized 2-C-Glucoside and an O-Glucoside. *J Org Chem,* 2018, 83, 13765-13775. https://doi.org/10.1021/acs.joc.8b02066.

[32] Zhao X, Zhang Z, Cai X, Ding B. Biological and Medical Applications of Materials and Interfaces Post-Synthetic Ligand Exchange of Metal Organic Framework for Photodynamic Therapy Post-Synthetic Ligand Exchange of Metal Organic Framework for Photodynamic Therapy. *ACS Appl Mater Interf,* 2019, 11, 7884-7892. https://doi.org/10.1021/acsami.9b00740.

[33] Liu B, Monro S, Li Z, Jabed MA, Ramirez D, Cameron CG, Colón K, Iii JR, Kilina S, Tian J, Mcfarland SA, Sun W, Colón K, Iii JR, Kilina S, Tian J, Sherri A. New Class of Homoleptic and Heteroleptic Bis(Terpyridine) Iridium(III) Complexes with Strong Photodynamic Therapy Effects. *ACS Appl Bio Mater*, 2019, 2, 2964-2977. https://doi.org/10.1021/ acsabm.9b00312.

[34] Liu B, Gao Y, Jabed MA, Kilina S, Liu G, Sun W. Lysosome Targeting Bis-terpyridine Ruthenium(II) Complexes: Photophysical Properties and In Vitro Photodynamic Therapy. *ACS Appl Bio Mater,* 2020, 3, 6025-6038. https://doi.org/10.1021/acsabm.0c00647.

[35] de Franca BM, Ghasemishahrestani Z, de Souza GFM, da Silva RN, Queiroz DD, Pierre MBR, Pereira MD, Forero JSB, Corrêa RJ. In vitro Studies of Antitumor Effect, Toxicity/Cytotoxicity and Skin Permeation/Retention of a Green Fluorescence Pyrene-based Dye for PDT Application. *Photochem Photobiol,* 2021, 97, 408-415. https://doi.org/ 10.1111/php.13335.

[36] Babu B, Ali TA, Ochappan T, Mack J, Nyokong T, Sethuraman MG. Photocytotoxicity of heavy-atom-free thiobarbituric acid functionalized pyrene derivatives against MCF-7 cancer cells Photodiagn. *Photo Ther,* 2021, 33, 102102. https://doi.org/10.1016/j.pdpdt.2020.102102.

[37] Atilgan A, Islamoglu T, Howarth AJ, Hupp JT, Farha OK. Detoxification of a Sulfur Mustard Simulant Using a BODIPY Functionalized Zirconium-Based Metal - Organic Framework. *ACS Appl Mater Interf,* 2017, 9, 24555-24560. https://doi.org/10.1021/acsami.7b05494.

[38] Ayoub G, Arhangelskis M, Zhang X, Son F, Islamoglu T, Friš T, Farha OK. Air Oxidation of Sulfur Mustard Gas Simulants Using a Pyrene-Based Metal - Organic Framework Photocatalyst. *Beilstein J Nanotechnol,* 2019, 10, 2422–2427. https://doi.org/10.3762/bjnano.10.232.

[39] Liu Y, Buru CT, Howarth AJ, Mahle JJ, Buchanan JH, Decoste JB, Hupp T, Farha OK. E Ffi Cient and Selective Oxidation of Sulfur Mustard Using Singlet Oxygen

Generated by a Pyrene-Based Metal - Organic Framework. *J Mater Chem A*, 2016, 4, 13809-13813 https://doi.org/10.1039/ C6TA05903A.

[40] Liu Y, Moon S, Hupp JT, Farha OK. Dual-Function Metal-Organic Framework as a Versatile Catalyst for Detoxifying Chemical Warfare Agent Simulants. *ACS Nano*, 2015, 9, 12358-12364. https://doi.org/10.1021/ acsnano.5b05660.

[41] Zhao N, Sun F, Zhu G, Accepted J. A Novel Pyrene-Based Anionic Metal-Organic Framework for Efficient Organic Dye Elimination. *Cryst Growth Des,* 2017, 17, 2453-2457. https://doi.org/10.1021/acs.cgd.6b01864.

[42] Zhang Y, Pang J, Li J, Yang X, Feng M, Cai P, Zhou H. Visible-light harvesting pyrene-based MOFs as efficient ROS generators. *Chem Sci,* 2019, 10, 8455-8460. https://doi.org/10.1039/C9SC03080H.

[43] Zhang T, Wang P, Gao Z, An Y, He C, Duan C. Pyrene-based metal–organic framework NU-1000 photocatalysed atom-transfer radical addition for iodoperfluoroalkylation and (Z)-selective perfluoroalkylation of olefins by visible-light irradiation. *RSC Adv,* 2018, 8, 32610-32620. https:// doi.org/10.1039/ C8RA06181E.

[44] Choi H, Peters AW, Noh H, Gallington LC, Platero-prats AE, Destefano MR, Rimoldi M, Chapman KW, Farha OK, Hupp JT, Farha OK, Hupp JT. Vapor-Phase Fabrication and Condensed-Phase Application of a MOF-Node-Supported Iron Thiolate Photocatalyst for Nitrate Conversion to Ammonium. *ACS Appl Energy Mater,* 2019, 2, 8695-8700. https://doi.org/ 10.1021/acsaem.9b01664.

[45] Chen Y, Li P, Zhou J, Buru CT, Đorđević L, Li P, Zhang X, Cetin MM, Stoddart JF, Stupp SI, Wasielewski MR, Farha OK. Integration of Enzymes and Photosensitizers in a Hierarchical Mesoporous Metal-Organic Framework for Light-Driven CO_2 Reduction. *J Am Chem Soc,* 2020, 142, 1768-1773. https://doi.org/ 10.1021/jacs.9b12828.

[46] Peters AW, Li Z, Farha OK, Hupp JT. Toward Inexpensive Photocatalytic Hydrogen Evolution: A Nickel Sul Fi de Catalyst Supported on a High-Stability Metal-Organic Framework. *ACS Appl Mater Interfaces,* 2016, 8, 20675-20681. https://doi. org/10.1021/acsami.6b04729.

[47] Bag PP, Wang X-S, Sahoo P, Xiong J, Cao R. Efficient photocatalytic hydrogen evolution under visible light by ternary composite CdS@NU-1000/RGO. *Catal Sci Technol*, 2017, 7, 5113-5119. https://doi.org/10.1039/ C7CY01254C.

[48] Cadiau A, Xie LS, Kolobov N, Shkurenko A, Qureshi M, Tchalala MR, Park SS, Bavykina A, Eddaoudi M, Dinca M, Hendon CH, Gascon J. Toward New 2D Zirconium-Based Metal-Organic Frameworks: Synthesis, Structures, and Electronic Properties. *Chem Mater,* 2020, 32, 97-104. https://doi.org/10.1021/acs.chemmater. 9b02462.

[49] Nasrallah H, Lyu P, Maurin G, El-Roz M, Nasrallah H, Lyu P, Maurin G, Highly ME. Highly Efficient CO_2 Reduction under Visible-Light on Noncovalent Ru•••Re Assembled Photocatalyst : Evidence on the Electron Transfer Mechanism. *J Catal*, 2021, 404, 46-55. https://doi.org/10.1016/ j.jcat.2021.09.007.

Chapter 12

Ruthenium Complexes in Dye-Sensitized Solar Cells: Experimental and Theoretical Perspectives

Smitha Roy[1], Bebeto G. Nair[2] and Aravind Krishnan[2,*]

[1]Union Christian College, Aluva, Kerala, India
[2]St. Berchmans College (Autonomous), Changanassery, Kottayam, Kerala, India

Abstract

"High efficiency with low-cost production" is the tag given to the 3rd generation of solar cells, viz. Dye Sensitized Solar Cells (DSSC). Ruthenium complexes serve as a major category belonging to the above class due to the advantages of high photostability and wide absorption range from visible to near IR. With the development of N3 dye, the godfather of ruthenium dyes, scientific research has developed remarkably in modifying the complexes with various functional groups. Improved binding of the dye to TiO_2 nano-surface, decreased back electron transfer, high absorbtion coefficient, resistance to water induced dye desorption, reversibility of ruthenium III/II couple, high absorption near the red region are some of the prerequsites for the production of efficient ruthenium dyes. An overall power conversion efficieny of 11% have been achieved using ruthenium polypyridyl complexes under standard illumination. The experimental modification includes the introduction of donor and acceptor moieties in the polypyridy ligands which is seen to achieve the above-mentioned requirements.

* Corresponding Author's Email: aravind@sbcollege.ac.in.

In: Photosensitizers and Their Applications
Editors: Davor Margetic and Renjith Thomas
ISBN: 978-1-68507-880-5

Computational methodology also played a major contribution in designing the dyes applicable for DSSC. With methods like DFT, an accurate knowledge of the electronic environment of the dye in both the ground and the excited states can be obtained. Fine tuning of the required properties has been achieved by careful design of the dye using computational knowledge.

This chapter, thus throws light on the development of experimental and computational methods involved in designing of ruthenium complexes for DSSC.

Keywords: solar cells, N3 dye, dye sensitized solar cells, Ruthenium bipyridine complexes

Introduction

We know that our dependency on natural gas or fossil fuels as energy sources are at stake and hence, alternative sources which are inexhaustible such as solar, geothermal, biomass etc. are required[1-3].

With this excessive demand for renewable, clean and green energy, significant research is being undertaken on photovoltaic cells (solar cells) to convert solar energy in to electrical energy [4-7]. The journey of solar cells began in 19th century when it was first noticed that solar energy can be converted to electrical energy. Solar cells are divided into three broad categories:

1. First generation solar cells - lower efficiency and huge cost of production.
2. Second generation solar cells - lower efficiency but cheaper cost of production than first generation type.
3. Third generation solar cells - highly efficient which are being used now by combining multiple devices like quantum dot solar cells (QDSCs), perovskite solar cells, and dye-sensitized solar cells (DSSCs) [8-9]

DSSC is based on the sensitization of mesoporous nanocrystalline metal oxide film to visible light. The sensitization process takes place by the principle of adsorption. In DSSC, the photosensitive dyes provide photoelectrons and semiconductor act as charge transporter. With the progress of time, better materials were discovered, reported, invented and developed with an idea of commercialization. DSSC belongs to a group of solar cells of thin film undergoing massive research in the past three decades. The properties that focused research on DSSC are: 1) High incident solar light-to-electricity conversion efficiency, 2) Colourful and decorative nature and 3) Low weight and low cost. [10-14].

Components and Assembly of DSSCs [15-20, 53]

DSSC is assembled from a working electrode with a sensitizer or a dye and sealed to the counter electrode having a thin layer of electrolyte coating with the help of a hot melt tape. This tape prevents the leakage of the electrolyte.

The following represents the various components of DSSC:

A. Transparent and conductive substrate - Two sheets of conductive transparent materials act as a current collector and are used for helping the substrate for the deposition of the semiconductor and catalyst. Usually, FTO and ITO is applied as substrate having transparency above 80%, higher electrical conductivity to prevent energy loss and proper charge transfer.
B. Working electrode - A thin layer of oxide semiconductor with wide energy band gap is deposited on the FTO or ITO glass and then it is dipped in a mixture of solvent and photosensitive molecular sensitizer allowing the dye to make a covalent bond with the oxide semiconductor's surface therefore the working electrode will have a better ability to absorb light.
C. Photosensitizer/Dye - It is considered as the heart of DSSC. The photosensitizer is placed in between the redox couple and semiconductor film. Because of this dye or photosensitizer, the ability of absorption of light is maximum.
D. Electrolyte (redox mediator) - In DSSC, three types of electrolytes are used: liquid electrolytes, solid-state electrolytes and quasi-solid gel electrolytes. Much care has to be given while selecting an electrolyte as it should not react with the dye and there should be no overlapping of spectra of absorption of electrolyte and dye, show great stability and should be non-corrosive, fast charge diffusion, good conductivity, proper contact between electrodes and should be transparent to visible light.

It has five main components: redox couple (I^-/I^{3-}), solvent, additives, ionic liquids, and cations.

E. Counter electrode – Platinum or graphite is used as counter electrode for catalyzing the electrolyte's reduction process.

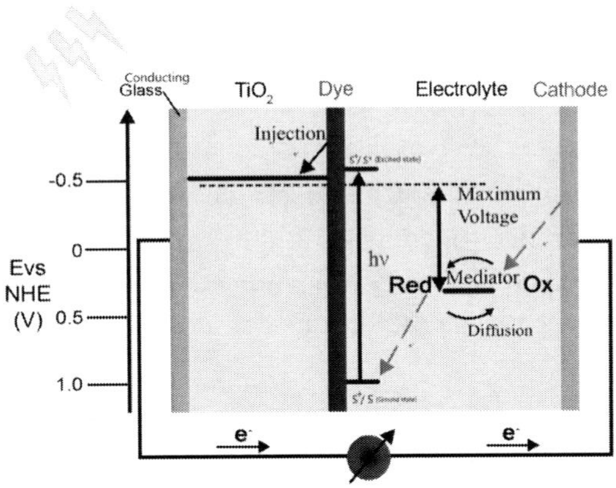

Figure 1. Schematic diagram of a dye sensitized solar cell.

Working Principle [17, 21-22]

The working steps are as follows:

1. The electron of the dye moves from the ground state to the first excited singlet state when photon is absorbed by the photosensitizer.

   (Dye molecule) + hν → (Dye molecule)*

2. The excited electron gets injected into conduction band of TiO_2 nano-crystalline particles. In this step, the semiconductor also has the ability to absorb photons therefore the dye gets oxidized.

   (Dye molecule)* → (Dye molecule)$^+$ + e^- (TiO_2)

3. Transportation of the injected electron followed by diffusion to the external circuit takes place.

   e^- (TiO_2) → e^- (wire)

4. The electron reaches the counter electrode

   $I^{3-} + 2e^{-}$ (counter electrode) $\rightarrow 3I^{-}$

5. Reduction of the electrolyte takes place followed by regeneration of the dye, wherein the reduced electrolyte is oxidized.

   $(\text{Dye molecule})^{+} + 3I^{-} \rightarrow 2(\text{Dye molecule}) + I^{3-}$ [17,21-22].

Incident Photon to Current Conversion (IPCC)

IPCE represents the term used to check the efficiency of DSSC, which, in turn is a combination of many factors. The first among them is Fill Factor. The ratio of maximum power output (Jmp × Vmp) to the product (VOC × JSC) gives FF:

$$FF = \frac{Area\ A}{Area\ B} = \frac{Jmp * Vmp}{Jsc * Voc}$$

At a constant light level exposure FF is the fill factor, short circuit current (JSC, $mAcm^{-2}$), open circuit voltage (VOC, V), overall efficiency is the percentage of solar energy being converted to electrical energy and is the intensity of incident light.

$$\eta\ (\%) = \frac{Jsc * Voc * FF}{Pin}$$

External quantum efficiency (also known as IPCE) is the ratio of number of electrons flowing through the external circuit to the number of photons incident on the cells surface at any wavelength λ (Pin). It is mathematically expressed as:

$$IPCE\ \%\ (\lambda) = 1240 * \left(\frac{Jsc}{Pin\lambda}\right)$$

Also,

$$IPCE\ (\lambda, nm) = LHE.\phi E1.\eta EC$$

where *LHE* is the light harvesting efficiency, *φE1* is electron injection quantum efficiency, and *ηEC* is the efficiency of collecting electrons in the external circuit [17,23-25].

Development of Dyes

In DSSC, dyes play an important role in the conversion of solar energy to electrical energy and hence major researches where focused on the development for better absorbing materials [17]. Shalini et al. provided the information on different types of sensitizers which include:

1. Metal complex inorganic dyes
2. Metal-free organic dyes
3. Quantum-dot sensitizer
4. Perovskite-based sensitizer
5. Mordant dyes
6. Natural dyes

Some examples of dyes which were developed and used as sensitizers are Mangosteen, D102, K68 etc. Dye sensitizer's general structure is $ML_2(X)_2$ where M represents metal, L represents ligand and X represents cyanide, thiocarbamate, thiocyanate, halide, acetyl acetonate, or water substituent group [17].

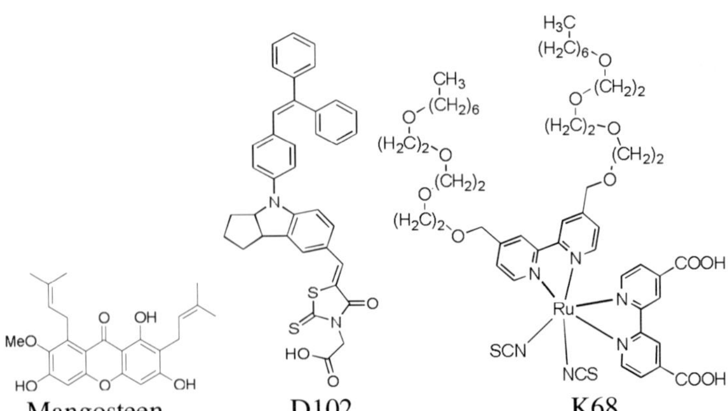

Mangosteen D102 K68

Figure 2. Structures of Mangosteen, D102 and K68.

Metal Complex Dyes

The inorganic dyes produced from the complex of heavy transition metals such as ruthenium, osmium, and iridium are called metal complex dyes. Properties which them good sensitizers include a long excited lifetimes, efficient charge transfer spectra, high redox properties and efficient metal to ligand charge transfer [17].

Ruthenium Metal Complex Dyes

The first ruthenium dye (N3 dye) (based DSSC was reported in 1991 by O'Regan and Grätzel having efficiency of 7.12% and later they reported N749 dye (black dye) with an efficiency of 10% [17]. After the initial report of 1991, there has been much research interest on DSSC leading to the synthesis of thousands of sensitizers of several kinds [5]. Nazeeruddin et al. reported efficiency of 10.4% with Ru dye (black dye). In the review by Anandan et al., which highlights the improvements and the arising challenges in dye sensitized solar cells till 2007, it was mentioned that maximum IPCE of 7% for naphthyridine coordinated Ru complex have been observed which is almost half to the efficiencies shown in later work after 2007. Mostly, ruthenium complexes consist of Ru(ll) atoms. Polypyridyl ligands coordinate to these ruthenium atoms along with thiocyanate moieties in octahedral geometry. Ruthenium polypyridyl complexes have shown better efficiencies and hence are being mostly used due to their chemical stability, thermal stability and having wide absorption range. With the use of Ru(II) dyes, the maximum efficiency of 12% could be calculated which when compared with the efficiency of first and second generation is very less i.e., almost half. When bipyridyl moieties is replaced by polynuclear bipyridyl Ru dyes, phosphate Ru dyes or carboxylate polypyridine Ru dyes, then there is improvement in the absorption, emission and electrochemical properties.

N3/N719/N712 Dyes

N3 dye is considered as a good photosensitize due to the following properties it possesses: broad range of visible light absorption, longer lived excited states, higher thermal stability and stable anchoring of dye due to terminal acidic carboxyl group [18]. The limitations include relatively low molar extinction coefficient and low absorption in the red region of visible spectrum [17].

N3 dye viz {*cis*-di(thiocyanato)bis(2,2-bipyridine-4,4-dicarboxylate) ruthenium} or *cis*-$[Ru(L)_2\text{-}(NCS)_2]$ in which L = 4,4'-dicarboxy-2,2'-bipyridine exhibits conversion efficiency of 10.3% [29]. Nazeeruddin et al. reported 10.3% efficiency in a system where N3 dye covered film with TBP [17]. Maximum absorption spectra were obtained at 518 nm and 380 nm wavelengths which can be extended to red and NIR region by ligand substitution. The excited state lifetime of 60 ns was observed and at 518 nm the extinction coefficient was 1.3 x 104 $M^{-1}cm^{-1}$ and at 380 nm it was 1.33 x 10^4 $M^{-1}cm^{-1}$. The activity of the device was found to be affected by protonation of N3 and N719 in the following ways: a) Due to the increase in the number of protons, there is positive charge in the surface of the semiconductor. Semiconductor's Fermi level has a shift downwards, therefore a positive shift in the conduction band takes place due to drop in VOC; b) The electron injection to CB of the semiconductor from dye's excited state is incurred due to the enhancement in the anionic Ru(ll) complex's absorption. This is because of the surface dipole being associated by the electric field [17].

Nazeeruddin et al. reported the N719 dye when N3's four counter ions were replaced by three TBA^+ and one H^+ therefore attaining 11.2% efficiency. When the counter ions are changed, there is alteration in the adsorption speed. The adsorption speed of N3 (3h) is fast but N719 is slow (24h) onto semiconductor electrode [17]. On investigating the levels of protonation of N3 and N719, Grätzel and group reported 9.3% PCE for monoprotonated sensitizer N3 and later doubly protonated salt of N3 or N719 showed higher PCE values of 8.4%. During esterification, there is a chemical binding between dye and semiconductor. Therefore, this results in partial transformation of protons of anchoring group to the semiconductor's surface [17].

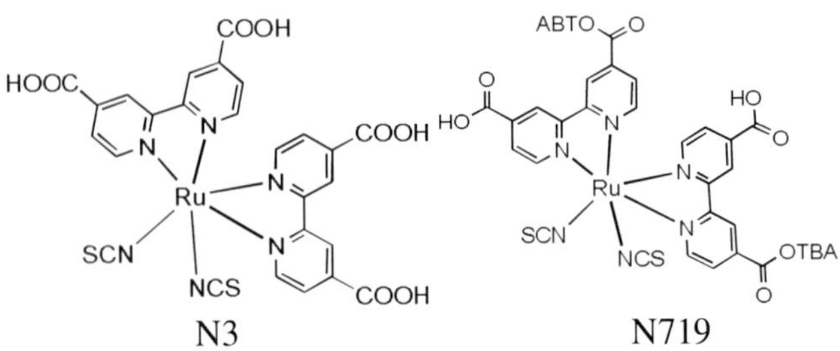

Figure 3. Structures of N3 and N719.

Ruthenium Dye (N3) Adsorbed on a TiO_2 Brookite Cluster

The chemisorption of N3 dye molecules on TiO_2 has the following benefits: 1) The light is harnessed with an efficiency of 20.1% and 2) Photoexcited electrons are injected into CB of semiconductor [31-33]. Absorption peaks for simulated spectra are at 311, 388 and 480 nm. HOMO of N3 lies at -5.03 eV and is centered on NCS moiety whereas LUMO lies at -3.01 eV and is centered on the 4,4-dicarboxy-2,2-bipyridine moiety. The *light-harvesting efficiency* measures the fraction of *light intensity absorbed* by the *dye* and from this absorption strength studies we can infer whether the dye's grafting on the semiconductor's surface is stable or not.

All these computational techniques were done by using Avogadro programs, Gaussian 03 quantum chemical package, DFT with PBE functional through GPAW, CASTEP module in Materials Studio BIOVIA, DFT by hybrid B3LYP in the gas phase and LANL2DZ pseudopotential for Ti atom. For C, N, H, O atoms, the basis set 6-311+G (d,p) was used. UV-Vis simulated absorption spectrum was calculated using TD-DFT [35-39].

From the results obtained we can infer that the separation of charges is efficient upon adsorption and electron injection from excited state of dye to CB of semiconductor [34].

When Ruthenium (II) polypyridyl complex (N3) dye was used as a sensitizer with TiO_2 nanostructured semiconductor then an efficiency as high as 13% have been reported based on DFT studies [40-43].

Ruthenium Phenanthroline Complexes

Another family of bidentate ligands is based on phenanthroline. It is used in Ru-DSSC to prepare sensitizer *cis*-$[Ru(L)_2(NCS)_2]$ where L = 4,7-dicarboxy-1,10-phenanthroline [18]. Eight ruthenium complexes of the formula *cis*-$[Ru(L)(L')(NCS)_2]$, have been reported by Gao et al. with the following bidentate ligands: 1,10-phenanthroline (P or phen); 4-monocarboxy-1,10-phenanthroline (MCP or mcphen); 4,7-dicarboxy-1,10-phenanthroline (DCP or dcphen); *cis*-$[Ru(P)_2(NCS)_2]$ labelled as P2; *cis*-$[Ru(MCP)_2(NCS)_2]$ labelled as MCP2; *cis*-$[Ru(DCP)_2(NCS)_2]$ labelled as DCP2; *cis*-$[Ru(MCP)(P)(NCS)_2]$ labelled as MCPP; *cis*- $[Ru(DCP)(P)(NCS)_2]$ labelled as DCPP and *cis*-$[Ru(MCP)(DCP)(NCS)_2]$ labelled as MCPDCP (Figure 4).

Structural optimisations were conducted using Becke's 3 parameter hybrid functional using the Lee-Yang-Parr correlation functional (B3LYP) and unrestricted B3LYP (UB3LYP) [44] of the 6 dyes having 0-4 carboxyls coordinating phenanthroline ligands (P2, MCPP, MCP2, DCPP, MCPDCP and DCP2) in the ground and triplet excited states. Structural and spectroscopic properties were determined by DFT and TD-DFT with CPCM solvent model (methanol) and were comparable to experimental values [45-52]. The GAUSSIAN03 program package was used for all calculations [38].

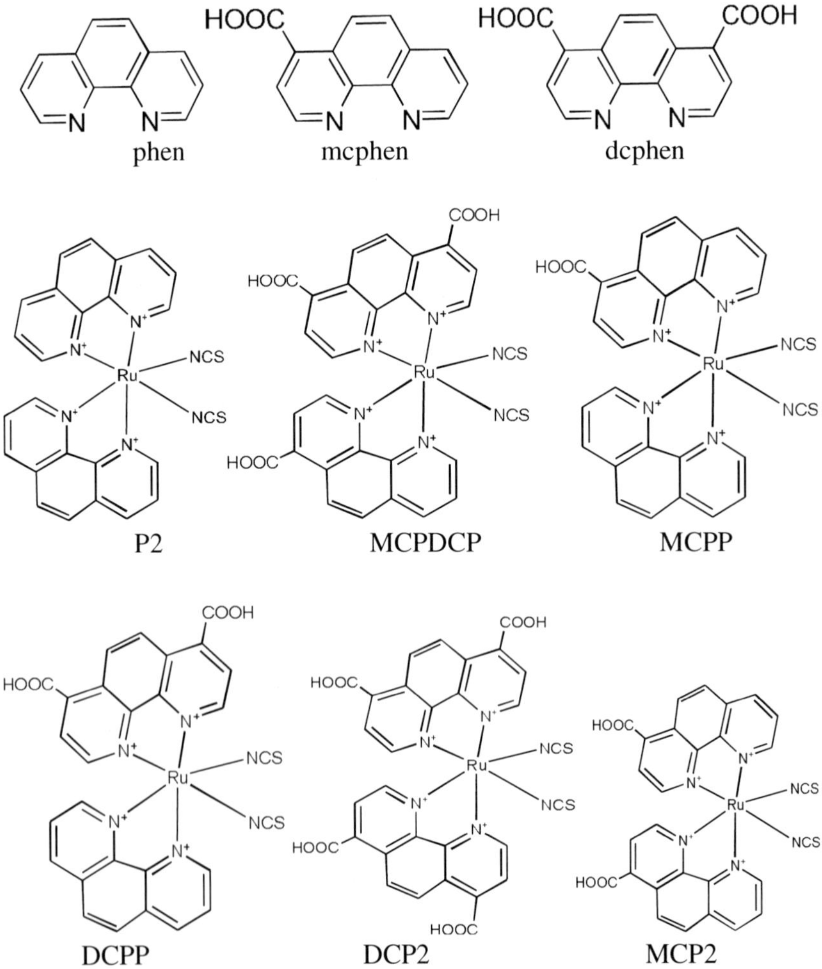

Figure 4. Structures of bidentate ligands and its corresponding ruthenium complexes.

The number and position of these carboxyl groups improve the electron accepted capability of phenanthroline. With the variation in the carboxyl substituent, there is difference in the absorption spectra. Gao *et al.* reported that all carboxylic sensitizers have wide and intense absorption bands in visible region therefore they are good choices for DSSC [18].

π-System Extension (N945, Z910, K19, N719, Z907, K8, K9)

The absorption and cell efficiency of organic dyes was expected to increase by replacing ligands with other π conjugated ligands. On increasing the π extension, sensitizers with higher molar extinction coefficients can be generated. In these kind of dyes with higher molar extinction coefficient, the LUMO can be changed to excited state and can also induce hydrophobic side chain to repel water. To increase the ε, thick layered TiO_2 can be used [54]. Efficiency rates are: Z910 dye (10.2%) [72] and N945 (10.8%) [55-56]. K8 and K9 dyes which were synthesised by Nazeeruddin et al. showed better results, broad and intense absorption band which lies in the range of 370-570 nm with an efficiency of 8.46% for K8 and 7.81% for K9 [57]. Z907 was the first amphiphilic dye having conversion efficiency of 6.0% and the second amphiphilic dye K19 was synthesised by Grätzel group showed 7.0% conversion efficiency. N719 dye reported an overall PCE of 6.7% [57-58]. But simple extension of π was not found to contribute for improvement in efficiency compared to N3 dye.

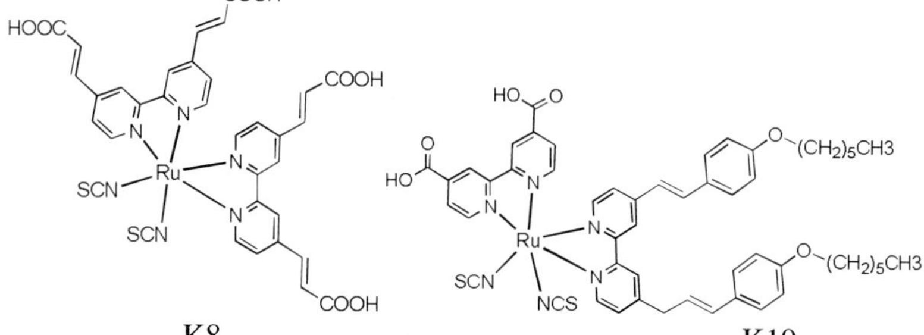

K8 K19

Figure 5. (Continued).

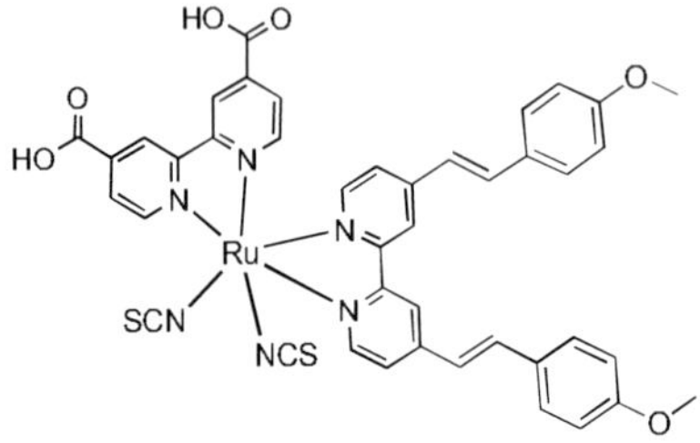

Z910

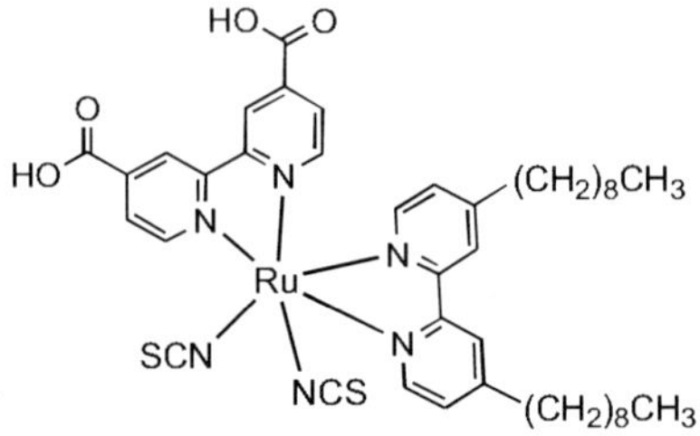

Z907

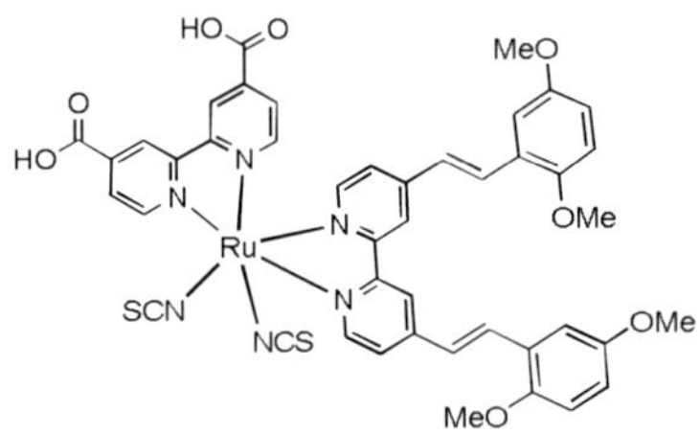

N945

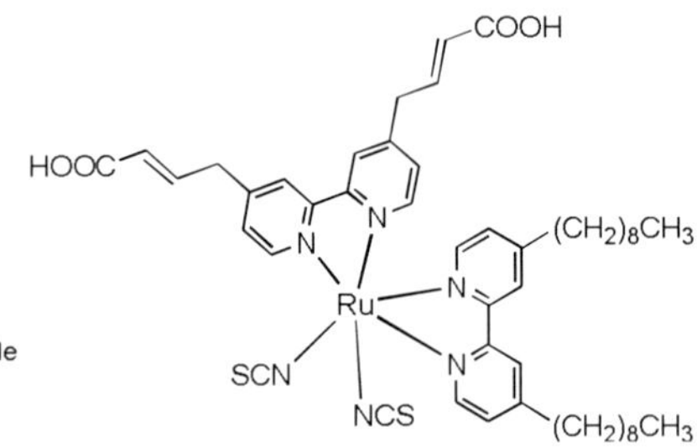

K9

Figure 5. Structures of dyes with extended π-system.

RuHDA1, RuHDA2, RuHDA3

Pashaei and Shahroosvand designed three ruthenium complex molecules shown in Figure 5. The electron transfer kinetics obtained by transient absorption spectroscopy (TAS) proved larger lifetime of regeneration of RuHDA3. Correlation between the structures and recombination process lifetime was also investigated. Amount of ancillary and anchoring groups have a direct relation with the lifetime of regeneration or regeneration dynamics and recombination loss therefore it is important to have a proper balance with regard to the amount of recombination loss and regeneration dynamics [59]. A good light harvesting photosensitizer must have ε higher than 5000 $M^{-1}cm^{-1}$. The value of ε for these dyes are as follows: RuHDA3 with 6650 $M^{-1}cm^{-1}$ > RuHDA1 with 6590 $M^{-1}cm^{-1}$ > RuHDA2 with 6545 $M^{-1}cm^{-1}$. The order of intensity and photoluminescence quantum yield (PLQY) is: RuHDA2 > RuHDA1 > RuHDA3. IPCE value of these dyes: RuHDA1 (58% at 465 nm), RuHDA2 (56% at 485 nm) and RuHDA3 (62% at 508 nm). From the IPCE and PCE results, it can be inferred that the performance level of RuHDA3 is better due to one ancillary and anchoring ligand along with two NCS group, one ancillary ligand with two carboxylic group has far better efficiency than two ancillary ligands with four carboxylic group, NCS makes the absorption band extend to NIR therefore improves light harvesting ability and also about the structure which is hybrid [59-62].

By replacing the anchoring ligand such as dcbpy in N3 with highly conjugated ancillary ligands is a strategy to extend the absorption band to NIR and also to increase the ε [63]. The experimentally calculated electron transfer properties were comparable with DFT and TD-DFT results.

Conclusions from the DFT and TD-DFT reports:

1. For RuHDA1 and RuHDA2, HOMO is localized on the d-orbitals of the metal centre.
2. For RuHDA3, charge density is localized on Ru center.
3. For RuHDA3, NCS contributes for the charge density's extension.
4. Electronic spectral predictions using computation where in good agreement with experimental values.
5. From MLCT and LLCT transitions, low energy excitation band arises therefore better charge separation and electron injection takes place easily
6. Efficient transition from LUMO to the conduction band takes place due to the delocalization of LUMO on π* of dcbpy ligand
7. Due to the presence of more electron rich ancillary phenanthroline ligand there is a raise in HOMO energy level.
8. When there is raise in the HOMO energy level, red shift of absorption band takes place and there may be delay in recombination [59-67].

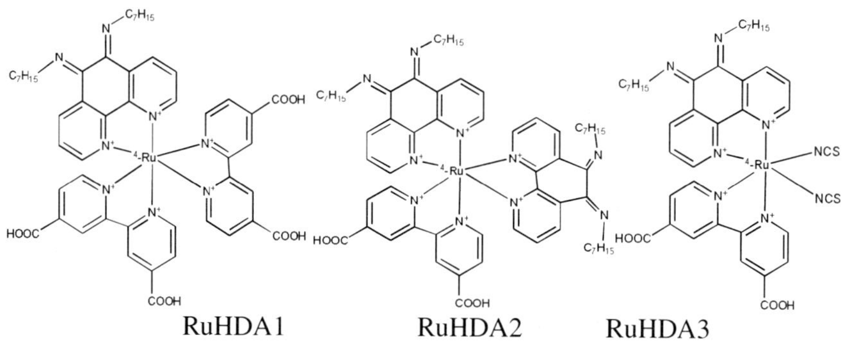

Figure 6. Structures of ruthenium complexes.

Amphiphilic Dyes with Alkyl Chains

The ester linkage of the dye to the semiconductor is prone to hydrolysis. Therefore, alkyl group was used to replace two carboxylic groups of N3 dye as there is more chance for lowering of absorption spectrum when water gets adsorbed on the semiconductor's surface due to bipyridyl-ligand's smaller conjugated π-system. Complex Z907 having 9 carbon atoms attached on the side shows appreciable PCE of 7.3%. It is high stable and moreover the setup could remain in 1000 h at 80 °C in dark and under light at 55 °C with undergoing no degradation. The nonyl group attached to the bipyridine creates a hydrophobic environment preventing desorption of the dye due to moisture [17]. Complex N621 having 13 carbon atom chains (Figure 6) shows appreciable PCE of 9.6% and high stability.

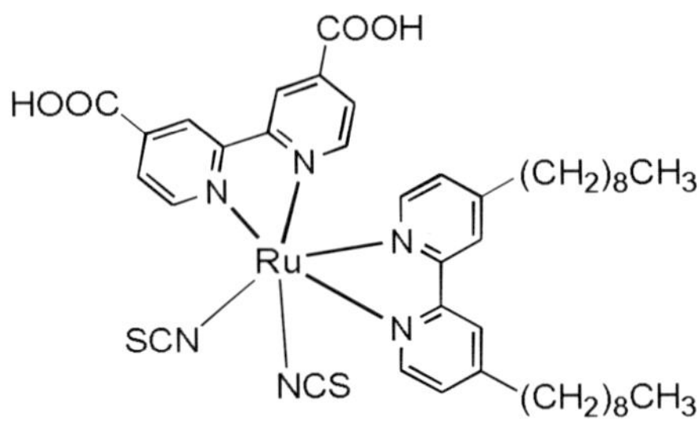

Figure 7. Structure of Z907.

Different Anchoring Groups

The efficiency of DSSC is found to vary depending on the nature of anchoring group attached to the dye [68]. Carboxylic acid groups are the most common anchoring group used for dye attachment to TiO_2. The decreasing order of binding strength is as follows: phosphonic acid > carboxylic acid > ester > acid chloride > carboxylate salts > amides. Many researches are focused on phosphonic acid due to binding strength to metal oxide and Z955 is an example of the same containing Ru complex [17]. In Ru sensitizers, the best anchoring ligand is 4,4'-dicarboxylic acid-2,2'-bipyridine (dcbpy) [69]. Grätzel and coworkers reported that when a dcbpy anchoring ligand is replaced with an ancillary ligand which is highly conjugated then the absorption coefficient increases and also the density of photocurrent [66].

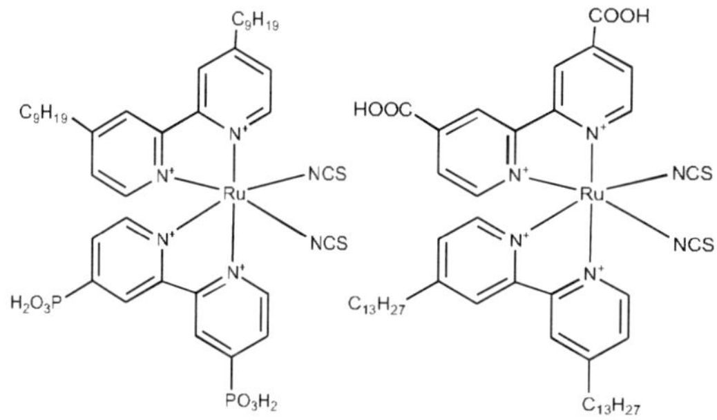

Z955 N621

Figure 8. Structures of ruthenium complexes.

CYC-B1

In the search for the best DSSC photosensitizer with the best stability, Chen et al. reported the preparation of a new Ru photosensitizer by one-pot synthesis [71, 78] in which a dcbpy ligand in N3 is replaced with a bipyridine ligand with alkyl bithiophene substitution (abtpy) [66]. This alkyl substituent has the ability to prevent desorption of the photosensitizer from the semiconductor's surface being induced by water [70]. It has the highest absorption coefficient among the group of ruthenium based photosensitizers and efficiency of 10% was measured more than that of N3. The absorption maximum of abtpy is red shifted by 76 nm. This is attributed to the longer length of conjugation on comparing with dcbpy [66]. Absorption spectrum of CYC-B1 shows three bands [66]: 1) 313 nm - intraligand π-π^* transitions of dcbpy; 2) 400 nm - π-π^* transitions of abtpy ligand and MLCT for CYC-B1 and 3) 553 nm - characteristic MLCT transition which is used for calculating efficiency.

The strategy to enhance MLCT band's absorption coefficient with the help of increase in the ancillary ligand's conjugation length is very well known [58, 71-74]. Absorption maxima of both MLCT bands in CYC-B1 are larger than those of N3 and since the binding of the two dcbpy ligand on N3 to TiO_2 particles at the same time is not possible therefore CYC-B1 is considered better than N3. For a sensitizer to be efficient, its LUMO should match properly to the conduction band's lower limit of the semiconductor whereas its HOMO must be able to accept electrons from the electrolyte for which HOMO should have low energy. Lower energy MLCT bands of CYC-B1's molar absorption coefficient is very high. It is higher than the rest of Ru dyes used [66] and this was confirmed by calculating locations of HOMOs and LUMOs by ZINDO/1 method [73, 75-77].

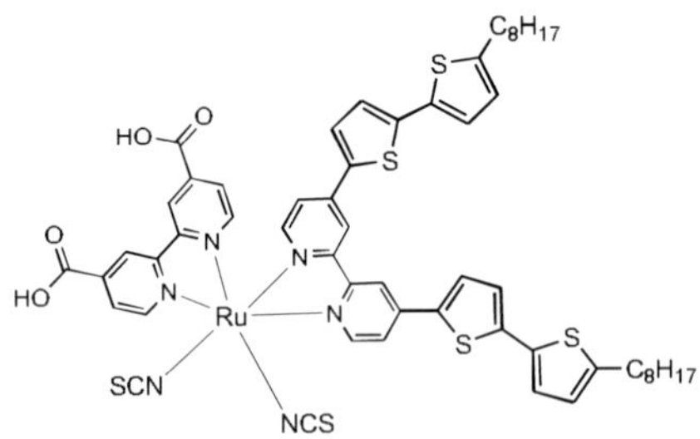

CYC-B1

Figure 9. Structure of CYC-B1.

π-Extended Ferrocene

Ferrocene with different anchoring group have been used as photosensitizers in DSSC [17]. Chauhan and co-workers have synthesised new compounds $C_5H_4FcC_5H_5CH{=}NC_6H_4COOH$ and $C_5H_4FcC_5H_5CH{=}NCH_2CH_2OH$ using starting material $C_5H_4FcC_5H_5CHO$ and characterisation have been done by cyclic voltammetry and DFT. The characterisation results explain the dye's quasi-reversible redox behaviour.

In 2017, from the studies on cyano group in the anchoring part of the organic dye, it was understood that on its addition, the adsorption's stability increased when adsorption is done via CN with the surface and the photovoltaic characteristics decreased when there is no involvement of binding [17].

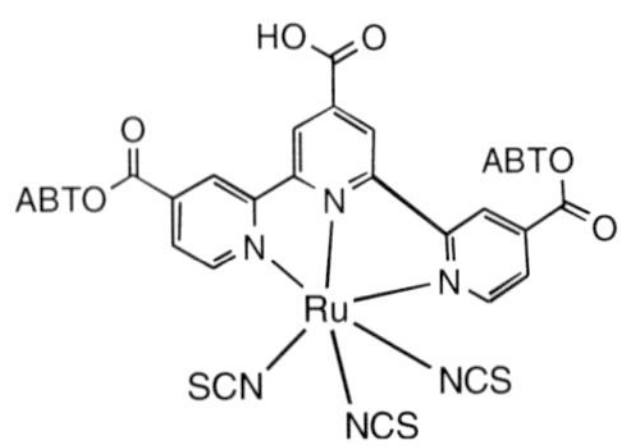

N749

Figure 10. Structure of N749 (Black dye).

Two important Ru dyes are N749 (Figure 8) and Z907 (described earlier in Figure 4) Complex N749 (black dye) has broad absorption, high efficiency of 10.5 - 11.1% when it was used in different devices have been reported by Wang et al. and Chiba et al., but the stability is not good when compared to other Ru sensitizers. At the National Renewable Energy Laboratory (NREL), black dye (N749)-sensitized DSSC showed efficiency of 10.4% with JSC = 20.53 mA/cm^2, VOC = 0.721 V, and FF = 0.704, where the active area of the cell was 0.186 cm^2. The response of N749 dye was observed to be extended 100 nm further into the IR compared to that of N3 [17].

Porphyrin-Based Sensitizers

The electrochemical, spectral, thermal, electronic and photovoltaic properties make porphyrins worthy application in DSSC as they play an important role in achieving high photovoltaic performance. L^4 complex which is $[Ru(NCS)_2L^4]^{2-}$ shows extra intense absorption peaks at 500 and 582 nm, allowing for its promising application in dye-sensitized solar cells (DSSCs) [78-79].

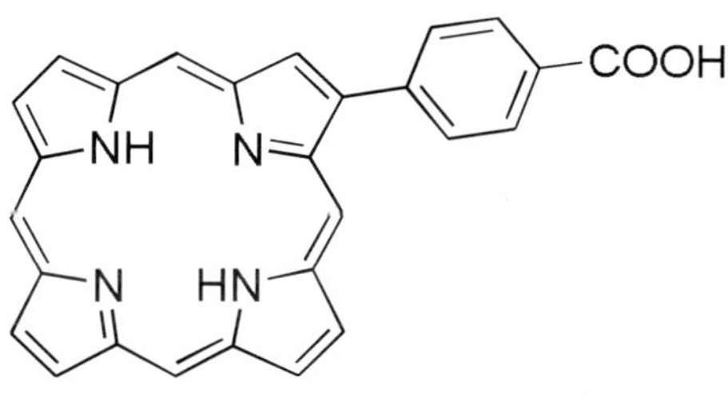

Figure 11. Structure of L^4.

N3 dye Adsorption onto TiO_2

TiO_2 is widely used as a photoanode because of its proven high efficiency in DSSC. Recent studies has shown that that doping titania transition metal ions can enhance optical and electrical properties by changing its band gap and trap states [80]. Theoretical studies were conducted to find the perfect dopand which can improve the efficiency. It has been noticed that there is improvement in efficiency with respect to conventional ruthenium dyes in

DSSC when cyclometalated Ru dyes were used and the geometrical, electronic, structural, photophysical and density studies on dye (both doped and undoped) in presence and absence of acetonitrile solvent were performed by DFT. There was about 20 nm red shift absorption showed by the dyes adsorbed on TiO_2 [81]. To understand the electronic properties of Ru dye@TiO_2 (1 0 0) in which brookite was used, the structural optimizations, band structure, and electronic density of states were estimated by the first-principles calculations based on DFT employing a plane-wave pseudo-potential method. For describing the exchange correlation functional, generalized gradient approximation (GGA) was used in the scheme of Perdew-Burke-Ernzerhof (PBE). CASTEP code was used for calculations. When the TiO_2 brookite (1 0 0) is doped with ruthenium dye there is a reduction in band gap [82]. DFT have been used to study the optical properties also. The energy levels were studied by using computational methods and compared with the experimental data. The chemisorption of $ZnTiO_3$ on N3 is by the carboxylic group and the calculated adsorption energy is 241.50 kcal/mol which is higher than ZnO as well as TiO_2. The adsorption energy was computed by taking Van der Waals dispersion based on Grimme scheme and the higher adsorption energy is due to the higher electron density on (101) surface. N3 adsorption onto $ZnTiO_3$ by bidentate bridging mode is higher on (101) surface and the large surface area leads to enhanced light harvesting efficiency. Also, the electron injection efficiency of Ru dye on (101) surface is higher [83]. Due to large surface area, high electron mobility and chemical stability $ZnTiO_3$ perovskite films are used as photo-anodes in DSSC with N3 dye [83]. Furthermore, TiO_2@heteroleptic Ru(II) dye was used for DFT and TD-DFT studies. From the analysis it was found that if Ru dye coordinated to 2,5-bis(*N*-pyrazolyl)pyridine and bipyridine ligands then alkaline electrolyte was needed [84].

There are some of the benefits of good modification of acceptor ligand (4-carboxyl-pyridine fragment): It expands absorption coverage and improves photon capturing within effective absorption bands [84]. Three Ru complexes a-c shown in Figure 9 were synthesized. The electron injection ability was ensured to be good by the structure of a@$(TiO_2)_5(OH)$ which has a higher average thermodynamic driving force. In b@$(TiO_2)_5(OH)$, higher short-circuit current density was noted due to the larger number of photons that is able to capture by the structure. This can be known from the extra absorption bands in the lower energy region. In c@$(TiO_2)_5(OH)$, even though there is an expand of absorption coverage, the light harvesting capacity is reduced due to the weak molar absorption coefficient [84].

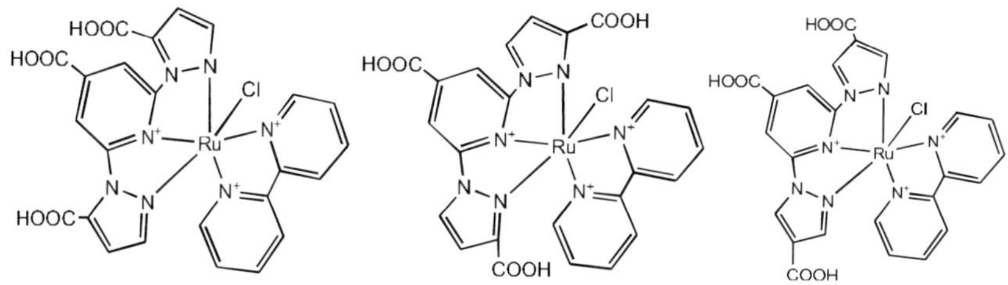

a b c

Figure 12. Structures of ruthenium complexes a, b and c.

Ru dye (N3 and chlorinated forms) and X^-, X_2^- (X = Br, I, At)

Structural differences and relative stabilities can be studied from the electronic structure's qualitative findings. Quantum chemistry especially DFT and continuum solvation model serves the purpose. These computations were used for the interaction studies taking different bond length and increasing reducing power X-/X_3^- redox couple of ruthenium dyes depicted in Figure 10. Various outer sphere complexes were found for each dye-redox couple combination except for inner sphere seven coordinate complex which does not exist. Clifford et al. put forward a mechanism in which [dye^+X^-] intermediate are probed. Intermediate and elementary steps have been confirmed. From the results it is understood that N3 has better dye regeneration thermodynamically than the chlorinated forms and also N3 is easily regenerated compared to 4^+. Regeneration of neutral dyes are in the order of At > I > Br [85-86].

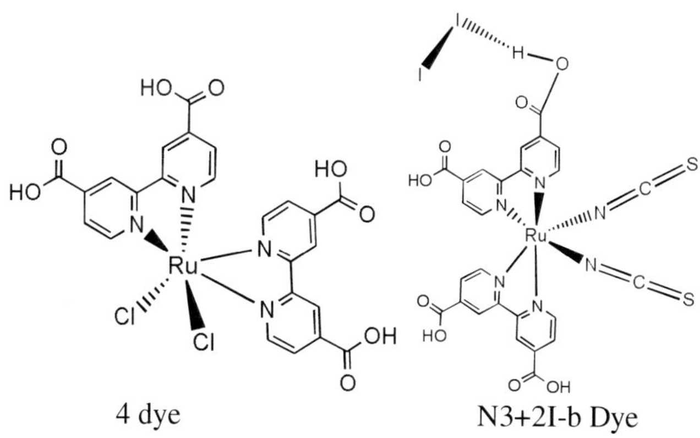

4 dye N3+2I-b Dye

Figure 13. (Continued).

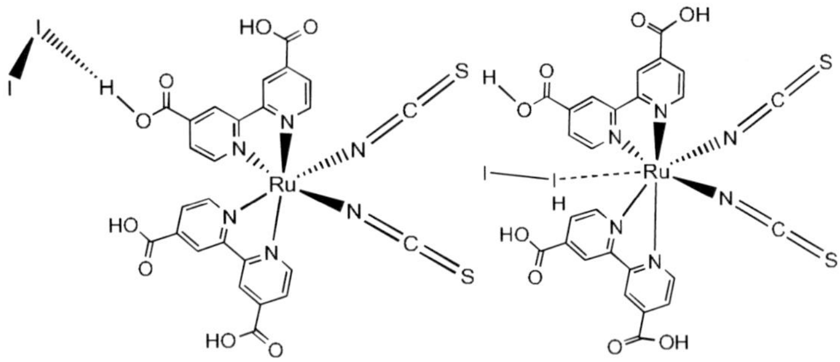

N3+2I-eDye N3+2I-z Dye

Figure 13. Structures of Ru dyes.

Anatase TiO_2 (1 0 1) Surface Doped with Ruthenium

Doping TiO_2 has proved to be a way for the improvement in DSSC's efficiency due to the change in energy band gap and adsorption on different wavelength region. The reaction mechanism of Ru-doped with anatase TiO_2 (1 0 1) was investigated by first principle DFT. With the use of generalized gradient approximation implemented within the CASTEP CODE, geometrical optimization of surface was done. From the results of band structures and calculated density of states, it can be said that impurities of Ru at titanium's lattice sites resulted in sharp band gap narrowed from 3.163 eV to 0.345 eV which was due to synergistic effect of ruthenium. The electronic properties were studied by implementing energy plane-wave pseudopotential method in DFT and studies confirmed that the bonds between Ru and Ti were weak [87].

An integrated computational tool for the studies on optical properties of nanoscale devices using computation which is an application to solar cells and molecular wires. On the basis of DFT and TD-DFT using several different codes computations were conducted to study the optical properties of nano scale systems, geometry optimizations of complex systems, TD-DFT excitation energies, solvation effects by continuum solvation models. Car-Parinello molecular dynamics simulations were particularly used along with available quantum chemistry packages. Prototypical applications have also been reviewed with the help of these studies. Ru complex of formula $[Ru_n(TPPZ)_{n+1})^{2n+}$ was studied using computational methods to understand its electronic structure, geometry, optical absorption spectra of monomer and oligomer with n ranging from 1 to 4. From computational results of oligomers of n = 1-4 in the visible region, the results have a good agreement with experimental results i.e., the red-shift and the increase in intensity of main feature with the increase in n. When n = 4 then the species had 237 atoms and dimensions more than 3 nm length. Experimental as well as computational data showed no saturation effect with the increase in "n" which is because of the extended charge delocalization in excited states which was noticed on going from monomer to oligomers, which are related to inter-metal communication through the pyrazine bridge [88].

Stability of Ruthenium/Organic DSSC

A DSSC was constructed using black dye (N749) and Y1 organic co-adsorbent to be compared (Figure 14). The co-sensitized solar cell has shown good efficiency under light soaking of 1100 h at 1 sun, at 55°C. Due to the partial desorption of Y1, there is a reduction of 20% in the photovoltaic efficiency and also upon oxidation of Y1, there is a reduction in the binding energy. This suggests that there is a possibility of dye desorption and therefore, this possibility explains the lower stability under stress conditions. Durability is shown to be lower in those devices which employ only Y1 organic dye. Having double carboxylic anchoring groups on N749 dye accounts for the higher energy compared to Y1 dye which have been proved by DFT computational studies [89].

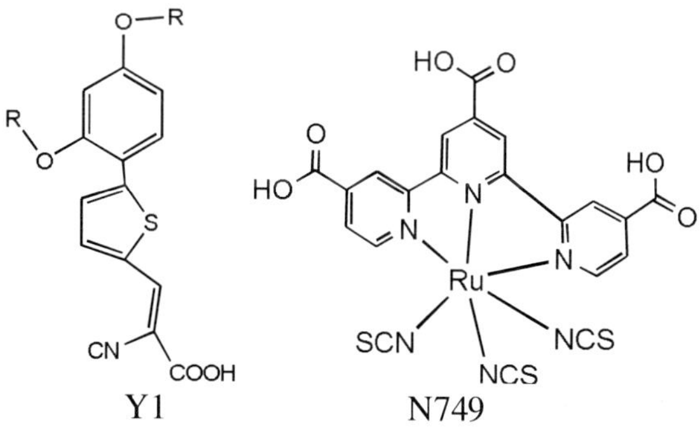

Figure 14. Structures of Y1 and N749.

New Ru Complex DSSC by BCL Approach

ADF 2010 package was used for the calculations of dyes given in Figure 12. The geometry optimization was done with PBE exchange correlation functional in conjunction with DZP quality basis set for first row elements and ZORA-TZP basis set for transition metals (Ru and Ti) and local minima was checked if it is present or not. ACN was taken as a solvent to conduct TDDFT. Solvent effect was done by COSMO. SAOP exchange correlation was put forward along with SZ basis set for C, H, N, O and a DZ basis set for F, S, I, and ZORA-TZP for transition metals. For the first 30 allowed transitions Hirschfeld fragment molecular decomposition was done. Mulliken's method was employed for the population analysis of ground and first triplet state. Excitations with oscillator strength greater than or equal to 0.05 was considered for absorbed energy calculation. From all these studies, we can increase the overall yield with the help of new ligand. There has been a doubt with regard to the anchoring ligand dcbp request two carboxy or one which could be clarified with the employment of non-symmetric mcbp ligand and synthesis of the complex N3mc which was under process. Studies were done with anchoring of various dcbp ligand on the two $Ti(OH)_3$ of N3 complex. With the replacement of dcbp with mcbp, F1 and F2 index response increased which caused the enhancement of global efficiency index. This enhancement was from 0.69 to 1.15 which was closer to BD [90].

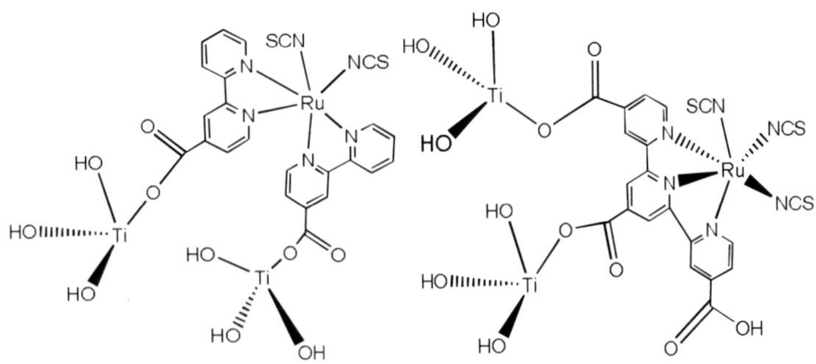

Figure 15. Structures of N3mc and BD.

Conclusion

Even though DSSC acts as an alternative source of energy, it was found to have a number of limitations. First and foremost, the overall efficiency of the dye is by far less, compared to the silicon solar cells. Many factors contribute to the above observation, the stability of the dye in the electrolyte and its regeneration after each cycle plays a very important role. On exposure to sunlight, there is always a possibility for the dye to desorb from TiO_2 and leaching can be expected in the presence of the electrolyte. The stability of the electrolyte, again, is also another important criterion for efficiency. At low temperatures the electrolyte can freeze, halting power production and potentially leading to physical damage and at high temperatures, leakage of electrolyte can be expected. The space between the electrodes plays another important role in the de-excitation (recombination) of the dye to ground state which adversely affects efficiency. These limitations, again, pose a challenge and at the same time stress for improved research in this area.

In conclusion, we have summarized selected very promising ruthenium complexes used for DSSC application, emphasizing on the reasons behind the choice of each entity providing a deep insight in to how the choice can affect the final efficiency of the system. The scope of the area is still open and new research, both experimental and computational are required for the production of new ruthenium based solar cells with exceptional efficiency.

References

[1] Seger, B. (2016). Global Energy Consumption: The Numbers for Now and in the Future. https://www.linkedin.com/pulse/global-energy-consumption-numbers-now-future-brian-seger

[2] Al Mogren, M. M., Ahmed, N. M., and Hasanein, A. A. (2020). Molecular modeling and photovoltaic applications of porphyrin-based dyes: A review. *Journal of Saudi Chemical Society*, 24(3), 303-320.

[3] Demirbaş, A. (2006). Global renewable energy resources. *Energy Sources*, 28(8), 779-792.

[4] Al Mogren, M. M., Ahmed, N. M., and Hasanein, A. A. (2020). Molecular modeling and photovoltaic applications of porphyrin-based dyes: A review. *Journal of Saudi Chemical Society*, 24(3), 303-320.

[5] Chang, W. C., Chen, H. S., Li, T. Y., Hsu, N. M., Tingare, Y. S., Li, C. Y., and Li, W. R. (2010). Highly Efficient N-Heterocyclic Carbene/Pyridine-Based Ruthenium

Sensitizers: Complexes for Dye-Sensitized Solar Cells. *Angewandte Chemie*, 122(44), 8337-8340.

[6] Lior, N. (2008). Energy resources and use: The present situation and possible paths to the future. *Energy*, 33(6), 842-857.

[7] Grätzel, M. (2001). Photoelectrochemical cells. *Nature*, 414, 338-344.

[8] Kibria, M. T., Ahammed, A., Sony, S. M., Hossain, F., and Islam, S. U. (2014, September). A Review: Comparative studies on different generation solar cells technology. In *Proc. of 5th International Conference on Environmental Aspects of Bangladesh* (pp. 51-53).

[9] Yun, S., Qin, Y., Uhl, A. R., Vlachopoulos, N., Yin, M., Li, D., and Hagfeldt, A. (2018). New-generation integrated devices based on dye-sensitized and perovskite solar cells. *Energy and Environmental Science*, 11(3), 476-526.

[10] Natarajan, P., Ramaprabha, R., and Ranganath, M. (2012). Application of Circuit Model for Photovoltaic Energy Conversion System, *International Journal of Photoenergy,* 2012, Article ID 410401, 14 pages.

[11] O'regan, B., and Grätzel, M. (1991). A low-cost, high-efficiency solar cell based on dye-sensitized colloidal TiO 2 films. *Nature*, 353(6346), 737-740.

[12] Gong, J., Sumathy, K., Qiao, Q., and Zhou, Z. (2017). Review on dye-sensitized solar cells (DSSCs): Advanced techniques and research trends. *Renewable and Sustainable Energy Reviews*, 68, 234-246.

[13] Tributsch, H., and Calvin, M. (1971). Electrochemistry of excited molecules: photo-electrochemical reactions of chlorophylls. *Photochemistry and Photobiology*, 14(2), 95-112.

[14] Grätzel, M. (2014). The light and shade of perovskite solar cells. *Nature Materials*, 13(9), 838-842.

[15] Wu, M., and Ma, T. (2012). Platinum-free catalysts as counter electrodes in dye-sensitized solar cells. *Chem Sus Chem*, 5(8), 1343-1357.

[16] Veerappan, G., Bojan, K., and Rhee, S. W. (2011). Sub-micrometer-sized graphite as a conducting and catalytic counter electrode for dye-sensitized solar cells. *ACS Applied Materials and Interfaces*, 3(3), 857-862.

[17] Sharma, K., Sharma, V., and Sharma, S. S. (2018). Dye-sensitized solar cells: fundamentals and current status. *Nanoscale Research Letters*, 13(1), 1-46.

[18] Gao, X. Q., Pan, Q. J., Li, L., Guo, Y. R., Zhang, H. X., and Fu, H. G. (2011). Structures and spectroscopic properties of ruthenium phenanthroline solar-cell sensitizers: A computational study. *Chemical Physics Letters*, 506(4-6), 146-151.

[19] Grätzel, M. (2003). Dye-sensitized solar cells. *Journal of photochemistry and photobiology C: Photochemistry Reviews*, 4(2), 145-153.

[20] Hagfeldt, A., Boschloo, G., Sun, L., Kloo, L., and Pettersson, H. (2010). Dye-sensitized solar cells. *Chemical Reviews*, 110(11), 6595-6663.

[21] Hardin, B. E., Hoke, E. T., Armstrong, P. B., Yum, J. H., Comte, P., Torres, T., and McGehee, M. D. (2009). Increased light harvesting in dye-sensitized solar cells with energy relay dyes. *Nature Photonics*, 3(7), 406-411.

[22] Jiao, Y., Zhang, F., Meng, S., and Kosyachenko, L. A. (2011). Dye sensitized solar cells principles and new design. Solar Cells-Dye-Sensitized Devices, *IntechOpen,* doi: 10.5772/21393.

[23] Zhang, C. R., Liu, Z. J., Chen, Y. H., Chen, H. S., Wu, Y. Z., Feng, W., and Wang, D. B. (2010). DFT and TD-DFT study on structure and properties of organic dye sensitizer TA-St-CA. *Current Applied Physics*, 10(1), 77-83.

[24] Murakami, T. N., and Grätzel, M. (2008). Counter electrodes for DSC: application of functional materials as catalysts. *Inorganica Chimica Acta*, 361(3), 572-580.

[25] Murakami Iha, N. Y., Garcia, C. G., and Bignozzi, C. A. (2003). Dye-sensitized photoelectrochemical solar cells. *Handbook of photochemistry and photobiology*. American Scientific Publishers, Stevenson Ranch, 49-82.

[26] Anthonysamy, A., Lee, Y., Karunagaran, B., Ganapathy, V., Rhee, S. W., Karthikeyan, S., and Kim, J. K. (2011). Molecular design and synthesis of ruthenium (II) sensitizers for highly efficient dye-sensitized solar cells. *Journal of Materials Chemistry*, 21(33), 12389-12397.

[27] Hagfeldt, A., and Graetzel, M. (1995). Light-induced redox reactions in nanocrystalline systems. *Chemical Reviews*, 95(1), 49-68.

[28] Mohmeyer, N., Kuang, D., Wang, P., Schmidt, H. W., Zakeeruddin, S. M., and Grätzel, M. (2006). An efficient organogelator for ionic liquids to prepare stable quasi-solid-state dye-sensitized solar cells. *Journal of Materials Chemistry*, 16(29), 2978-2983.

[29] Duncan, W. R., and Prezhdo, O. V. (2007). Theoretical studies of photoinduced electron transfer in dye-sensitized TiO_2. *Annu. Rev. Phys. Chem.*, 58, 143-184.

[30] Ooyama, Y., and Harima, Y. (2012). Photophysical and electrochemical properties, and molecular structures of organic dyes for dye-sensitized solar cells. *Chem Phys Chem*, 13(18), 4032-4080.

[31] Nazeeruddin, M. K., De Angelis, F., Fantacci, S., Selloni, A., Viscardi, G., Liska, P., and Grätzel, M. (2005). Combined experimental and DFT-TDDFT computational study of photoelectrochemical cell ruthenium sensitizers. *Journal of the American Chemical Society*, 127(48), 16835-16847.

[32] Liang, J., Zhu, C., and Cao, Z. (2013). Electronic and optical properties of the triphenylamine-based organic dye sensitized TiO_2 semiconductor: insight from first principles calculations. *Physical Chemistry Chemical Physics*, 15(33), 13844-13851.

[33] Anselmi, C., Mosconi, E., Pastore, M., Ronca, E., and De Angelis, F. (2012). Adsorption of organic dyes on TiO_2 surfaces in dye-sensitized solar cells: interplay of theory and experiment. *Physical Chemistry Chemical Physics*, 14(46), 15963-15974.

[34] Elegbeleye, I. F., Maluta, N. E., and Maphanga, R. R. (2018, September). Density functional theory studies of ruthenium dye (N3) adsorbed on a TiO_2 brookite cluster for application in dye sensitized solar cells. In :*International Workshop on Quantum Systems in Chemistry, Physics, and Biology* (pp. 143-155). Springer, Cham.

[35] Hanwell, M. D., Curtis, D. E., Lonie, D. C., Vandermeersch, T., Zurek, E., and Hutchison, G. R. (2012). Avogadro: An advanced semantic chemical editor, visualization, and analysis platform. *J. Cheminform*, 4(1), 17 pages.

[36] Mortensen, J. J., Hansen, L. B., and Jacobsen, K. W. (2005). Real-space grid implementation of the projector augmented wave method. *Physical Review B*, 71(3), 035109.

[37] Clark, S. J., Segall, M. D., Pickard, C. J., Hasnip P. J. Probert MIJ Refson K., Payne M. C. (2005). First principles methods using CASTEP. *Z. Kristallogr. Cryst. Mater*, 220, 567-570.

[38] Frisch, M. J., G. W. Trucks, H. B. Schlegel, G. E. Scuseria, M. A. Robb, J. R. Cheeseman, J. A. Montgomery Jr. et al. *Gaussian 03, Revision D. 01*, Gaussian, Inc., Wallingford CT, 2004.

[39] Corneliu, I. O., Petre, P., Fanica, C., Marilena, F., and Mihai, A. G. (2013). (DFT) study of coumarin-based dyes adsorbed on TiO_2 nanoclusters-applications to dye-sensitized solar cells. *Materials (Basel)*, 6, 2372-2392.

[40] Yella, A., Lee, H. W., Tsao, H. N., Yi, C., Chandiran, A. K., Nazeeruddin, M. K., and Grätzel, M. (2011). Porphyrin-sensitized solar cells with cobalt (II/III)-based redox electrolyte exceed 12 percent efficiency. *Science*, 334(6056), 629-634.

[41] Zhang, C. R., Liu, Z. J., Chen, Y. H., Chen, H. S., Wu, Y. Z., Feng, W., and Wang, D. B. (2010). DFT and TD-DFT study on structure and properties of organic dye sensitizer TA-St-CA. *Current Applied Physics*, 10(1), 77-83.

[42] Jungsuttiwong, S., Tarsang, R, Pansay, S., Yakhantip, T., Promarak, V., Sudyoadsuk, T., Kaewin, T. Saengsuwan, S., Namuangrak, S. (1999) *Int J Chem Mol Nucl Mater and Metallur Eng* 77, 561-567.

[43] Elegbeleye, I. F., Maluta, N. E., and Maphanga, R. R. (2018, September). Density functional theory studies of ruthenium dye (N3) adsorbed on a TiO_2 brookite cluster for application in dye sensitized solar cells. In *International Workshop on Quantum Systems in Chemistry, Physics, and Biology* (pp. 143-155). Springer, Cambridge.

[44] Axel D. Becke, (1993) Density-functional thermochemistry. III. The role of exact exchange, *J. Chem. Phys.* 98, 5648-5652.

[45] Baranoff, E., Collin, J. P., Furusho, J., Furusho, Y., Laemmel, A. C., and Sauvage, J. P. (2002). Photochemical or thermal chelate exchange in the ruthenium coordination sphere of complexes of the Ru (phen) 2L family (L= diimine or dinitrile ligands). *Inorg Chem*, 41(5), 1215-1222.

[46] Bonneson, P. J. L. W., Walsh, J. L., Pennington, W. T., Cordes, A. W., and Durham, B. (1983). Six-coordinate complexes with 1,10-phenanthroline ligands in the trans configuration. Preparation of trans-bis(1,10-phenanthroline) ruthenium(II) complexes and crystal structure of trans-bis(1,10-phenanthroline)bis(pyridine) ruthenium(II)hexafluorophosphate. *Inorganic Chemistry*, 22(12), 1761-1765.

[47] Pan, Q. J., Zhou, X., Zhang, H. X., and Fu, H. G. (2008). A theoretical probe on the ground-and excited-state properties of heterobinuclear Au-Pt complex with phosphine ligands: Comparison with analogous homobinuclear Au-Au and Pt-Pt complexes. *Chemical Physics Letters*, 453(1-3), 7-12.

[48] Pan, Q. J., Zhou, X., Guo, Y. R., Fu, H. G., and Zhang, H. X. (2009). Theoretical Studies on Metal- Metal Interaction, Excited States, and Spectroscopic Properties of Binuclear Au- Au, Au- Rh, and Rh- Rh Complexes with Diphosphine Ligands: Buildup of Complexity from Monomers to Dimers. *Inorganic Chemistry*, 48(7), 2844-2854.

[49] Pan, Q. J., Guo, Y. R., and Zhang, H. X. (2010). Theoretical Study of Metallophilic Interactions and Excited States of Heterobimetallic d^{10}-d^8 Complexes with Bridging

Ligands: The Tuning of Electronic Spectroscopy. *Organometallics*, 29(15), 3261-3270.

[50] Schüürmann, G., Cossi, M., Barone, V., and Tomasi, J. (1998). Prediction of the pKa of carboxylic acids using the ab initio continuum-solvation model PCM-UAHF. *J Phys Chem A*, 102(33), 6706-6712.

[51] Wadt, W. R., and Hay, P. J. (1985). Ab initio effective core potentials for molecular calculations. Potentials for main group elements Na to Bi. *The Journal of Chemical Physics*, 82(1), 284-298.

[52] Hay, P. J., and Wadt, W. R. (1985). Ab initio effective core potentials for molecular calculations. Potentials for K to Au including the outermost core orbitals. *The Journal of Chemical Physics*, 82(1), 299-310.

[53] Grätzel, M., & Moser, J. E. (2001). Solar energy conversion. *Electron transfer in chemistry*, *5*, 589-644.

[54] Al-Rawashdeh, N. A., Albiss, B. A., and Mo'Ath, H. I. (2018, February). Graphene-Based Transparent Electrodes for Dye Sensitized Solar Cells. In *IOP Conference Series: Materials Science and Engineering* (Vol. 305, No. 1, p.012019). IOP Publishing.

[55] Kuang, D., Ito, S., Wenger, B., Klein, C., Moser, J. E., Humphry-Baker, R., & Grätzel, M. (2006). High molar extinction coefficient heteroleptic ruthenium complexes for thin film dye-sensitized solar cells. *Journal of the American Chemical Society*, *128*(12), 4146-4154.

[56] Nazeeruddin, M. K., Bessho, T., Cevey, L., Ito, S., Klein, C., De Angelis, F., and Grätzel, M. (2007). A high molar extinction coefficient charge transfer sensitizer and its application in dye-sensitized solar cell. *Journal of Photochemistry and Photobiology A: Chemistry*, 185(2-3), 331-337.

[57] Nazeeruddin, M. K., Klein, C., Liska, P., and Grätzel, M. (2005). Synthesis of novel ruthenium sensitizers and their application in dye-sensitized solar cells. *Coordination Chemistry Reviews*, 249(13-14), 1460-1467.

[58] Wang, P., Klein, C., Humphry-Baker, R., Zakeeruddin, S. M., and Grätzel, M. (2005). A high molar extinction coefficient sensitizer for stable dye-sensitized solar cells. *Journal of the American Chemical Society*, 127(3), 808-809.

[59] Pashaei, B., and Shahroosvand, H. (2020). Molecularly engineered ruthenium polypyridyl complexes for using in dye-sensitized solar cell. *Inorganic Chemistry Communications*, 112, 107737.

[60] Bomben, P. G., Robson, K. C., Koivisto, B. D., and Berlinguette, C. P. (2012). Cyclometalated ruthenium chromophores for the dye-sensitized solar cell. *Coordination Chemistry Reviews*, 256(15-16), 1438-1450.

[61] De Angelis, F., Fantacci, S., Selloni, A., Grätzel, M., and Nazeeruddin, M. K. (2007). Influence of the sensitizer adsorption mode on the open-circuit potential of dye-sensitized solar cells. *Nano Letters*, 7(10), 3189-3195.

[62] Shahroosvand, H., Najafi, L., Mohajerani, E., Janghouri, M., and Nasrollahzadeh, M. (2013). Unusual electroluminescence in ruthenium(II) tetrazole complexes. *RSC Advances*, 3(18), 6323-6326.

[63] Jang, S. R., Lee, C., Choi, H., Ko, J. J., Lee, J., Vittal, R., and Kim, K. J. (2006). Oligophenylenevinylene-functionalized Ru(II)-bipyridine sensitizers for efficient

dye-sensitized nanocrystalline TiO_2 solar cells. *Chemistry of Materials*, 18(23), 5604-5608.

[64] Yen, Y. S., Chen, Y. C., Hsu, Y. C., Chou, H. H., Lin, J. T., and Yin, D. J. (2011). Heteroleptic Ruthenium Sensitizers That Contain an Ancillary Bipyridine Ligand Tethered with Hydrocarbon Chains for Efficient Dye-Sensitized Solar Cells. *Chemistry-A European Journal*, 17(24), 6781-6788.

[65] Cao, Y., Zhang, J., Bai, Y., Li, R., Zakeeruddin, S. M., Grätzel, M., and Wang, P. (2008). Dye-sensitized solar cells with solvent-free ionic liquid electrolytes. *The Journal of Physical Chemistry C*, 112(35), 13775-13781.

[66] Chen, C. Y., Wu, S. J., Wu, C. G., Chen, J. G., and Ho, K. C. (2006). A Ruthenium complex with superhigh light-harvesting capacity for dye-sensitized solar cells. *Angewandte Chemie*, 118(35), 5954-5957.

[67] Giribabu, L., Kumar, C. V., Rao, C. S., Reddy, V. G., Reddy, P. Y., Chandrasekharam, M., and Soujanya, Y. (2009). High molar extinction coefficient amphiphilic ruthenium sensitizers for efficient and stable mesoscopic dye-sensitized solar cells. *Energy and Environmental Science*, 2(7), 770-773.

[68] Cai, N., Moon, S. J., Cevey-Ha, L., Moehl, T., Humphry-Baker, R., Wang, P., and Grätzel, M. (2011). An organic D-π-A dye for record efficiency solid-state sensitized heterojunction solar cells. *Nano Letters*, 11(4), 1452-1456.

[69] Yanagida, M., Singh, L. P., Sayama, K., Hara, K., Katoh, R., Islam, A., and Grätzel, M. (2000). A new efficient photosensitizer for nanocrystalline solar cells: synthesis and characterization of cis-bis(4,7-dicarboxy-1,10-phenanthroline) dithiocyanato ruthenium(II). *Journal of the Chemical Society, Dalton Transactions*, (16), 2817-2822.

[70] Wang, P., Zakeeruddin, S. M., Moser, J. E., Nazeeruddin, M. K., Sekiguchi, T., and Grätzel, M. (2003). A stable quasi-solid-state dye-sensitized solar cell with an amphiphilic ruthenium sensitizer and polymer gel electrolyte. *Nature Materials*, 2(6), 402-407.

[71] Sugihara, H., Singh, L. P., Sayama, K., Arakawa, H., Nazeeruddin, M. K., and Grätzel, M. (1998). Efficient photosensitization of nanocrystalline TiO_2 films by a new class of sensitizer: cis-dithiocyanato bis (4,7-dicarboxy-1,10-phenanthroline) ruthenium (II). *Chemistry Letters*, 27(10), 1005-1006.

[72] Wang, P., Zakeeruddin, S. M., Moser, J. E., Humphry-Baker, R., Comte, P., Aranyos, V., and Grätzel, M. (2004). Stable new sensitizer with improved light harvesting for nanocrystalline dye-sensitized solar cells. *Advanced Materials*, 16(20), 1806-1811.

[73] Renouard, T., Fallahpour, R. A., Nazeeruddin, M. K., Humphry-Baker, R., Gorelsky, S. I., Lever, A. B. P., and Grätzel, M. (2002). Novel ruthenium sensitizers containing functionalized hybrid tetradentate ligands: synthesis, characterization, and INDO/S analysis. *Inorganic Chemistry*, 41(2), 367-378.

[74] Klein, C, Nazeeruddin, Md. K, Liska P, Di Censo D, Hirata, N. Palomares, E Durrant JR, and Grätzel M. (2005). Engineering of a Novel Ruthenium Sensitizer and Its Application in Dye-Sensitized Solar Cells for Conversion of Sunlight into Electricity. *Inorg. Chem*, 44, 178-180.

[75] Rensmo, H., Södergren, S., Patthey, L., Westermark, K., Vayssieres, L., Kohle, O., and Siegbahn, H. (1997). The electronic structure of the cis-bis (4, 4′-dicarboxy-2,2′-bipyridine)-bis (isothiocyanato) ruthenium (II) complex and its ligand 2,2′-bipyridyl-4,4′-dicarboxylic acid studied with electron spectroscopy. *Chemical Physics Letters*, 274(1-3), 51-57.

[76] Nazeeruddin, M. K., Zakeeruddin, S. M., Humphry-Baker, R., Gorelsky, S. I., Lever, A. B. P., and Grätzel, M. (2000). Synthesis, spectroscopic and a ZINDO study of cis-and trans-(X_2) bis (4,4′-dicarboxylic acid-2,2′-bipyridine) ruthenium (II) complexes (X= Cl-, H_2O, NCS-). *Coordination Chemistry Reviews*, 208(1), 213-225.

[77] Figgemeier, E., Aranyos, V., Constable, E. C., Handel, R. W., Housecroft, C. E., Risinger, C., ... and Mukhtar, E. (2004). Modification of electron transfer properties in photoelectrochemical solar cells by substituting $\{Ru(terpy)_2\}^{2+}$ dyes with thiophene. *Inorganic Chemistry Communications*, 7(1), 117-121.

[78] Birel O., Nadeem S., Duman H. (2017). Porphyrin-Based Dye-Sensitized Solar Cells (DSSCs): a Review. *Journal of Fluorescence*, 27 (3), pp. 1075-1085.

[79] Zhang, M. J., Guo, Y. R., Fang, G. Z., and Pan, Q. J. (2013). DFT/TD-DFT studies on structural and spectroscopic properties of metalloporphyrin complexes: a design of ruthenium porphyrin photosensitizer. *Computational and Theoretical Chemistry*, 1019, 94-100.

[80] Anbarasan, P. M., Priyadharsini, C. I., Sathiyapriya, R., Hariharan, V., Parabakaran, K., and Aroulmoji, V. (2019). Development of TiO_2 Nanomaterials and Dyes Selection (using DFT) for DSSC Applications-A Stepwise Review. *International Journal of Advanced Science and Engineering*, 6(02), 1326-1350.

[81] Chitumalla, R. K., and Jang, J. (2018). Density functional theory study on ruthenium dyes and dye@TiO_2 assemblies for dye sensitized solar cell applications. *Solar Energy*, 159, 283-290.

[82] Maluta, N. E., Dima, R. S., Nemudzivhadi, H., and Maphanga, R. R. (2017, October). Computational study of dye adsorption onto Brookite TiO_2 surfaces for the applications in dye sensitized solar cells. In *Journal of Physics: Conference Series* Vol. 905, No. 1, p. 012034, IOP Publishing.

[83] Cherifi, K., Cheknane, A., Benghia, A., Hilal, H. S., Rahmoun, K., Benyoucef, B., and Goumri-Said, S. (2019). Exploring N3 ruthenium dye adsorption onto $ZnTiO_3$ (101) and (110) surfaces for dye sensitized solar cell applications: Full computational study. *Materials Today Energy,* 13, 109-118.

[84] He, L. J., Chen, J., Bai, F. Q., Jia, R., Wang, J., and Zhang, H. X. (2016). The influence of a dye-TiO_2 interface on DSSC performance: a theoretical exploration with a ruthenium dye. *RSC Advances*, 6(85), 81976-81982.

[85] Clifford, J. N., Palomares, E., Nazeeruddin, M. K., Grätzel, M., and Durrant, J. R. (2007). Dye dependent regeneration dynamics in dye sensitized nanocrystalline solar cells: evidence for the formation of a ruthenium bipyridyl cation/iodide intermediate. *The Journal of Physical Chemistry C*, 111(17), 6561-6567.

[86] Hu, C. H., Asaduzzaman, A. M., and Schreckenbach, G. (2010). Computational studies of the interaction between ruthenium dyes and X- and X2-, X= Br,I,At.

Implications for dye-sensitized solar cells. *The Journal of Physical Chemistry C*, 114(35), 15165-15173.

[87] Nemudzivhadi, H., Maluta, N. E., Maphanga, R. R., and Sankaran, V. (2017, October). DFT calculations of Anatase TiO_2 (101) Surface Doped with Ruthenium for Application in Dye Sensitised Solar Cell. In *Journal of Physics: Conference Series* (Vol. 905, No. 1, p. 012013). IOP Publishing.

[88] De Angelis, F., Fantacci, S., and Sgamellotti, A. (2007). An integrated computational tool for the study of the optical properties of nanoscale devices: Application to solar cells and molecular wires. *Theoretical Chemistry Accounts*, 117(5), 1093-1104.

[89] Salvatori, P., Agrawal, S., Barreddi, C., Malapaka, C., de Borniol, M., and De Angelis, F. (2014). Stability of ruthenium/organic dye co-sensitized solar cells: A joint experimental and computational investigation. *RSC Advances*, 4(101), 57620-57628.

[90] Barrera, M. (2017). The BCL Approach for the Molecular Design of New Ruthenium Complexes for Dye Sensitized Solar Cells. *J Anal Pharm Res,* 4(1), 00090.

About the Editors

Dr. Davor Margetic is head of Division of Organic Chemistry and biochemistry at Rudjer Boskovic Institute in Zagreb, Croatia. He obtained his PhD degree in the field of theoretical and physical organic chemistry at Zagreb University and carried out postdoctoral research studies at the Centre for Molecular Architecture, Central Queensland University, Australia (working with Professors Ronald Warrener and Doug Butler) for nine years. Now he holds the position of senior research scientist (an equivalent of full professor at the university).

During his scientific career, Dr. Margetic has published more than 110 research papers, seventeen book chapters and four books (*Microwave Assisted Cycloaddition Reactions*, *Mechanochemical Organic Synthesis*, *Cycloaddition Reactions: Advances in Research and Applications*, and *High Pressure Organic Synthesis*.)

His research interests span areas which include the synthesis and investigation of theoretically interesting molecules, the study of reaction mechanisms, computational organic chemistry, supramolecular chemistry and the development of environmentally friendly organic reactions using novel techniques (extremely high pressures, microwave irradiation and mechanochemistry).

Dr Renjith Thomas is currently a senior faculty member at the Department of Chemistry and the Director of Centre for Interdisciplinary Research, St. Berchmans College, Mahatma Gandhi University, Kerala, India. He is a theoretical and computational chemist and has contributed much to the scientific literature.

Dr. Renjith was born in 1980 at Changanassery, India, studied BSc chemistry from St. Berchmans College, MSc from Gandhigram University with gold medal and received his PhD in theoretical chemistry from Bharathidasan University. His research is concentrated on chemical physics, theoretical evaluation of the reaction mechanism, molecular structure, spectroscopy, weak interactions, drug design and machine learning approaches. He is a regular reviewer to many journals of repute and author of more than 90 research papers in different well-known journals. He is a member of the editorial board of BMC Chemistry and topic editor for Frontiers in Chemistry and Coatings. Recently he was listed in the World's top 2% scientists in chemical physics by Stanford University and Elsevier. He enjoys membership in ICS, RSC and ACS.

Index

D

E

F

Q

R

S